Essential Mathematics

Essential Mathematics

A Worktext
Second Edition

D. FRANKLIN WRIGHT

Cerritos College

D. C. Heath and Company

Lexington, Massachusetts Toronto

To the memory of Don Wright

Cover illustration:
Marvin D. Cone
Patterns of Rectangles, 1957
Oil on Canvas, 30" × 18"
Cedar Rapids Museum of Art
Gift of Winnifred Cone and Family, and Museum Purchase, 1983
All Rights Reserved.

International Standard Book Number: 0-669-19912-5

10   9   8   7

Preface

PURPOSE AND STYLE

The purpose of this text is to provide students with a learning tool that will help them to understand arithmetic and to achieve some satisfaction in learning so that they will be encouraged to continue their education in mathematics. The style of writing gives carefully worded explanations that are direct and easy to understand. Students are expected to participate actively. Each section begins with a short list of important terms or rules to be used, followed by examples that are completely worked out and explained. Next, students are to complete partially worked out examples and then work similar exercises in the margins before proceeding to the next topic. The answers to the completion examples appear at the bottom of the page and the answers to the margin exercises appear at the end of each section. This approach provides students with immediate reinforcement and the instructor with immediate feedback.

The text is designed for both lecture and independent study classes. The intended teaching style of the text for either type of class is to allow time for working completion examples and margin exercises before proceeding to the next idea or to the homework assignment. Space is provided in the text for students to write answers on the text pages themselves.

SPECIAL FEATURES

1. The important terms, rules, and procedures in each section are provided in list form and are highlighted with boldface print and shading. The explanations are brief and easy to read.

2. Meaningful word problems are provided throughout the text.

3. Completion examples lead students through the proper formats to use when working problems, while providing step-by-step explanations.

4. Margin exercises provide immediate reinforcement before new ideas are introduced.

5. Students are encouraged to use calculators in the application problems involving several operations in Chapters 7, 8, and 9.

6. Each chapter contains a chapter test similar to a test students can expect to take in class. If the instructor prefers a multiple-choice format, multiple-choice tests are supplied in the Instructor's Guide.

7. Answers to the odd-numbered section exercises and all chapter test questions appear in the back of this text.

NEW IN THE SECOND EDITION

1. The discussion of whole numbers has been condensed into one chapter (Chapter 1) so students can progress through this beginning material more quickly and easily.

2. Exponents are now introduced in Chapter 2 and used throughout the text.

3. Methods for estimating answers are introduced in the Chapter 1 discussion of whole numbers. Similar concepts for estimation with decimals and in word problems involving decimals are an integral part of Chapter 5.

4. A new section on comparing and converting fractions and decimals has been added to the end of Chapter 3.

5. Chapter 7 now uses a single formula to solve the three basic types of percent problems. A table of common percent–decimal–fraction equivalents has been added to Section 7.3.

6. Chapter 9, a new chapter, covers basic statistics and reading graphs and histograms.

7. Measurement and geometry are now covered in separate chapters. Chapter 10 discusses both the metric and the US customary systems of measurement.

8. In Chapter 11, the coverage of geometry has been expanded to include angle measure; properties of triangles, including similar triangles; the Pythagorean Theorem; and square roots.

9. Most sections now begin with short paragraphs that introduce the topics to be covered.

10. The number of exercises has been increased by more than 25% to over 3700 problems.

CONTENT

There is sufficient material for a three- or four-semester hour course. Many topics are presented in an effort to provide flexibility in the course content and in the order that topics may be taught.

Chapter 1 (Whole Numbers) is designed to develop the basic skills of addition, subtraction, multiplication, and division with whole numbers. Estimation is used to aid understanding and word problems are used to reinforce these skills.

Chapter 2 (Prime Numbers) now includes exponents. With the topics of prime factorization and least common multiples, Chapter 2 forms the foundation for the work with fractions and mixed numbers in Chapters 3 (Fractions) and 4 (Mixed Numbers). Many students have indicated that fractions "make more sense" with this approach, and that working with fractions is no longer a series of mysterious steps that they cannot remember or understand.

Chapter 5 (Decimals) now discusses estimating as an integral part in developing an understanding of operations with decimal numbers. It also emphasizes reading and writing decimal numbers in the form of English words, and contains a new section relating decimals and fractions.

POSSIBLE COURSE OFFERINGS

Short Course (Chapters 1–7)	Longer Course (Chapters 1–9)	Optional Course (Chapters 2–9 with selected topics from Chapters 10, 11, and 12)
Whole Numbers	Whole Numbers	Prime Numbers
Prime Numbers	Prime Numbers	Fractions
Fractions	Fractions	Mixed Numbers
Mixed Numbers	Mixed Numbers	Decimals
Decimals	Decimals	Ratio and Proportion
Ratio and Proportion	Ratio and Proportion	Percent
Percent	Percent	Applications Using Calculators
	Applications Using Calculators	Statistics
	Statistics	Selected Topics from Measurement, Geometry, and Algebra

Chapter 6 (Ratio and Proportion) develops an understanding of equations, along with techniques for solving equations for unknown terms in proportions. With these techniques as a foundation, a new approach to solving percent problems is covered in Chapter 7. This approach uses the formula $R \times B = A$, where R is in decimal form.

Chapter 8 (Consumer Applications) is made up of practical applications that are specifically designed to involve several calculations, as in real-life mathematical situations. Except for Sections 8.1 and 8.2, the topics in Chapter 8 are independent; the instructor may use any part or all of this chapter as time permits.

In Chapters 7 and 8 the use of calculators is recommended. The calculator should be regarded as a tool to aid in speed and accuracy of calculations, however, and not as a replacement for basic skills and knowledge. In addition, students should find the work with estimation from Chapters 1 and 5 particularly helpful in avoiding obvious errors.

Chapter 9 (Statistics) discusses the mean, median, and range for given data, as well as how to read information from various types of graphs and make calculations using this information.

Chapter 10 (Measurement) introduces the metric system and discusses equivalent measures between the metric system and US customary system. There is also a special section designed for nursing students that covers the metric and household measures used in medical applications.

Chapters 11 (Geometry) and 12 (Introduction to Algebra) should be particularly helpful for those students who plan to continue in mathematics. In Chapter 11, the measures introduced in Chapter 10 are applied in calculating the perimeter, area, and volume of various geometric figures. The properties of triangles are analyzed and, through the Pythagorean Theorem, square roots are introduced. In Chapter 12 students are introduced to positive and negative numbers, and the basic skills needed for solving first-degree equations. Word problems involving number phrases and some geometric concepts conclude the chapter.

EXERCISES

There are more than 3700 exercises, carefully chosen and graded, proceeding from easy exercises to more difficult ones, plus more than 370 margin exercises. Each chapter includes a chapter test similar to length and content to the tests provided in the instructor's guide. The answers to the odd-numbered exercises and to all the chapter test questions are provided in the back of the book. The answers to the margin exercises appear at the end of each section, just before the section exercises. I have made a special effort to develop word problems that are both interesting and meaningful.

This text is designed for the students to work through all the examples, completion examples, and margin exercises as well as the section exercises. The students should learn by the active participation of working many problems.

ADDITIONAL AIDS

Instructor's Guide. As well as the answers to the even-numbered section exercises in the textbook, this supplement contains a selection of tests. Each chapter test appears in four forms: two are tests similar to those in the text and two are composed of multiple-choice questions. Four comprehensive tests—one covering Chapters 1–2, another covering Chapters 1–5, and another covering Chapters 1–7, and a final exam covering Chapters 1–12—appear in four forms each. Answers to all test questions are included in this supplement.

Student Solutions Guide. This manual contains the solutions to all the odd-numbered section exercises and chapter test questions in the text.

HeathTest Plus. This new, versatile test-generating program allows instructors to customize tests for their own classes. It contains over 1500 multiple-choice and open-ended test items and offers full graphics capability (including mathematical symbols).

With this program, instructors can preview questions on-screen and then add each item to a test with one keystroke. Random generation of test items by chapter is possible. In addition, instructors may edit existing items or add new items either to the database or to individual tests. Tests may be saved and then printed in multiple scrambled versions. Answer keys are automatically generated.

Hardware requirements: IBM PC or compatible, two disk drives, IBM graphics-compatible dot-matrix printer or laser printer.

Items included: one program disk, test items disks, User Manual/ Printed Test Item File.

ACKNOWLEDGMENTS

I would like to thank Kathy Sessa, developmental editor, and Anne Starr, production editor, for their hard work and consistent attention to detail throughout the development of this second edition. Special thanks to Carol L. Johnson for her excellent work on the Instructor's Guide and to Helen Medley for her writing and typing of the Student Solutions Guide. Pat Wright has again done a superb job of typing the manuscript and proofreading, making all the related work just that much easier.

Many thanks to the following reviewers for their constructive and critical comments for the second edition: David E. Conroy, Northern Virginia Community College; Ray Emett, Salt Lake City Community College; James Magliano, Union County College; Roger Melton, Central Wyoming College; and Howard Penn, El Centro College.

I would also like to thank the reviewers of the first edition. They are: Martha Breitweiser, Kellogg Community College; Reginald Luke, Middlesex County College; Steven M. Mahoney, Shasta College; and Claude S. Moore, Danville Community College.

D. Franklin Wright

Contents

12

INTRODUCTION TO ALGEBRA   563

APPENDIX   623

ANSWER KEY   625

INDEX   645

Whole Numbers

1.1
READING AND WRITING WHOLE NUMBERS

The Hindu-Arabic system that we use was invented about A.D. 800. It is called the **decimal system** (**deci** means **ten** in Latin). This system allows us to add, subtract, multiply, and divide more easily and faster than any of the ancient number systems.

Whole Numbers: The numbers used for counting and the number 0:

0, 1, 2, 3, 4, 5, 6, 7, 8, 9, 10, 11, 12, 13, . . .

Digits: The symbols used to write whole numbers:

0, 1, 2, 3, 4, 5, 6, 7, 8, 9

Decimal System (Base 10 System): A place value system used to write any whole number

The decimal system depends on three things:

1. the **ten digits:** 0, 1, 2, 3, 4, 5, 6, 7, 8, 9.
2. the **placement** of each digit.
3. the **value** of each place.

Decimal Point: The beginning point for writing digits in the decimal system

Powers of Ten: 1, 10, 100, 1000, and so on. The values of the places in the decimal system

Memorize the material in Figure 1.1.

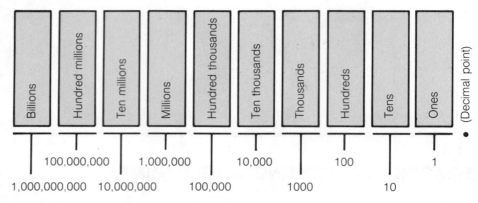

FIGURE 1.1

Standard Notation: Digits are written in places corresponding to values of powers of ten:

$$\text{six hundred ninety-three} = 693$$

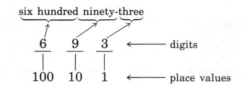

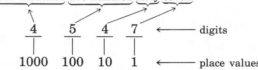

Expanded Notation: The values represented by each digit in standard notation are added:

$$693 = 600 + 90 + 3$$

six hundred   ninety   three

This notation allows the number to be read easily from the sum.

Write each number in Examples 1 and 2 in expanded notation.

Example 1   843

$843 = 800 + 40 + 3$ Expanded notation

$843 = 800 + 40 + 3$

Example 2   4956

   $4956 = 4000 + 900 + 50 + 6$ Expanded notation

   $4956 = 4000 + 900 + 50 + 6$

Fill in the blanks in Examples 3 and 4.

Completion Example 3

   $532,081 = 500,000 + 30,000 + \underline{2000} + \underline{0} + \underline{80} + 1$

Completion Example 4

   $497,500 = 400,000 + \underline{90,000} + \underline{7,000} + \underline{500} + 0 + 0$

☑ **NOW WORK EXERCISES 1–3 IN THE MARGIN.**

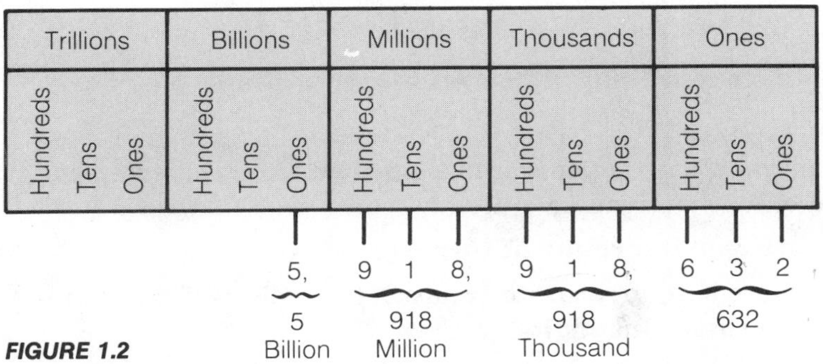

FIGURE 1.2

> **English Word Equivalents:** A whole number written in English words
>
>   268:   two hundred sixty-eight
>
> Notice that the word **and** is not used when writing the English words for whole numbers.

Write each number in Examples 5–8 in expanded notation and in its English word equivalent.

Example 5   653

   $653 = 600 + 50 + 3$

   six hundred fifty-three

   $653 = 600 + 50 + 3$

   SIX HUNDRED FIFTY-THREE

Completion Example Answers

3. $500,000 + 30,000 + 2000 + 0 + 80 + 1$

4. $400,000 + 90,000 + 7000 + 500 + 0 + 0$

1. Write 4635 in expanded notation.

   $4635 = 4000 + 600 + 30 + 5$

2. Write 73,428 in expanded notation.

   $= 70,000 + 3,000 + 400 + 20 + 8$

3. Write $500 + 20 + 1$ in standard notation.

   521

4. Write 9567 in expanded notation and in its English word equivalent.

= 9000 + 500 + 60 + 7

NINE THOUSAND FIVE HUNDRED SIXTY-SEVEN

5. Write 25,400 in expanded notation and in its English word equivalent.

= 20,000 + 5,000 + 400 + 0 + 0

TWENTY FIVE THOUSAND FOUR HUNDRED

6. Write 6000 + 700 + 90 + 2 in standard notation and in its English word equivalent.

6792

SIX THOUSAND SEVEN HUNDRED NINTY-TWO

Example 6   75,900

$$75,900 = 70,000 + 5000 + 900 + 0 + 0$$

seventy-five thousand, nine hundred

75,900 = 70,000 + 5000 + 900 + 0 + 0

SEVENTY-FIVE THOUSAND, NINE HUNDRED

Completion Example 7   8407

$8407 = 8000 + \underline{400} + \underline{0} + \underline{7}$

eight thousand, *FOUR HUNDRED SEVEN*

Completion Example 8   15,352

$15,352 = 10,000 + \underline{5000} + \underline{300} + \underline{50} + \underline{2}$

fifteen thousand, *THREE HUNDRED FIFTY-TWO*

☑ **NOW WORK EXERCISES 4–6 IN THE MARGIN.**

> You should note four things when reading or writing whole numbers:
>
> 1. The word **and** does not appear in English word equivalents. **And** is said only when reading a decimal point. (See Chapter 5.)
> 2. Digits are read in **groups of three.** (See Figure 1.2.)
> 3. Commas are used to separate groups of three digits **if a number has more than four digits.**
> 4. Hyphens (-) are used to write English words for the two-digit numbers from 21 to 99 except for those that end in 0.

Example 9   Write four hundred eighty thousand, five hundred thirty-three in standard notation.

480,533   *480,533*

Example 10   Write five hundred seventy-two thousand, six hundred in standard notation.

572,600   *572,600*

Completion Example Answers

7. 8000 + 400 + 0 + 7
   eight thousand, four hundred seven
8. 10,000 + 5000 + 300 + 50 + 2
   fifteen thousand, three hundred fifty-two

Completion Example 11 Write two million, eight hundred thousand, thirty-five in standard notation.

2,800, <u>035</u>

Completion Example 12 Write five million, three hundred fifty thousand in standard notation.

5, <u>350</u> , <u>000</u>

☑ **NOW WORK EXERCISES 7–10 IN THE MARGIN.**

7. Write twenty-three thousand, six hundred forty-two in standard notation.

23,642

8. Write three hundred sixty-three thousand, nine hundred seventy-five in standard notation.

363,975

9. Write 6,300,500 in its English word equivalent.

SIX MILLION, THREE HUNDRED THOUSAND, FIVE HUNDRED

10. Write 4,875,000 in its English word equivalent.

FOUR MILLION, EIGHT HUNDRED SEVENTY FIVE THOUSAND

Completion Example Answers

11. 2,800,035   **12.** 5,350,000

Answers to Marginal Exercises

1. 4000 + 600 + 30 + 5   **2.** 70,000 + 3000 + 400 + 20 + 8   **3.** 521   **4.** 9000 + 500 + 60 + 7; nine thousand, five hundred sixty-seven   **5.** 20,000 + 5000 + 400 + 0 + 0; twenty-five thousand, four hundred   **6.** 6792; six thousand, seven hundred ninety-two   **7.** 23,642   **8.** 363,975   **9.** six million, three hundred thousand, five hundred   **10.** four million, eight hundred seventy-five thousand

NAME _____ SECTION _____ DATE _____

EXERCISES **1.1** _____

1. List the ten digits used to write whole numbers.

2. The beginning point for writing digits in the decimal system is called a

   _____ .

3. The powers of ten are _____ .

4. The word _____ does not appear in English word equivalents

   for whole numbers.

5. Use a comma to separate groups of three digits if a number has more

   than _____ digits.

6. Hyphens are used to write English words for numbers with

   _____ .

Write the following whole numbers in expanded notation.

  7. 37   **8.** 84   **9.** 56   **10.** 821

11. 1892   **12.** 2059   **13.** 25,658   **14.** 32,341

Write the following whole numbers in their English word equivalents.

15. 83   **16.** 122   **17.** 10,500   **18.** 683,100

19. 592,300   **20.** 16,302,590   **21.** 71,500,000

Answers:

1. 0, 1, 2, 3, 4 5 6 7 8 9
2. DECIMAL POINT
3. 1, 10, 100, 1000, 10,000
4. AND
5. FOUR
6. TWO DIGITS (21–99)
7. $= 30 + 7 = 37$
8. $= 80 + 4 = 84$
9. $50 + 6 = 56$
10. $800 + 20 + 1 = 821$
11. $1000 + 800 + 90 + 2 = 1892$
12. $2000 + 0 + 50 + 9 = 2059$
13. $20,000 + 5000 + 600 + 50 + 8$
14. $30,000 + 2,000 + 300 + 40 + 1$
15. EIGHTY-THREE
16. ONE HUNDRED, TWENTY-TWO
17. TEN THOUSAND, FIVE HUNDRED
18. SIX HUNDRED, EIGHTY THREE THOUSAND, ONE HUNDRED
19. FIVE HUNDRED NINETY TWO THOUSAND, THREE HUNDRED
20. SIXTEEN MILLION, THREE HUNDRED TWO THOUSAND FIVE HUNDRED NINETY
21. SEVENTY ONE MILLION, FIVE HUNDRED THOUSAND

Answers

22. ___75___

23. ___98___

24. ___142___

25. ___573___

26. ___3,834___

27. ___10,011___

28. ___400,736___

29. ___537,082___

30. ___63,251,065___

31. ONE, TEN, TEN MILLION

32. TWO MILLION, EIGHT HUNDRED SIXTEEN THOUSAND

33. NINETY THREE MILLION

Write the following whole numbers in standard notation.

22. seventy-five

23. ninety-eight

24. one hundred forty-two

25. five hundred seventy-three

26. three thousand, eight hundred thirty-four

27. ten thousand, eleven

28. four hundred thousand, seven hundred thirty-six

29. five hundred thirty-seven thousand, eighty-two

30. sixty-three million, two hundred fifty-one thousand, sixty-five

31. Name the position of each nonzero digit in the following number: 2,403,189,500.

Write the English word equivalent for the number in each sentence.

32. The population of Los Angeles is 2,816,000.

33. The distance from the earth to the sun is about 93,000,000 miles.

1.2
ADDITION

> **Addition:** Counting the total represented by two or more whole numbers
>
> **Sum:** Result of addition
>
> **Addends:** Numbers that are added

Example 1   5   +   3   =   8
                 ↑   ↑   ↑
              addend addend   sum

Or, writing the numbers vertically,

$$
\begin{array}{r}
5 \longleftarrow \text{addend} \\
+3 \longleftarrow \text{addend} \\
\hline
8 \longleftarrow \text{sum}
\end{array}
$$

You should **memorize** the basic addition facts in Table 1.1 to be able to add larger numbers with speed and accuracy.

TABLE 1.1 Basic Addition Facts

+	0	1	2	3	4	5	6	7	8	9	
0	0	1	2	3	4	5	6	7	8	9	
1	1	2	3	4	5	6	7	8	9	10	
2	2	3	4	5	6	7	8	9	10	11	
3	3	4	5	6	7	8	9	10	11	⑫	3 + 9 = 12
4	4	5	6	7	8	9	10	11	12	13	
5	5	6	7	8	9	10	11	12	13	14	
6	6	7	8	9	10	11	12	13	14	15	
7	7	8	9	10	11	12	13	14	15	16	
8	8	9	10	11	12	⑬	14	15	16	17	
9	9	10	11	12	13	14	15	16	17	18	

8 + 5 = 13

For Exercises 1 and 2, find both sums and tell what property of addition is illustrated.

1.   8   5
    +5   +8
    ‾‾‾   ‾‾‾
    13   13

2. 9 + 6 = __15__

    6 + 9 = __15__

☑ **NOW WORK PRACTICE PROBLEMS 1–55.**

PRACTICE PROBLEMS
Find the following sums mentally as quickly as you can. If you make any errors, practice those sums until you have them memorized.

1. 1 +3	2. 4 +6	3. 7 +9	4. 0 +1	5. 2 +8	6. 5 +5	7. 3 +4
4	10	16	1	10	10	7

8. 6 +7	9. 8 +8	10. 0 +3	11. 8 +1	12. 2 +2	13. 5 +0	14. 9 +3
13	16	3	9	4	5	12

15. 0 +0	16. 6 +3	17. 7 +5	18. 4 +2	19. 7 +7	20. 8 +4	21. 7 +0
0	9	12	6	14	12	7

22. 6 +1	23. 8 +3	24. 9 +9	25. 4 +7	26. 2 +0	27. 1 +4	28. 4 +9
7	12	18	11	2	5	13

29. 8 +5	30. 6 +6	31. 4 +4	32. 8 +0	33. 3 +3	34. 2 +6	35. 0 +9
12	12	8	8	6	8	9

36. 9 +6	37. 8 +7	38. 1 +1	39. 3 +2	40. 5 +4	41. 9 +8	42. 1 +2
15	15	2	5	9	17	3

43. 3 +5	44. 5 +6	45. 9 +1	46. 2 +9	47. 5 +1	48. 6 +0	49. 8 +6
8	11	10	11	6	6	14

50. 3 +7	51. 5 +2	52. 0 +4	53. 7 +1	54. 9 +5	55. 2 +7	
10	7	4	8	14	9	

By looking at Table 1.1, we can see that reversing the **order** of any two addends does not change their sum. This fact is called the **commutative property of addition.** We state this property using letters for whole numbers.

COMMUTATIVE PROPERTY OF ADDITION
For whole numbers a and b, $a + b = b + a$

Example 2 $5 + 4 = 9$ and $4 + 5 = 9$

Thus, $5 + 4 = 4 + 5$.

Example 3   6   5
                 + 5   + 6
                 ‾‾‾‾   ‾‾‾‾
                  11   11

☑ **NOW WORK EXERCISES 1 AND 2 IN THE MARGIN.**

Another property, called the **associative property of addition,** states that numbers may be **grouped** (or associated) differently and still have the same sum.

ASSOCIATIVE PROPERTY OF ADDITION
For whole numbers a, b, and c,
$$a + b + c = (a + b) + c = a + (b + c)$$

Example 4 $6 \ + \ 3 \ + \ 5$ $6 \ + \ 3 \ + \ 5$

$= \ 9 \quad + \ 5$ $= \ 6 \ + \ 8$
$= 14$ $= 14$

Thus, $(6 + 3) + 5 = 6 + (3 + 5)$.

Example 5 $7 + 1 + 4 = (7 + 1) + 4 = 8 + 4 = 12$ and
$7 + 1 + 4 = 7 + (1 + 4) = 7 + 5 = 12$

☑ **NOW WORK EXERCISES 3 AND 4 IN THE MARGIN.**

Another property of addition with whole numbers is addition with 0. Whenever 0 is added to a number, the result is the original number. Zero (0) is called the **additive identity** or the **identity element for addition.**

0 IS THE ADDITIVE IDENTITY
For any whole number a, $a + 0 = a$

Example 6 $7 + 0 = 7$

Example 7 $\begin{array}{r} 8 \\ +0 \\ \hline 8 \end{array}$

For Exercises 3 and 4, find both sums and tell what property of addition is illustrated.

3. $6 + (5 + 4) =$ _15_
   $(6 + 5) + 4 =$ _15_
   ASSOCIATIVE

4. $(9 + 1) + 7 =$ _17_
   $9 + (1 + 7) =$ _17_

ASSOCIATIVE

Show that each of the following statements is true by doing the addition both ways vertically.

5. 9 + 1 = 1 + 9

$$\begin{array}{r} 9 \\ +1 \\ \hline 10 \end{array} \qquad \begin{array}{r} 1 \\ +9 \\ \hline 10 \end{array}$$

6. (8 + 4) + 3 = 8 + (4 + 3)

$$\begin{array}{r} 8 \\ 4 \\ +3 \\ \hline 15 \end{array} \qquad \begin{array}{r} 8 \\ 4 \\ +3 \\ \hline 15 \end{array}$$

7. 3 + 0 = 0 + 3

$$\begin{array}{r} 3 \\ +0 \\ \hline 3 \end{array} \qquad \begin{array}{r} 0 \\ +3 \\ \hline 3 \end{array}$$

8. 5 + 7 = 7 + 5

$$\begin{array}{r} 5 \\ +7 \\ \hline 12 \end{array} \qquad \begin{array}{r} 7 \\ +5 \\ \hline 12 \end{array}$$

In Examples 8–10, find the sums and tell what property of addition is illustrated.

Completion Example 8 1 + (6 + 5) = (1 + 6) + 5 = _12_

ASSOCIATIVE

Completion Example 9 8 + 3 = 3 + 8 = _11_

COMMUTATIVE

Completion Example 10 5 + 0 = _5_

ADDITIVE IDENTITY

You should understand the following distinction between the associative and commutative properties of addition:

> In the associative property, the order of the numbers is unchanged. (The grouping is changed.)

> In the commutative property, the order of the numbers is changed.

▢ **NOW WORK EXERCISES 5–8 IN THE MARGIN.**

> To add whole numbers with more than one digit,
>
> 1. write the numbers vertically so that the place values are **lined up** in columns.
> 2. add only the digits with the same place value.

Example 11   Add 5623 + 3172.

$$\begin{array}{r} 5\ 6\ 2\ \boxed{3} \\ +\ 3\ 1\ 7\ \boxed{2} \\ \hline 5 \end{array}$$ ← addend
← addend
Add ones.

$$\begin{array}{r} 5\ 6\ \boxed{2}\ 3 \\ +\ 3\ 1\ \boxed{7}\ 2 \\ \hline 9\ 5 \end{array}$$ Add tens.

$$\begin{array}{r} 5\ \boxed{6}\ 2\ 3 \\ +\ 3\ \boxed{1}\ 7\ 2 \\ \hline 7\ 9\ 5 \end{array}$$ Add hundreds.

$$\begin{array}{r} \boxed{5}\ 6\ 2\ 3 \\ +\ \boxed{3}\ 1\ 7\ 2 \\ \hline 8\ 7\ 9\ 5 \end{array}$$ Add thousands.

Completion Example Answers

8. 12, associative property

9. 11, commutative property

10. 5, additive identity

```
    5 6 2 3   ← You do not write all the steps. You
  + 3 1 7 2       write only the addends and the sum.
    8 7 9 5   ← sum
```

☐ **NOW WORK EXERCISES 9–12 IN THE MARGIN.**

> **CARRYING**
>
> If the sum of the digits in one place-value column is more than 9, then
>
> 1. write the ones digit in that column;
> 2. **carry** the other digit to the next column to the left.

Example 12 Add 28 + 66.

```
  2 8        8
 +6 6      + 6    You think: 8 + 6 = 14.
            14
```

```
  1
  2 8          Write only the digit 4 and carry 1
 +6 6          to the next column.
    4
```

```
  1
  2 8          Now add all the digits in the tens
 +6 6          column, including the 1 that was
  9 4          carried.
```

Example 13 Example 2 can be explained using expanded notation.

```
 28 = 20 + 8 = 2 tens + 8 ones
+66 = 60 + 6 = 6 tens + 6 ones
               8 tens  14 ones

             1 ten     4 ones
             9 tens    4 ones = 90 + 4 = 94
```

Example 14 Add 347 + 295.

```
    1 ←
    347       ┐— 12 ones
   +295       │
      2 ←─────┘
```

```
   ↓
   11
   347       ┐— 14 tens
  +295       │
   42 ←──────┘
```

Find each sum.

9.   4361
       +2528
       6889

10. 3590
       +4207
       7797

11. 6781
       + 213
       6994

12. 5307
       + 401
       5708

Find the following sums.

13.
$$
\begin{array}{r}
359 \\
+647 \\
\hline
\end{array}
$$
1006

14.
$$
\begin{array}{r}
8793 \\
+4595 \\
\hline
\end{array}
$$
13388

15.
$$
\begin{array}{r}
53,895 \\
+69,146 \\
\hline
\end{array}
$$
123041

16.
$$
\begin{array}{r}
64 \\
58 \\
+38 \\
\hline
\end{array}
$$
160

17.
$$
\begin{array}{r}
196 \\
357 \\
492 \\
804 \\
+621 \\
\hline
\end{array}
$$
2470

18.
$$
\begin{array}{r}
21,452 \\
32,551 \\
364,625 \\
+\;75,807 \\
\hline
\end{array}
$$
494435

$$
\begin{array}{r}
11 \\
347 \\
+295 \\
\hline
642 \;\leftarrow\; \text{sum}
\end{array}
$$

☐ **NOW WORK EXERCISES 13–15 IN THE MARGIN.**

ADDITION WITH MORE THAN TWO NUMBERS

1. You can add the digits in one column in any order or from the top down or from the bottom up because of the commutative and associative properties of addition.

2. With several digits in a column, look for combinations of digits that total 10. This will increase your speed.

3. **Carry** digits just as before.

Example 15

$$
\begin{array}{r}
2 \\
217 \\
389 \\
634 \\
+536 \\
\hline
6
\end{array}
$$

Add $7 + 9 + 4 + 6$
$= 7 + 9 + 10$
$= 26$

Carry the 2.

$$
\begin{array}{r}
12 \\
217 \\
389 \\
634 \\
+536 \\
\hline
76
\end{array}
$$

Add $②+ 1 + ⑧ + 3 + 3$
$= 10 + 1 + 3 + 3$
$= 17$

$$
\begin{array}{r}
12 \\
217 \\
389 \\
634 \\
+\;536 \\
\hline
1776
\end{array}
$$

Add $①+ 2 + ③ + ⑥ + 5$
$= 10 + 2 + 5$
$= 17$

☐ **NOW WORK EXERCISES 16–18 IN THE MARGIN.**

Answers to Marginal Exercises

1. 13, 13; commutative property of addition   **2.** 15, 15; commutative property of addition   **3.** 15, 15; associative property of addition   **4.** 17, 17; associative property of addition   **5.** 10, 10   **6.** 15, 15   **7.** 3, 3   **8.** 12, 12   **9.** 6889   **10.** 7797   **11.** 6994   **12.** 5708   **13.** 1006   **14.** 13,388   **15.** 123,041   **16.** 160   **17.** 2470   **18.** 494,435

Answers to Practice Problems

1. 4   **2.** 10   **3.** 16   **4.** 1   **5.** 10   **6.** 10   **7.** 7   **8.** 13   **9.** 16   **10.** 3
11. 9   **12.** 4   **13.** 5   **14.** 12   **15.** 0   **16.** 9   **17.** 12   **18.** 6   **19.** 14
20. 12   **21.** 7   **22.** 7   **23.** 11   **24.** 18   **25.** 11   **26.** 2   **27.** 5   **28.** 13
29. 13   **30.** 12   **31.** 8   **32.** 8   **33.** 6   **34.** 8   **35.** 9   **36.** 15   **37.** 15
38. 2   **39.** 5   **40.** 9   **41.** 17   **42.** 3   **43.** 8   **44.** 11   **45.** 10   **46.** 11
47. 6   **48.** 6   **49.** 14   **50.** 10   **51.** 7   **52.** 4   **53.** 8   **54.** 14   **55.** 9

NAME _____   SECTION _____   DATE _____

EXERCISES **1.2** _____

Add mentally and write only the answers.

1.   6
    3
  +7
	2. 4
	    5
	  +6
		3. 2
		    3
		  +8

4.   9
    2
  +8
	5. 8
	    7
	  +2
		6. 4
		    3
		  +6

7. $(2 + 6) + (4 + 5)$ **8.** $(5 + 3) + (7 + 2)$

9. $5 + 4 + 6$ **10.** $7 + 3 + 5$

Find each sum and state the property of addition that is illustrated.

11. $9 + 3 = 3 + 9$ **12.** $8 + 7 = 7 + 8$

13. $4 + (5 + 3) = (4 + 5) + 3$ **14.** $4 + 8 = 8 + 4$

Answers

1. _16_
2. _15_
3. _13_
4. _19_
5. _17_
6. _13_
7. _17_
8. _17_
9. _15_
10. _15_
11. _12 COMMUTATIVE_
12. _15 " " "_
13. _12 ASSOCIATIVE_
14. _12 COMMUTATIVE_

Answers

15. 9 ASSOCIATIVE

16. 18

17. 9 ADDITIVE IDENTITY

18. 9 ASSOCIATIVE

19. 16 ASSOCIATIVE

20. 13 ASSOCIATIVE

21. _____

15. $2 + (1 + 6) = (2 + 1) + 6$

16. $(8 + 7) + 3 = 8 + (7 + 3)$

17. $9 + 0 = 9$

18. $(2 + 3) + 4 = 2 + (3 + 4)$

19. $7 + 6 + 3 = 7 + 9$

20. $8 + 1 + 4 = 8 + 5$

21. Complete the table.

+	5	8	7	9
3	8	11	10	12
6	11	14	13	15
5	10	13	12	14
2	7	10	9	11

NAME _____ SECTION _____ DATE _____

Find the following sums.

22. 56
      +95
      151

23. 37
      +88
      125

24. 156
      +285
      441

25. 816
      +736
      1552

26. 1076
      +3095
      4171

27. 7328
      +5996
      13324

28. 65
       43
      +54
      162

29. 24
       78
      +95
      197

30. 73
       68
      +98
      239

31. 165
       276
      +394
      835

32. 876
       279
      +143
      1298

33. 268
       93
      +192
      553

34. 981
       146
       92
      + 17
      1256

35. 2112
       147
       904
      +1005
      4168

36. 114
       5402
       710
      + 643
      6869

37. 1403
       7010
       622
      + 29
      9064

38. 213,116
       116,018
       722,988
       24,336
      +526,968
      1603426

39. 21,442
       32,462
       564,792
       801,801
      + 43,433
      1483930

40. 438,966
       1,572,486
       327,462
       181,753
      + 90,000
      2610667

41. 123,456
       456,123
       879,282
       617,500
      +740,765
      2817126

Answers

22. 151

23. 125

24. 441

25. 1552

26. 4171

27. 13,324

28. 162

29. 197

30. 239

31. 835

32. 1298

33. 553

34. 1256

35. 4168

36. 6869

37. 9064

38. 1,603,426

39. 1,483,930

40. 2610,667

41. 2817,126

Answers

42. _2762_

43. _6,313,323_

44. _20,121_

45. _493,154_

46. _1,818_

---

42. Mr. Jones kept the mileage records indicated in the table shown here. How many miles did he drive during the six months?

Month	Mileage
Jan.	546
Feb.	378
Mar.	496
Apr.	357
May	503
June	482

_2762_

43. The Modern Products Corp. showed profits as indicated in the table for the years 1982–1985. What were the company's total profits for the years 1982–1985?

Year	Profits
1982	$1,078,416
1983	1,270,842
1984	2,000,593
1985	1,963,472

_6313323_

44. During six years of college, including two years of graduate school, Fred estimated his expenses each year as: $2035, $2786, $3300, $4000, $3500, $4500. What were his total expenses for six years of schooling? (**Note:** He had some financial aid.)

45. Apple County has the following items budgeted: highways, $270,455; salaries, $95,479; maintenance, $127,220. What is the county's total budget for these three items?

46. The following numbers of students at South Junior College are enrolled in mathematics courses: 303 in arithmetic, 476 in algebra, 293 in trigonometry, 257 in college algebra, and 189 in calculus. Find the total number of students taking mathematics.

1.3
SUBTRACTION

Subtraction: Taking away one amount from another (the opposite of addition)

Difference: Result of subtraction

Minus sign (−): Used to indicate subtraction

Example 1 $7 - 5 = 2$ or
$$\begin{array}{r} 7 \\ -5 \\ \hline 2 \end{array} \text{ difference}$$

Or, in terms of addition, think:

5 plus what number gives 7?

$$5 \;+\; \boxed{?} \;=\; 7$$

is the same as $\boxed{?} \;=\; 7 \;-\; 5$

SUBTRACTION	*ADDITION*

$$\begin{array}{r} 7 \\ -5 \\ \hline \boxed{?} \end{array} \qquad \begin{array}{r} 5 \\ +\boxed{?} \\ \hline 7 \end{array} \text{ missing addend}$$

difference $\boxed{?}$

Example 2 $10 \;-\; 7 \;=\; \boxed{?}$ or
$$\begin{array}{r} 10 \\ -\;7 \\ \hline \boxed{?} \end{array} \text{ difference}$$

     ↑   ↑   ↑

sum   addend   missing addend (or difference)

Think: What number added to 7 will give 10?

Since $7 + \boxed{3} = 10$, we have $10 - 7 = \boxed{3}.$

Some basic subtraction facts are shown in Table 1.2. (Read the number in the left column and subtract the number in the top row.)

TABLE 1.2 Basic Subtraction Facts

−	0	1	2	3	4	5	6	7	8	9	
18	18	17	16	15	14	13	12	11	10	9	
17	17	16	15	14	13	12	11	10	⑨	8	17 − 8 = 9
16	16	15	14	13	12	11	10	9	8	7	
15	15	14	13	12	11	10	9	8	7	6	
14	14	13	12	11	10	9	8	7	6	5	
13	13	12	11	10	9	8	7	6	5	4	
12	12	11	10	9	8	7	6	5	4	3	
11	11	10	9	8	7	6	5	4	3	2	
10	10	9	8	7	6	5	4	3	2	1	

☐ **NOW WORK PRACTICE PROBLEMS 1–28.**

PRACTICE PROBLEMS
Find the following differences mentally as quickly as you can.

1. $\begin{array}{r} 12 \\ -\ 7 \\ \hline 5 \end{array}$
2. $\begin{array}{r} 11 \\ -\ 5 \\ \hline 6 \end{array}$
3. $\begin{array}{r} 6 \\ -6 \\ \hline 0 \end{array}$
4. $\begin{array}{r} 9 \\ -2 \\ \hline 7 \end{array}$
5. $\begin{array}{r} 13 \\ -\ 4 \\ \hline 9 \end{array}$
6. $\begin{array}{r} 13 \\ -\ 9 \\ \hline 4 \end{array}$
7. $\begin{array}{r} 10 \\ -\ 8 \\ \hline 2 \end{array}$

8. $\begin{array}{r} 11 \\ -\ 9 \\ \hline 2 \end{array}$
9. $\begin{array}{r} 11 \\ -\ 7 \\ \hline 4 \end{array}$
10. $\begin{array}{r} 16 \\ -\ 7 \\ \hline 9 \end{array}$
11. $\begin{array}{r} 16 \\ -\ 9 \\ \hline 7 \end{array}$
12. $\begin{array}{r} 15 \\ -\ 8 \\ \hline 7 \end{array}$
13. $\begin{array}{r} 15 \\ -\ 9 \\ \hline 6 \end{array}$
14. $\begin{array}{r} 15 \\ -\ 6 \\ \hline 9 \end{array}$

15. $\begin{array}{r} 14 \\ -\ 7 \\ \hline 7 \end{array}$
16. $\begin{array}{r} 14 \\ -\ 5 \\ \hline 9 \end{array}$
17. $\begin{array}{r} 16 \\ -\ 7 \\ \hline 9 \end{array}$
18. $\begin{array}{r} 17 \\ -\ 8 \\ \hline 9 \end{array}$
19. $\begin{array}{r} 18 \\ -\ 9 \\ \hline 9 \end{array}$
20. $\begin{array}{r} 13 \\ -\ 8 \\ \hline 5 \end{array}$
21. $\begin{array}{r} 9 \\ -0 \\ \hline 9 \end{array}$

22. $\begin{array}{r} 11 \\ -\ 8 \\ \hline 3 \end{array}$
23. $\begin{array}{r} 14 \\ -\ 9 \\ \hline 5 \end{array}$
24. $\begin{array}{r} 17 \\ -\ 9 \\ \hline 8 \end{array}$
25. $\begin{array}{r} 10 \\ -\ 6 \\ \hline 4 \end{array}$
26. $\begin{array}{r} 9 \\ -4 \\ \hline 5 \end{array}$
27. $\begin{array}{r} 12 \\ -\ 4 \\ \hline 8 \end{array}$
28. $\begin{array}{r} 14 \\ -\ 6 \\ \hline 8 \end{array}$

> To subtract whole numbers with a difference of more than one digit,
>
> 1. write the numbers vertically so that the place values are **lined up** in columns.
> 2. subtract only the digits with the same place value.

Example 3   Subtract $496 - 342$.

$$
\begin{array}{r}
4\ 9\ 6 \\
-\ 3\ 4\ 2 \\
\hline
4
\end{array}
$$
Subtract ones.
$6 - 2 = 4$

$$
\begin{array}{r}
4\ 9\ 6 \\
-\ 3\ 4\ 2 \\
\hline
5\ 4
\end{array}
$$
Subtract tens.
$9 - 4 = 5$

$$
\begin{array}{r}
4\ 9\ 6 \\
-\ 3\ 4\ 2 \\
\hline
1\ 5\ 4
\end{array}
$$
Subtract hundreds.
$4 - 3 = 1$

$\leftarrow$ difference

Example 4   Example 3 can be shown using expanded notation.

$$
\begin{array}{rcl}
496 & = & 400 + 90 + 6 \\
-342 & = & 300 + 40 + 2 \\
\hline
& & 100 + 50 + 4 = 154 \quad \leftarrow \text{difference}
\end{array}
$$

In Example 5, find the difference using expanded notation.

Completion Example 5   Subtract $897 - 364$.

$$
\begin{array}{rcl}
897 & = & 800 \ + \ 90 \ + \ \underline{7} \\
-364 & = & 300 \ + \ \underline{60} \ + \ \underline{4} \\
\hline
& & \underline{500} + \underline{30} + \underline{3} = \underline{533}
\end{array}
$$

Completion Example Answer

5. $\begin{array}{rcl}
897 & = & 800 + 90 + 7 \\
-364 & = & 300 + 60 + 4 \\
\hline
& & 500 + 30 + 3 = 533
\end{array}$

Find the differences. Use expanded notation only if you find it helpful.

1. 654
   −421

   233

2. 1857
   −346

   1511

3. 2469
   −1125

   1344

Find the following differences. Use expanded notation to show any borrowing.

4. 867
   −328

   539

5. 426
   −388

   38

☑ **NOW WORK EXERCISES 1–3 IN THE MARGIN.**

> BORROWING
>
> 1. **Borrowing** is necessary when a digit is smaller than the digit being subtracted. We cannot add two whole numbers and get a smaller number.
> 2. The process starts from the rightmost digit. **Borrow** from the digit to the left.

In Examples 6–7 subtract by borrowing. Expanded notation is helpful in explaining the technique.

Example 6 $65 = 60 + 5$
$$\underline{-28 = 20 + 8}$$

Starting from the right, 5 is smaller than 8. We cannot subtract 8 from 5, so borrow 10 from 60.

Borrow 10 from 60. →   10 borrowed from 60, plus 5. ↓

$$65 = 60 + 5 = 50 + 10 + 5 = 50 + 15$$ Now subtract.
$$\underline{-28 = 20 + 8 = 20 + \qquad 8 = 20 + \ 8}$$
$$30 + \ 7 = 37$$

Borrowing can be done more than once.

Example 7 $736 = 700 + 30 + 6$
$$\underline{-258 = 200 + 50 + 8}$$

Since 6 is smaller than 8, borrow 10 from 30.   Since 20 is smaller than 50, borrow 100 from 700.

$$736 = 700 + 30 + 6 = 700 + 20 + 16 = 600 + 120 + 16$$
$$\underline{-258 = 200 + 50 + 8 = 200 + 50 + \ 8 = 200 + \ 50 + \ 8}$$
$$400 + \ 70 + \ 8 = 478$$

☑ **NOW WORK EXERCISES 4 AND 5 IN THE MARGIN.**

A common practice is to indicate borrowing by crossing out digits and writing new digits instead of expanded notation.

Example 8

$$65 = 60 + 5 = 50 + 15$$
$$-28 = 20 + 8 = 20 + \ 8$$

can be written

50 in place of 60.   Borrowed 10

$$
\begin{array}{cc}
5 & 1 \\
\cancel{6} & 5 \\
-2 & 8 \\
\hline
3 & 7
\end{array}
$$

Example 9   This example uses the same numbers as in Example 7 to illustrate the different techniques.

$$
\begin{array}{r}
736 \\
-258
\end{array}
$$

STEP 1: Since 6 is smaller than 8, borrow 10 from 30. This leaves 20, so cross out 3 and write 2.

$$
\begin{array}{ccc}
 & 2 & 1 \\
7 & \cancel{3} & 6 \\
-2 & 5 & 8
\end{array}
$$

STEP 2: Since 2 is smaller than 5, borrow 100 from 700. This leaves 600, so cross out 7 and write 6.

$$
\begin{array}{ccc}
6 & 12 & 1 \\
\cancel{7} & \cancel{3} & 6 \\
-2 & 5 & 8
\end{array}
$$

STEP 3: Now subtract.

$$
\begin{array}{ccc}
6 & 12 & 1 \\
\cancel{7} & \cancel{3} & 6 \\
-2 & 5 & 8 \\
\hline
4 & 7 & 8
\end{array}
$$

Example 10
$$
\begin{array}{r}
8000 \\
- \ 657
\end{array}
$$

STEP 1: Trying to borrow from 0 each time, we end up borrowing 1000 from 8000. Cross out 8 and write 7.

$$
\begin{array}{cccc}
7 & & & \\
\cancel{8} & 0 & 0 & 0 \\
- & 6 & 5 & 7
\end{array}
$$

STEP 2: Now borrow 100 from 1000. Cross out 10 and write 9.

$$
\begin{array}{cccc}
 & 9 & & \\
7 & \cancel{1} & & \\
\cancel{8} & \cancel{0} & 0 & 0 \\
- & & 6 & 5 & 7
\end{array}
$$

STEP 3: Borrow 10 from 100. Cross out 10 and write 9.

$$
\begin{array}{cccc}
 & 9 & 9 & \\
7 & \cancel{1} & \cancel{1} & \\
\cancel{8} & \cancel{0} & \cancel{0} & 0 \\
- & & 6 & 5 & 7
\end{array}
$$

STEP 4: Now subtract.

$$
\begin{array}{cccc}
 & 9 & 9 & \\
7 & \cancel{1} & \cancel{1} & 1 \\
\cancel{8} & \cancel{0} & \cancel{0} & 0 \\
- & & 6 & 5 & 7 \\
\hline
7 & 3 & 4 & 3
\end{array}
$$

☑ **NOW WORK EXERCISES 6–8 IN THE MARGIN.**

Find the following differences without using expanded notation.

6.
$$
\begin{array}{r}
537 \\
-249 \\
\end{array}
$$

$$
\begin{array}{r}
537 \\
-249 \\
\hline
288
\end{array}
$$

7.
$$
\begin{array}{r}
6545 \\
-2687 \\
\end{array}
$$

$$
\begin{array}{r}
6545 \\
-2687 \\
\hline
3858
\end{array}
$$

8.
$$
\begin{array}{r}
4000 \\
-3946 \\
\end{array}
$$

$$
\begin{array}{r}
4000 \\
-3946 \\
\hline
054
\end{array}
$$

Example 11   Two painters bid on painting the same house. The first painter bid $1685 and the second painter bid $1827. What was the difference between the two bids?

Solution

This is a subtraction problem because you are asked for a difference.

$$
\begin{array}{lr}
 & \overset{7}{}\ \overset{1}{} \\
\text{Higher Bid} & \$\ 1\ \cancel{8}\ 2\ 7 \\
-\ \text{Lower Bid} & -1\ 6\ 8\ 5 \\
\hline
\text{Difference} & \$\quad 1\ 4\ 2 \quad \text{difference}
\end{array}
$$

Answers to Marginal Exercises

1. 233   **2.** 1511   **3.** 1344   **4.** 539   **5.** 38   **6.** 288   **7.** 3858   **8.** 54

Answers to Practice Problems

1. 5   **2.** 6   **3.** 0   **4.** 7   **5.** 9   **6.** 4   **7.** 2   **8.** 2   **9.** 4   **10.** 9
11. 7   **12.** 7   **13.** 6   **14.** 9   **15.** 7   **16.** 9   **17.** 9   **18.** 9   **19.** 9
20. 5   **21.** 9   **22.** 3   **23.** 5   **24.** 8   **25.** 4   **26.** 5   **27.** 8   **28.** 8

NAME _____ SECTION _____ DATE _____

EXERCISES **1.3** _____

Find the following differences.

1. 15
    − 9

2. 14
    − 6

3. 12
    − 8

4. 15
    − 7

5. 14
    − 9

6. 12
    − 6

7. 6
    −5

8. 7
    −7

9. 8
    −0

10. 14 − 8

11. 5 − 5

12. 15 − 7

13. 16 − 9

14. 10 − 9

15. 14 − 5

16. 15 − 6

17. 17 − 9

18. 12 − 3

19. 7 − 0

Answers

1. _____6_____
2. _____8_____
3. _____4_____
4. _____8_____
5. _____5_____
6. _____6_____
7. _____1_____
8. _____0_____
9. _____8_____
10. _____6_____
11. _____0_____
12. _____5_____
13. _____7_____
14. _____1_____
15. _____9_____
16. _____9_____
17. _____8_____
18. _____9_____
19. _____7_____

Answers

20. _0_
21. _11_
22. _13_
23. _20_
24. _69_
25. _5_
26. _54_
27. _126_
28. _218_
29. _376_
30. _593_
31. _395_
32. _1569_
33. _966_
34. _1568_
35. _1424_
36. _694_
37. _5871_
38. _2806644_
39. _3800559_
40. _7352439_
41. _7671011_
42. _140_
43. _398_
44. _559_
45. _815_

Subtract. Use expanded notation only if you find it helpful.

20. $\begin{array}{r} 17 \\ -17 \end{array}$	21. $\begin{array}{r} 42 \\ -31 \end{array}$	22. $\begin{array}{r} 89 \\ -76 \end{array}$	23. $\begin{array}{r} 53 \\ -33 \end{array}$
24. $\begin{array}{r} 96 \\ -27 \end{array}$	25. $\begin{array}{r} 23 \\ -18 \end{array}$	26. $\begin{array}{r} 126 \\ -\ 32 \end{array}$	27. $\begin{array}{r} 174 \\ -\ 48 \end{array}$
28. $\begin{array}{r} 347 \\ -129 \end{array}$	29. $\begin{array}{r} 543 \\ -167 \end{array}$	30. $\begin{array}{r} 900 \\ -307 \end{array}$	31. $\begin{array}{r} 603 \\ -208 \end{array}$

32. $\begin{array}{r} 7843 \\ -6274 \end{array}$	33. $\begin{array}{r} 6793 \\ -5827 \end{array}$	34. $\begin{array}{r} 4376 \\ -2808 \end{array}$
35. $\begin{array}{r} 4900 \\ -3476 \end{array}$	36. $\begin{array}{r} 5070 \\ -4376 \end{array}$	37. $\begin{array}{r} 8007 \\ -2136 \end{array}$
38. $\begin{array}{r} 7,085,076 \\ -4,278,432 \end{array}$	39. $\begin{array}{r} 6,543,222 \\ -2,742,663 \end{array}$	40. $\begin{array}{r} 8,000,000 \\ -\ \ 647,561 \end{array}$

41. $\begin{array}{r} 6,000,000 \\ -\ \ 328,989 \end{array}$

42. What number should be added to 860 to get a sum of 1000?

43. If the sum of two numbers is 537 and one of the numbers is 139, what is the other number?

44. If you had $834 in your checking account and you wrote a check for $275, what would be the balance?

45. The selling price of a refrigerator is $1200. What is still owed if a down payment of $385 is made?

1.4
ROUNDING OFF AND ESTIMATING

Rounding Off a given number means finding another number close to the given number. The desired place of accuracy must be stated.

Rounded-off answers are often acceptable. For example, we might say that the distance across the United States from east coast to west coast is 3000 miles. This number is rounded off and approximate and, with this understood, acceptable in most discussions.

Numbers resulting from measurement can only be approximate and, therefore, involve some form of rounding off. There are several methods for rounding off. For example, in retail stores, the price per unit is always raised to the next higher cent. In this text, we will use a method based on looking at the digit just to the right of the desired place of accuracy.

We begin by using number lines as visual aids in understanding the rounding-off process.

Example 1   **Round** 37 to the nearest ten.

Solution

A number line gives a visual reference.

37 is closer to 40 than to 30. So, 37 rounds off to 40 (to the nearest ten).

Example 2   **Round** 253 to the nearest ten.

Solution

Use a number line.

253 is closer to 250 than to 260. So, 253 rounds off to 250 (to the nearest ten).

Round off each number to the place indicated. Draw a number line as an aid.

1. 57 (nearest ten)

60

2. 345 (nearest hundred)

300

3. 2345 (nearest thousand)

2000

Example 3   **Round** 278 to the nearest hundred.

Solution

Again, a number line is helpful.

278 is closer to 300 than to 200. So, 278 rounds off to 300 (to the nearest hundred).

NOW WORK EXERCISES 1–3 IN THE MARGIN.

The following rule serves in place of a number line.

> ROUNDING-OFF RULE
>
> 1. Look at the single digit just to the right of the place of desired accuracy.
> 2. If this digit is 5 or greater, make the digit in the desired place of accuracy one larger and replace all digits to the right with zeros. All digits to the left remain unchanged.
> 3. If this digit is less than 5, leave the digit in the desired place of accuracy as it is and replace all digits to the right with zeros. All digits to the left remain unchanged.

Example 4   **Round** 5749 to the nearest hundred.

Solution   5749   5749   5700
                  ↑   ↑   ↑
            place of desired   One digit to the right;   Leave 7 and
            accuracy   4 is less than 5.   fill in zeros.

So, 5749 rounds off to 5700 (to the nearest hundred).

Example 5   **Round** 6500 to the nearest thousand.

Solution   6500   6500   7000
                  ↑   ↑   ↑
            place of desired   Look at 5; 5 is 5 or   Increase 6 to 7
            accuracy   greater.   (one larger) and
                                                     fill in zeros.

So, 6500 rounds off to 7000 (to the nearest thousand).

Example 6 **Round** 397 to the nearest ten.

Solution   397   397   400

          ↑   ↑   ↑

   place of desired   Look at 7; 7 is 5   Increase 9 to 10
   accuracy   or greater.   (this affects two
                                                 digits, both 3 and 9).

So, 397 rounds off to 400 (to nearest ten).

☐ **NOW WORK EXERCISES 4–10 IN THE MARGIN.**

One very good use for rounded-off numbers is to **estimate** an answer before any calculations are actually made. Thus, answers that are not at all reasonable, due to pushing a wrong button on a calculator or some other large calculator error, can be spotted. Usually, we simply repeat the calculations and the error is found. There are also times, particularly in working word problems, when the wrong operation has been performed. For example, someone may have subtracted when they should have added.

> **Estimating an answer** means to use rounded-off numbers in a calculation to get some idea of what the size of the actual answer should be. This is a form of checking your work before you do it.

> To estimate a sum (or difference):
> 1. Round off each number to the place of the last digit on the left.
> 2. Add (or subtract) the rounded-off numbers.

Use the **Rounding-Off Rule** to round off the following numbers as indicated.

4. 576 (nearest ten)

580

5. 839 (nearest hundred)

800

6. 1500 (nearest thousand)

2000

7. 2589 (nearest hundred)

2600

8. 43,610 (nearest thousand)

44,000

9. 1983 (nearest hundred)

2000

10. 1938 (nearest hundred)

1900

Example 7   Add: 568
                       934
                      +712

Solution

(a) Estimate the sum first by rounding off each number to the nearest hundred, then adding. In actual practice, many of these steps can be done mentally.

$$
\begin{array}{rcr}
568 & \rightarrow & 600 \\
934 & \rightarrow & 900 \\
+712 & \rightarrow & +\ 700 \\
\hline
 & & 2200
\end{array}
$$

(b) Now, we find the sum with the knowledge that the answer should be close to 2200.

$$
\begin{array}{r}
11 \\
568 \\
934 \\
+\ 712 \\
\hline
2214
\end{array}
$$

10 → 934   10

This sum is quite close to 2200.

Example 8   Subtract: 658
                           −189

Solution

(a) Estimate the difference first by rounding off each number to the nearest hundred, then subtracting.

$$
\begin{array}{rcr}
658 & \rightarrow & 700 \\
-189 & \rightarrow & -200 \\
\hline
 & & 500
\end{array}
$$

(b) Now we find the difference with the knowledge that the answer should be approximately 500.

$$
\begin{array}{r}
1 \\
5\quad 4\ 1 \\
\cancel{6}\ \cancel{5}\ 8 \\
-1\ 8\ 9 \\
\hline
4\ 6\ 9
\end{array}
$$

This difference rounds off to 500, the **estimated** value.

Completion Example 9 Add: 5483
                                  2321
                                 +6857

Solution

(a) Estimate the sum by rounding off each number to the nearest thousand, then adding.

$$
\begin{array}{rr}
5483 & 5000 \\
2321 & 2000 \\
+6857 & +7000 \\
\hline
 & 14000 \quad \leftarrow \text{estimated sum}
\end{array}
$$

(b) Find the sum and compare your answer with the estimated sum. Are they close?

$$
\begin{array}{r}
5483 \\
2321 \\
+6857 \\
\hline
14661
\end{array}
$$

Estimating answers is particularly useful when problems involve multiplication and/or division. In any case, major errors are easily detected if you have some idea of a reasonable answer before performing the actual calculations.

Completion Example Answer

9. (a) 5000   (b)   5483
         2000   2321
      + 7000   + 6857
      ─────────   ───────
       14,000 ← estimated sum   14,661

   The numbers are close.

Answers to Marginal Exercises

1.
   Answer: 60

2.
   Answer: 300

3.
   Answer: 2000

4. 580   5. 800

6. 2000   7. 2600   8. 44,000   9. 2000   10. 1900

NAME _____ SECTION _____ DATE _____

EXERCISES **1.4** _____

Round off as indicated.

To the nearest ten:

1. 763	**2.** 31	**3.** 82	**4.** 503
5. 296	**6.** 722	**7.** 987	**8.** 347

To the nearest hundred:

9. 4163	**10.** 4475	**11.** 495	**12.** 572
13. 637	**14.** 3789	**15.** 76,523	**16.** 7007

To the nearest thousand:

17. 6912	**18.** 5500	**19.** 7500	**20.** 7499
21. 13,499	**22.** 13,501	**23.** 62,265	**24.** 47,800

To the nearest ten thousand:

25. 78,419	**26.** 125,000	**27.** 256,000	**28.** 62,200
29. 118,200	**30.** 312,500	**31.** 184,900	**32.** 615,000

Answers

1. 760   2. 30
3. 80   4. 500
5. 300   6. 700
7. 990   8. 350
9. 4200   10. 4500
11. 500   12. 570
13. 600   14. 3800
15. 76,500   16. 7000
17. 7000   18. 6000
19. 8000   20. 7000
21. 13,000   22. 14,000
23. 62,000   24. 48,000
25. 80,000
26. 130,000
27. 260,000
28. 60,000
29. 120,000
30. 310,000
31. 180,000
32. 620,000

Answers

33. __100__

34. __0__

35. __1000__

36. __1__

37. __12__

38. __11__

39. __27__

40. __25__

41. __19__

42. __141__

43. __276__

44. __1,671__

45. __333__

46. __231__

47. __931__

48. __102,600__

49. __804__

50. __1,223__

33. 87 to the nearest hundred

34. 46 to the nearest hundred

35. 532 to the nearest thousand

First estimate the answers using rounded-off numbers, then find the following sums and differences.

36.   83 = 80
          62 = 60
        +78 = 80
        223 220

37.   97 = 100
          46 = 50
        +25 = 30
        168 180

38.   146 = 100
          259 = 300
        +384 = 400
        789   800

39.   475 = 500
          126 = 100
        +572 = 600
        1173 1200
              1173
               27

40.   600 = 600
          542 = 500
        +483 = 500
        1625 1600

41.   851 = 900
          736 = 700
        +294 = 300
        1881 1900
              1881
               19

42.   5742 = 6000
          6271 = 6000
          8156 = 8000
        + 972 = 1000
        21141 21000

43.   483 = 500
          1681 = 2000
          3054 = 3000
        +4006 = 4000
        9224   9500
               224
               276

44.   22,506 = 20,000
          38,700 = 40,000
        +10,465 = 10,000
        71,671   70,000

45.   8742 9000
         −3275 3000
         5467   6000
               5467
                333

46.   6421 = 7000
         −1652 = 2000
         4769   5000
                4769
                231

47.   10,531 = 10,000
        − 4,600   5,000
         5931   5000

48.   275,600 = 300,000
        − 94,300 = 90,000
         181300   210000
                  181 300
                  18,700

49.   63,504
        −42,700

50.   74,305
        −33,082

1.5
BASIC MULTIPLICATION AND POWERS OF 10 ____

Multiplication: Repeated addition

Product: Result of multiplication

Factors: Numbers that are multiplied (Each number multiplied is a **factor** of the product.)

SYMBOLS THAT INDICATE MULTIPLICATION		
·	raised dot	$4 \cdot 7$
×	cross sign	5×6
( )	parentheses	$6(8)$ or $(6)8$ or $(6)(8)$

Example 1　$6 + 6 + 6 + 6 =$　$\underset{\text{factor}}{4}$　$\cdot$　$\underset{\text{factor}}{6}$　$=$　$\underset{\text{product}}{24}$

4 and 6 are factors of 24.

Example 2　$9 + 9 + 9 + 9 + 9 =$　$\underset{\text{factor factor}}{5(9)}$　$=$　$\underset{\text{product}}{45}$

9 is used 5 times in repeated addition. The numbers can be written vertically as

$$
\begin{array}{r}
9 \quad \leftarrow \text{factor} \\
\times\ 5 \quad \leftarrow \text{factor} \\
\hline
45 \quad \leftarrow \text{product}
\end{array}
$$

Memorize the basic multiplication facts in Table 1.3. These basic multiplication facts are very important in mathematics. You **must** know all of them to be successful. Practice using these facts. You might try making flash cards to use at home. Write the various combinations of two digits on one side of a card and the answer on the other side. Then get someone to hold the cards up one at a time. Your helper can check your answer from the back of the card as you call it out.

TABLE 1.3 Basic Multiplication Facts

·	0	1	2	3	4	5	6	7	8	9
0	0	0	0	0	0	0	0	0	0	0
1	0	1	2	3	4	5	6	7	8	9
2	0	2	4	6	8	10	12	14	16	18
3	0	3	6	9	12	15	18	21	24	27
4	0	4	8	12	16	20	24	28	32	36
5	0	5	10	15	20	25	30	35	40	45
6	0	6	12	18	24	30	36	42	48	54
7	0	7	14	21	28	35	42	49	56	63
8	0	8	16	24	32	40	48	56	64	72
9	0	9	18	27	36	45	54	63	72	81

Note in Table 1.3 that 0 times any number is 0.

ZERO PROPERTY OF MULTIPLICATION

For any whole number a, $0 \cdot a = 0$

☐ **NOW WORK PRACTICE PROBLEMS 1–55.**

PRACTICE PROBLEMS

Find the following products mentally. Work as quickly as you can. If you make any errors (or even hesitate in writing an answer), practice those problems until you have them memorized.

1. 1   2. 4   3. 7   4. 0   5. 2   6. 5   7. 3
   ×3   ×6   ×9   ×1   ×8   ×5   ×4
   3   *24*   *63*   *0*   *16*   *25*   *12*

8. 6   9. 8   10. 0   11. 8   12. 2   13. 5   14. 9
   ×7   ×8   ×3   ×1   ×2   ×0   ×3
   42   *64*   *0*   *8*   *4*   *0*   *27*

15. 0   16. 6   17. 7   18. 4   19. 7   20. 8   21. 7
   ×0   ×3   ×5   ×2   ×7   ×4   ×0
   0   *18*   *35*   *8*   *49*   *32*   *0*

22. 6   23. 8   24. 9   25. 4   26. 2   27. 1   28. 4
   ×1   ×3   ×9   ×7   ×0   ×4   ×9
   6   *24*   *81*   *28*   *0*   *4*   *36*

29. 8   30. 6   31. 4   32. 8   33. 3   34. 2   35. 0
   ×5   ×6   ×4   ×0   ×3   ×6   ×9
   40   *36*   *16*   *0*   *9*   *12*   *0*

36. 9   37. 8   38. 1   39. 3   40. 5   41. 9   42. 1
   ×6   ×7   ×1   ×2   ×4   ×8   ×2
   54   *56*   *1*   *6*   *20*   *72*   *2*

43.	44.	45.	46.	47.	48.	49.
3	5	9	2	5	6	8
×5	×6	×1	×9	×1	×0	×6
15	*30*	*9*	*18*	*5*	*0*	*48*

50.	51.	52.	53.	54.	55.
3	5	0	7	9	2
×7	×2	×4	×1	×5	×7
21	*10*	*0*	*7*	*45*	*14*

As with addition, there are three basic properties of multiplication: The **commutative property** (dealing with the order of the numbers), the **associative property** (dealing with grouping), and the **multiplicative identity** (1).

> **COMMUTATIVE PROPERTY OF MULTIPLICATION**
>
> For whole numbers a and b, $a \cdot b = b \cdot a$

Example 3 $6 \cdot 3 = 18$ and $3 \cdot 6 = 18$

Thus, $6 \cdot 3 = 3 \cdot 6$.

Example 4

$$\begin{array}{cc} 9 & 7 \\ \times\ 7 & \times\ 9 \\ \hline 63 & 63 \end{array}$$

Thus, $9 \times 7 = 7 \times 9$.

☐ **NOW WORK EXERCISES 1 AND 2 IN THE MARGIN.**

> **ASSOCIATIVE PROPERTY OF MULTIPLICATION**
>
> For whole numbers a, b, and c,
>
> $$a \cdot b \cdot c = a(b \cdot c) = (a \cdot b)c$$

Example 5

$$6 \cdot 2 \cdot 3 \qquad 6 \cdot 2 \cdot 3$$
$$= 12 \cdot 3 \qquad = 6 \cdot 6$$
$$= 36 \qquad = 36$$

Thus, $(6 \cdot 2)3 = 6(2 \cdot 3)$.

Example 6 $7 \cdot 4 \cdot 1 = (7 \cdot 4)1 = 28 \cdot 1 = 28$
$7 \cdot 4 \cdot 1 = 7(4 \cdot 1) = 7 \cdot 4 = 28$

☐ **NOW WORK EXERCISES 3 AND 4 IN THE MARGIN.**

For Exercises 1 and 2, find both products and tell what property is illustrated.

1. $8 \times 7 =$ _*56*_

   $7 \times 8 =$ _*56*_

   COMMUTATIVE

2. $7 \cdot 5 =$ _*35*_

   $5 \cdot 7 =$ _*35*_

   COMMUTATIVE

For Exercises 3 and 4, find both products and tell what property is illustrated.

3. $(2 \cdot 3)7 =$ _*42*_

   $2(3 \cdot 7) =$ _*42*_

   ASSOCIATIVE

4. $4(4 \cdot 2) =$ _*32*_

   $(4 \cdot 4)2 =$ _*32*_

In Examples 7–9, find each product using the associative property of multiplication.

Completion Example 7 $5 \cdot 1 \cdot 8 =$ _____40_____

Completion Example 8 $2 \cdot 4 \cdot 6 =$ _____48_____

Completion Example 9 $(3)(0)(5) =$ _____0_____

> **1 IS THE MULTIPLICATIVE IDENTITY**
>
> For any whole number a, $a \cdot 1 = a$

Example 10 $8 \cdot 1 = 8$

Example 11
$$\begin{array}{r} 6 \\ \times 1 \\ \hline 6 \end{array}$$

In Examples 12–15, find the products and tell what property of multiplication is illustrated.

Completion Example 12 $1(2 \cdot 6) = (1 \cdot 2)6 =$ _____

Completion Example 13 $5 \cdot 9 = 9 \cdot 5 =$ _____

Completion Example Answers
  7. 40
  8. 48
  9. 0
 12. 12, associative property of multiplication
 13. 45, commutative property of multiplication

Completion Example 14 $6 \cdot 1 = $ _____

Completion Example 15 $0 \cdot 4 = $ _____

☐ **NOW WORK EXERCISES 5–8 IN THE MARGIN.**

▌**Powers of 10:** 1, 10, 100, 1000, 10,000, 100,000, 1,000,000 and so on

POWERS OF 10

$$1$$
$$10$$
$$10 \cdot 10 = 100$$
$$10 \cdot 10 \cdot 10 = 1000$$
$$10 \cdot 10 \cdot 10 \cdot 10 = 10,000$$
$$10 \cdot 10 \cdot 10 \cdot 10 \cdot 10 = 100,000$$

and so on.

Multiplication by powers of 10 is useful in explaining multiplication with whole numbers in general, as we will see in the next section. Such multiplication should be done mentally.

> To multiply a whole number
>
> by 10, write 0 to the right,
> by 100, write 00 to the right,
> by 1000, write 000 to the right,
> by 10,000, write 0000 to the right,
> and so on.

Example 16 $6 \cdot 1 = 6$

Example 17 $6 \cdot 10 = 60$

Show that each of the following statements is true by doing the multiplication both ways vertically.

5. $9 \cdot 1 = 1 \cdot 9$

$$\begin{array}{r} 9 \\ \times 1 \\ \hline \end{array} \qquad \begin{array}{r} 1 \\ \times 9 \\ \hline \end{array}$$

6. $0 \cdot 3 = 3 \cdot 0$

$$\begin{array}{r} 0 \\ \times 3 \\ \hline \end{array} \qquad \begin{array}{r} 3 \\ \times 0 \\ \hline \end{array}$$

7. $5(7) = 7(5)$

$$\begin{array}{r} 5 \\ \times 7 \\ \hline \end{array} \qquad \begin{array}{r} 7 \\ \times 5 \\ \hline \end{array}$$

8. $8 \cdot 9 = 9 \cdot 8$

$$\begin{array}{r} 8 \\ \times 9 \\ \hline \end{array} \qquad \begin{array}{r} 9 \\ \times 8 \\ \hline \end{array}$$

Completion Example Answers

14. 6, multiplicative identity

15. 0, zero property of multiplication

Find the following products.

9. $8 \cdot 1000 =$ _____

10. $15 \cdot 100 =$ _____

11. $100 \cdot 25 =$ _____

12. $10 \cdot 300 =$ _____

13. $27 \cdot 10,000 =$ _____

Example 18 $6 \cdot 100 = 600$

Example 19 $6 \cdot 1000 = 6000$

Example 20 $6 \cdot 10,000 = 60,000$

Example 21 $93 \cdot 10 = 930$

Example 22 $17 \cdot 1000 = 17,000$

Example 23 $262 \cdot 100 = 26,200$

☐ **NOW WORK EXERCISES 9–13 IN THE MARGIN.**

| A number ending with 0's has a power of 10 as a factor.

Example 24 $90 = 9 \quad \cdot \quad 10$
$$\uparrow \qquad \uparrow$$
$$\text{factor} \quad \text{factor}$$

Example 25 $500 = 5 \quad \cdot \quad 100$
$$\uparrow \qquad \uparrow$$
$$\text{factor} \quad \text{factor}$$

Example 26 $36,000 = 36 \quad \cdot \quad 1000$
$$\uparrow \qquad \uparrow$$
$$\text{factor} \quad \text{factor}$$

Write each number in Examples 27–29 as a product with a power of ten as one factor.

Completion Example 27 $800 = 8 \cdot$ _____

Completion Example 28 $2000 =$ _____ $\cdot 1000$

Completion Example 29 $6500 = 65 \cdot$ _____

☐ **NOW WORK EXERCISES 14–18 IN THE MARGIN.**

> To multiply two or more whole numbers that end with 0's,
> 1. multiply the nonzero digits,
> 2. count the zeros,
> 3. write the counted number of zeros at the end.

Example 30 $6 \cdot 90 = 6(9 \cdot 10) = (6 \cdot 9)10 = 54 \cdot 10 = 540$

Example 31 $3 \cdot 400 = 3(4 \cdot 100) = (3 \cdot 4)100 = 12 \cdot 100 = 1200$

Example 32 $2 \cdot 300 = (2 \cdot 3)100 = 600$

Example 33

   $50 \cdot 700 = (5 \cdot 10)(7 \cdot 100) = (5 \cdot 7)(10 \cdot 100) = 35 \cdot 1000 = 35,000$

Example 34 $200 \cdot 800 = (2 \cdot 8)(100 \cdot 100) = 16 \cdot 10,000 = 160,000$

Write each number as a product with a power of ten as one factor.

14. $5000 =$ _____

15. $300 =$ _____

16. $70 =$ _____

17. $630 =$ _____

18. $190,000 =$ _____

Completion Example Answers

27. $8 \cdot 100$ **28.** $2 \cdot 1000$ **29.** $65 \cdot 100$

Find each of the following products mentally.

19. $19 \cdot 10 = $ _____

20. $25 \cdot 10 = $ _____

21. $14 \cdot 100 = $ _____

22. $37 \cdot 100 = $ _____

23. $80 \cdot 60 = $ _____

24. $70 \cdot 30 = $ _____

25. $6 \cdot 300 = $ _____

26. $5 \cdot 7000 = $ _____

27. $200 \cdot 30 = $ _____

28. $40 \cdot 600 = $ _____

Note: Some intermediate steps are shown here to help clarify the technique. With practice, you should be able to find the products mentally.

Completion Example 35 $50 \cdot 60 = 5 \cdot 6 \cdot 100 = 30 \cdot 100 = $ _____

Completion Example 36 $80 \cdot 90 = 8 \cdot 9 \cdot $ _____ $= 72 \cdot $ _____ $= $ _____

☐ **NOW WORK EXERCISES 19–28 IN THE MARGIN.**

Completion Example Answers

35. 3000

36. $8 \cdot 9 \cdot 100 = 72 \cdot 100 = 7200$

Answers to Marginal Exercises

1. 56, 56; commutative property of multiplication   **2.** 35, 35; commutative property of multiplication   **3.** 42, 42; associative property of multiplication   **4.** 32, 32; associative property of multiplication   **5.** 9, 9   **6.** 0, 0   **7.** 35, 35   **8.** 72, 72
9. 8000   **10.** 1500   **11.** 2500   **12.** 3000   **13.** 270,000   **14.** $5 \cdot 1000$
15. $3 \cdot 100$ **16.** $7 \cdot 10$ **17.** $63 \cdot 10$ **18.** $19 \cdot 10,000$ **19.** 190   **20.** 250
21. 1400   **22.** 3700   **23.** 4800   **24.** 2100   **25.** 1800   **26.** 35,000
27. 6000   **28.** 24,000

Answers to Practice Problems

1. 3   **2.** 24   **3.** 63   **4.** 0   **5.** 16   **6.** 25   **7.** 12   **8.** 42   **9.** 64   **10.** 0
11. 8   **12.** 4   **13.** 0   **14.** 27   **15.** 0   **16.** 18   **17.** 35   **18.** 8   **19.** 49
20. 32   **21.** 0   **22.** 6   **23.** 24   **24.** 81   **25.** 28   **26.** 0   **27.** 4   **28.** 36
29. 40   **30.** 36   **31.** 16   **32.** 0   **33.** 9   **34.** 12   **35.** 0   **36.** 54   **37.** 56
38. 1   **39.** 6   **40.** 20   **41.** 72   **42.** 2   **43.** 15   **44.** 30   **45.** 9   **46.** 18
47. 5   **48.** 0   **49.** 48   **50.** 21   **51.** 10   **52.** 0   **53.** 7   **54.** 45   **55.** 14

NAME _____ SECTION _____ DATE _____

EXERCISES **1.5** _____

Do the following problems mentally and write only the answers.

1. $8 \cdot 9$ **2.** $7 \cdot 6$ **3.** $8(6)$ **4.** $6 \cdot 5$

5. $5 \cdot 9$ **6.** $3(9)$ **7.** $6(4)$ **8.** $7(4)$

9. $8(7)$ **10.** $0(3)$ **11.** $5(0)$ **12.** $1 \cdot 9$

13. $\begin{array}{r} 3 \\ \times 8 \\ \hline \end{array}$ **14.** $\begin{array}{r} 9 \\ \times 6 \\ \hline \end{array}$ **15.** $\begin{array}{r} 8 \\ \times 4 \\ \hline \end{array}$ **16.** $\begin{array}{r} 7 \\ \times 9 \\ \hline \end{array}$

17. $1 \cdot 1 \cdot 1 \cdot 1 \cdot 1 \cdot 1$ **18.** $1(2)(3)(5)(0)$

Find each product and tell what property of multiplication is illustrated.

19. $9 \cdot 3 = 3 \cdot 9$ **20.** $2(1 \cdot 6) = (2 \cdot 1)6$

21. $1 \cdot 7$ **22.** $8 \cdot 0$

23. Complete the table.   **24.** Complete the table.

$\cdot$	5	8	7	9
3				
6				
5				
2				

$\cdot$	4	1	9	6
5				
9				
6				
0				

Answers

1. _____ 2. _____

3. _____ 4. _____

5. _____ 6. _____

7. _____ 8. _____

9. _____ 10. _____

11. _____ 12. _____

13. _____ 14. _____

15. _____ 16. _____

17. _____ 18. _____

19. _____

20. _____

21. _____

22. _____

23. _____

24. _____

Answers

25. _____ 26. _____

27. _____ 28. _____

29. _____ 30. _____

31. _____ 32. _____

33. _____ 34. _____

35. _____ 36. _____

37. _____ 38. _____

39. _____ 40. _____

41. _____ 42. _____

43. _____ 44. _____

45. _____ 46. _____

47. _____ 48. _____

49. _____

50. _____

51. _____

52. _____

53. _____

54. _____

55. _____

56. _____

57. _____

58. _____

Write each number as a product with a power of 10 as one factor.

25. 80   **26.** 700   **27.** 6000   **28.** 3000

29. 5400   **30.** 63,000   **31.** 7,000,000

Use the technique of multiplying by the powers of 10 to find the following products mentally.

32. $25 \cdot 10$ **33.** $47 \cdot 1000$ **34.** $18 \cdot 10$

35. $13 \cdot 1$ **36.** $50 \cdot 60$ **37.** $20 \cdot 20$

38. $60 \cdot 60$ **39.** $70 \cdot 80$ **40.** $90 \cdot 70$

41. $200 \cdot 20$ **42.** $500 \cdot 700$ **43.** $120 \cdot 30$

44. $130 \cdot 140$ **45.** $200 \cdot 80$ **46.** $300 \cdot 600$

47. $100 \cdot 50$ **48.** $50 \cdot 200$ **49.** $2000 \cdot 400$

50. $4000 \cdot 4000$ **51.** $800 \cdot 4000$ **52.** $900 \cdot 3000$

53.   30   **54.**   40   **55.**   6000
      $\times\ 8$ $\times\ 9$ $\times\quad 6$

56.   90   **57.**   300   **58.**   700
      $\times 90$ $\times 500$ $\times\ 70$

1.6
MULTIPLICATION

We first illustrate multiplication by using expanded notation and our skills with multiplication by powers of 10. This technique helps in developing an understanding of the more familiar multiplication process that we will call the short method.

MULTIPLYING WHOLE NUMBERS USING PARTIAL PRODUCTS

1. Write the numbers vertically.
2. Write each number in expanded notation.
3. Find each partial product by using your knowledge of multiplying by powers of 10.
4. Add the partial products.

Example 1 Multiply $4 \cdot 68$.

$$
\begin{array}{r}
68 \\
\times\ 4 \\
\hline
\end{array}
$$

$$
\begin{array}{r}
60 + 8 \\
4 \\
\hline
240 + 32 = 272 \\
\end{array}
$$
partial products   product

or

$$
\begin{array}{r}
68 \\
4 \\
\hline
32 \quad \leftarrow 4 \cdot 8 \\
240 \quad \leftarrow 4 \cdot 60 \\
\hline
272 \quad \leftarrow \text{product}
\end{array}
$$
partial products

Example 2 Multiply $6 \cdot 39$.

$$
\begin{array}{r}
39 \\
\times\ 6 \\
\hline
\end{array}
$$

$$
\begin{array}{r}
30 + 9 \\
6 \\
\hline
180 + 54 = 234 \\
\end{array}
$$
partial products   product

or

$$
\begin{array}{r}
39 \\
6 \\
\hline
54 \quad \leftarrow 6 \cdot 9 \\
180 \quad \leftarrow 6 \cdot 30 \\
\hline
234 \quad \leftarrow \text{product}
\end{array}
$$
partial products

Example 3 Multiply $37 \cdot 42$.

$$
\begin{array}{r}
42 \\
\times 37 \\
\hline
\end{array}
$$

$$
\begin{array}{r}
40 + 2 \\
30 + 7 \\
\hline
280 + 14 \\
1200 + 60 \\
\hline
1200 + 340 + 14 = 1554
\end{array}
$$

or

$$
\begin{array}{rl}
42 & \text{factor} \\
37 & \text{factor} \\
\hline
14 & (7 \cdot 2 = 14) \\
280 & (7 \cdot 40 = 280) \\
60 & (30 \cdot 2 = 60) \\
1200 & (30 \cdot 40 = 1200) \\
\hline
1554 & \text{product}
\end{array}
$$

Find each product using partial products.

1.   256
       × 7

2.   83
       ×49

3.   372
       × 64

Find the missing partial products in Examples 4 and 5; then add the partial products to find the product.

Completion Example 4   468   400   +   60   +   8
                           × 4   _____
                                                              4
                                    _____ + _____ + 32 = _____
                                    _____ partial products _____/   ___ product ___/

Completion Example 5   75
                             ×93
                             ____
                               15   ← 3·5   ⎫
                             ____   ← 3·70   ⎪
                                              ⎬ partial products
                             ____   ← 90·5   ⎪
                             _____ ← 90·70 ⎭
                             _____
                             _____ ← product

☐ **NOW WORK EXERCISES 1–3 IN THE MARGIN.**

The method of partial products is recommended only as an aid to understanding the faster **short method** that we use most of the time.

> **SHORT METHOD**
>
> Faster and recommended most of the time
>
>   1. Digits are carried to be added to the next partial product.
>   2. Sums of some partial products are found mentally.
>
> (Study Examples 6 and 7 carefully for understanding.)

Example 6

    WRITING PARTIAL PRODUCTS   *SHORT METHOD*

   5 6   1   ← 1 carried from 18
×   3   5 6
 1 8 (3·6 = 18)   ×   3
15 0 (3·50 = 150) ⑯8   From 6·3 = 18, write 8 below the
⑯8 product   3 and carry 10 by writing 1 above
                       the 5. Then multiply 3·5 and
                       add 1: 3·5 = 15
                            15 + 1 = 16

Completion Example Answers

4. 1600 + 240 + 32 = 1872

5.   75
           ×93
           ____
             15
            210
            450
           6300
           ____
           6975

Example 7

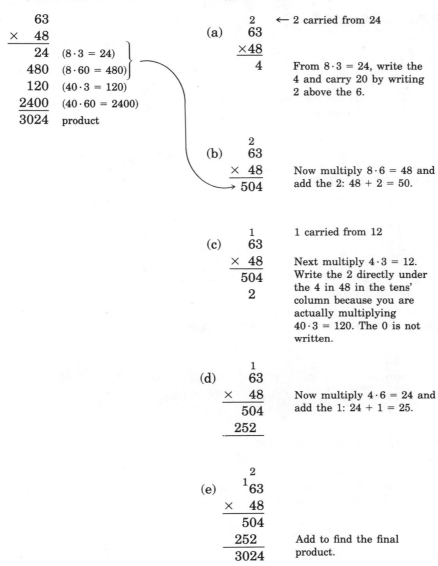

WRITING PARTIAL PRODUCTS   SHORT METHOD

```
      63
   ×  48
      24   (8·3 = 24)
     480   (8·60 = 480)
     120   (40·3 = 120)
    2400   (40·60 = 2400)
    3024   product
```

(a)
```
       2      ← 2 carried from 24
      63
     ×48
       4
```
From 8·3 = 24, write the 4 and carry 20 by writing 2 above the 6.

(b)
```
       2
      63
    × 48
     504
```
Now multiply 8·6 = 48 and add the 2: 48 + 2 = 50.

(c)
```
       1      1 carried from 12
      63
    × 48
     504
       2
```
Next multiply 4·3 = 12. Write the 2 directly under the 4 in 48 in the tens' column because you are actually multiplying 40·3 = 120. The 0 is not written.

(d)
```
       1
      63
    × 48
     504
     252
```
Now multiply 4·6 = 24 and add the 1: 24 + 1 = 25.

(e)
```
       2
      ¹63
    × 48
     504
     252
    3024
```
Add to find the final product.

Steps (a), (b), (c), and (d) are shown for clarification of the Short Method. You will write only Step (e).

Find each product using the Short Method.

4.   142
       × 6

5.   37
       × 85

6.   273
       × 24

In Examples 8 and 9, tell what product or sum gives the circled digits.

Completion Example 8

$$
\begin{array}{r}
{}_1{}^3 \\
4\,5 \\
\times\ 2\,7 \\
\hline
③①5 \\
⑨0 \\
\hline
121\,5
\end{array}
$$

31 comes from $7 \cdot 4 = 28$ and $28 + 3 = 31$

9 comes from _____

Completion Example 9

$$
\begin{array}{r}
{}_1{}^4 \\
5\,6 \\
\times\ 3\,8 \\
\hline
④④⑧ \\
①⑥8 \\
\hline
212\,8
\end{array}
$$

8 is from the product $8 \cdot 6 = 48$

44 comes from _____

16 comes from _____

☐ **NOW WORK EXERCISES 4–6 IN THE MARGIN.**

> To multiply by a number that ends with 0's, write the 0's to the right, then multiply as shown in Examples 10, 11, and 12.

Example 10   Multiply $423 \cdot 2000$.

$$
\begin{array}{r}
423 \\
\times\ 2000 \\
\hline
846,\!000
\end{array}
$$

We know $2000 = 2 \cdot 1000$. We multiply $423 \cdot 2$, then multiply the result by 1000.

Example 11   Multiply $596 \cdot 3000$.

$$
\begin{array}{r}
596 \\
\times\ 3000 \\
\hline
1,\!788,\!000
\end{array}
$$

Completion Example Answer

8. 9 comes from $2 \cdot 4 = 8$ and $8 + 1 = 9$

9. 44 comes from $8 \cdot 5 = 40$ and
$40 + 4 = 44$

16 comes from $3 \cdot 5 = 15$ and
$15 + 1 = 16$

Example 12 Multiply $265 \cdot 1500$.

```
        265
×      1500
    132 500
    265
    397,500
```

☐ **NOW WORK EXERCISES 7–9 IN THE MARGIN.**

Estimating Products _____

Estimating the product of two numbers before actually performing the multiplication can be of help in detecting errors. Just as with addition and subtraction, estimations are done with rounded-off numbers.

> To estimate a product:
>
> 1. round off each number to the place of the last digit on the left.
> 2. multiply the rounded-off numbers using your knowledge of multiplying by powers of 10.

Example 13 Find the product $52 \cdot 38$, but first estimate the product by multiplying the numbers in rounded-off form.

Solution

(a)
```
  52   →        50     rounded-off value of 52
 ×38   →    ×   40     rounded-off value of 38
             2000      The actual product should be close to 2000.
```

(b)
```
     1
     52
 ×   38
    416
    156
   1976
```

Find the following products.

7.
```
     62
 ×  800
```

8.
```
      47
 ×  1300
```

9.
```
      354
 ×  24,000
```

Example 14 Estimate the product $257 \cdot 83$ using rounded-off values; then find the product.

Solution

(a) Estimation:

$$
\begin{array}{r}
257 \longrightarrow 300 \\
\times\ 83 \longrightarrow \times\ \ 80 \\
\hline
24{,}000
\end{array}
$$

(b) The product should be near 24,000.

$$
\begin{array}{r}
257 \\
\times\ \ \ 83 \\
\hline
771 \\
2056 \\
\hline
21{,}331
\end{array}
$$

In this case, the estimated value of 24,000 is considerably larger than 21,231. However, we do not expect precise estimates, only estimates that are reasonable.

NAME _____ SECTION _____ DATE _____

EXERCISES **1.6**

Find the following products. Estimate each answer before multiplying.

1. 56
   × 4

2. 27
   × 6

3. 48
   × 9

4. 65
   × 5

5. 43
   × 8

6. 72
   × 6

7. 91
   × 5

8. 39
   × 2

9. 84
   × 3

10. 95
   × 8

11. 42
   ×56

12. 25
   ×33

13. 15
   ×22

14. 29
   ×41

15. 67
   ×36

16. 54
   ×27

17. 48
   ×20

18. 93
   ×30

19. 83
   ×85

20. 96
   ×62

Answers

1. _____

2. _____

3. _____

4. _____

5. _____

6. _____

7. _____

8. _____

9. _____

10. _____

11. _____

12. _____

13. _____

14. _____

15. _____

16. _____

17. _____

18. _____

19. _____

20. _____

Answers

21. _____

22. _____

23. _____

24. _____

25. _____

26. _____

27. _____

28. _____

29. _____

30. _____

31. _____

32. _____

33. _____

34. _____

35. _____

36. _____

37. _____

38. _____

39. _____

40. _____

21. 17
     ×32

22. 28
     ×91

23. 20
     ×44

24. 16
     ×26

25. 25
     ×15

26. 93
     ×47

27. 24
     ×86

28. 72
     ×65

29. 12
     ×13

30. 81
     ×36

31. 126
     × 41

32. 232
     × 76

33. 114
     × 25

34. 72
     ×106

35. 207
     ×143

36. 420
     ×104

37. 200
     × 49

38. 849
     ×205

39. 673
     ×186

40. 192
     ×467

NAME _____ SECTION _____ DATE _____

Multiply using your knowledge of powers of 10.

41.   52
    × 600

42.   72
    ×930

43.   76
    × 5000

44.   500
    × 8000

45.   68
    ×7300

46.   320
    × 4700

47.   41
    ×5300

48.   157
    × 6000

49. If your salary is $1300 per month and you are to get a raise once a year of $130 per month and you love your work, what will you earn over a two-year period? Over a five-year period?

50. If you rent an apartment with three bedrooms for $650 per month and you know the rent will increase once a year by $30 per month, what will you pay in rent over a three-year period? Over a five-year period?

Answers

41. _____

42. _____

43. _____

44. _____

45. _____

46. _____

47. _____

48. _____

49. _____

50. _____

Answers

In Exercises 51–54, first make an estimate of the answers. Then, after you have performed the actual calculations, check to see that your estimate and your answer are reasonably close.

51. _____

51. If you drive at 55 miles per hour for 5 hours, how far will you drive? If you drive at 50 miles per hour in a new four-door car for 4 hours, how far will you drive?

52. _____

52. Your company bought 18 new cars at a price of $9750 per car. Each one had an air conditioner. What did the company pay for all the new cars?

53. _____

53. A rectangular lot for a house measures 208 feet long by 175 feet wide. Find the area of the lot in square feet. (Area is found by multiplying length times width.)

54. _____

54. Find the number of square feet of area for a rectangular swimming pool and deck if the plans call for a length of 215 feet and a width of 82 feet. (Area is found by multiplying length times width.)

1.7
DIVISION

We know that $6 \cdot 10 = 60$ and that 6 and 10 are factors of 60. They are also called **divisors.** The process of division can be thought of as the reverse of multiplication. Therefore, since $6 \cdot 10 = 60$, we know that there are 10 sixes in 60, and we say that 60 **divided by** 6 is 10 (or $60 \div 6 = 10$).

Division: Finding how many times one number is contained in another (the reverse of multiplication)

Quotient: Result of division

Divide Sign ($\div$): Used to indicate division

Example 1   How many 6's are in 18?

Since $6 \cdot 3 = 18$, there are three 6's in 18. We write $18 \div 6 = 3$.

DIVISION				*MULTIPLICATION*		

$$18 \div 6 = 3 \qquad 6 \cdot 3 = 18$$

dividend   divisor   quotient   factor   factor   product

Dividend: Number that is divided

Divisor: Number that divides into the dividend

Example 2   How many 7's are in 28?

Since $7 \cdot 4 = 28$, there are four 7's in 28.
We write

$$28 \div 7 = 4$$

dividend   divisor   quotient

We can also write

$$\text{divisor} \rightarrow 7\overline{)28} \quad \begin{array}{l} \leftarrow \text{quotient} \\ \leftarrow \text{dividend} \end{array}$$

Example 3

| | DIVISION | | | | MULTIPLICATION | |
DIVIDEND	DIVISOR	QUOTIENT			FACTORS	PRODUCT
21	÷ 7	= 3		since	$7 \cdot 3$ =	21
24	÷ 6	= 4		since	$6 \cdot 4$ =	24
36	÷ 4	= 9		since	$4 \cdot 9$ =	36
12	÷ 2	= 6		since	$2 \cdot 6$ =	12

Example 4 If one factor of 72 is 8, what is the corresponding factor?

We are looking for the quotient. $72 \div 8 = \square$.

Since $72 \div 8 = 9$, the corresponding factor is 9.

> **Division can be thought of as repeated subtraction.**

Example 5 To find how many 4's are in 20, subtract 4 repeatedly and count how many times you subtract.

$$
\begin{array}{ccccc}
20 & 16 & 12 & 8 & 4 \\
-4 & -4 & -4 & -4 & -4 \\
\hline
16 & 12 & 8 & 4 & 0
\end{array}
$$

subtraction 5 times

There are five 4's in 20.

Example 6 Find how many 15's there are in 60 by doing repeated subtraction.

$$
\begin{array}{r}
60 \\
-15 \\
\hline
45 \\
-15 \\
\hline
30 \\
-15 \\
\hline
15 \\
-15 \\
\hline
0
\end{array}
$$

subtraction 4 times

There are four 15's in 60.

We can shorten subtraction techniques considerably and, in so doing, are led to the familiar method of division known as the **division algorithm.** Sometimes there is an amount remaining after the repeated subtraction that is less than the divisor. This amount is called the **remainder.** We summarize the important terms in the following list.

Division: Finding how many times one number is contained in another

Division Algorithm: A process or pattern of steps used in division

Quotient: Result of division

Dividend: Number that is divided

Divisor: Number that divides into the dividend

Remainder: Number left over after division (must be less than the divisor)

Now we explain the division algorithm using the technique of repeated subtraction.

Example 7   Find $275 \div 6$ using repeated subtraction and check your work.

(a)
$$
\begin{array}{r}
6\overline{)275} \\
-180 \\
\hline
95 \\
-60 \\
\hline
35 \\
-18 \\
\hline
17 \\
-12 \\
\hline
5
\end{array}
$$

    Subtract 30 sixes. $(30 \cdot 6 = 180)$

    Subtract 10 sixes. $(10 \cdot 6 = 60)$

    Subtract 3 sixes. $(3 \cdot 6 = 18)$

    Subtract 2 sixes. $(2 \cdot 6 = 12)$

            45 sixes total

   ↑   ↑

remainder quotient

Note: You can subtract any number of sixes less than the quotient. But this will not lead to a good explanation of the shorter division algorithm. You should subtract the largest number of thousands, hundreds, tens, or units you can at each step.

(b)
$$
\begin{array}{r}
6\overline{)275} \\
-240 \\
\hline
35 \\
-30 \\
\hline
5
\end{array}
$$

    Subtract 40 sixes. $(40 \cdot 6 = 240)$
$(50 \cdot 6 = 300;$ too much since 300 is greater than 275$)$

    Subtract 5 sixes.

            45 sixes total

   ↑   ↑

remainder quotient

CHECK: Division is checked by multiplying the quotient and the divisor and then adding the remainder. The result should be the dividend.

$$
\begin{array}{r}
45 \\
\times\ 6 \\
\hline
270
\end{array}
\quad
\begin{array}{l}
\text{quotient} \\
\text{divisor}
\end{array}
\qquad
\begin{array}{r}
270 \\
+\ \ 5 \\
\hline
275
\end{array}
\quad
\begin{array}{l}
\\
\text{remainder} \\
\text{dividend}
\end{array}
$$

We could also write the quotient above the dividend as follows.

Divide using repeated subtraction, and check.

1. 9)849

2. 11)6038

STEP 1:

$$6)\overline{275}$$
$$-240$$
$$\overline{35}$$

STEP 2:

$$5$$
$$40$$
$$6)\overline{275}$$
$$-240$$
$$\overline{35}$$
$$\underline{30}$$

STEP 3:

$$\underline{45} \quad \text{quotient}$$
$$5 \leftarrow$$
$$40 \leftarrow$$
$$6)\overline{275}$$
$$-240 \leftarrow$$
$$\overline{35}$$
$$\underline{30} \leftarrow$$
$$5 \quad \text{R} \quad \text{remainder}$$

☐ **NOW WORK EXERCISES 1 and 2 IN THE MARGIN.**

Example 8   Find 2076 ÷ 8.

STEP 1:
$$2 \leftarrow$$
$$8)\overline{2076}$$
$$-1600$$
$$\overline{476}$$

Write 2 in the hundreds position.
200 eights (200 · 8 = 1600)

STEP 2:
$$25 \leftarrow$$
$$8)\overline{2076}$$
$$-1600$$
$$\overline{476}$$
$$-400$$
$$\overline{76}$$

Write 5 in the tens position.

50 eights (50 · 8 = 400)

STEP 3:
$$259 \leftarrow$$
$$8)\overline{2076}$$
$$-1600$$
$$\overline{476}$$
$$-400$$
$$\overline{76}$$
$$-72$$
$$\overline{4}$$

Write 9 in the units position.

9 eights (9 · 8 = 72)

SUMMARY: The process can be shortened by not writing all the 0's and **bringing down** only one digit at a time.

```
      259 R4
   8)2076
      16
       47   ← Bring down the 7 only, then divide 8 into 47.
       40
        76  ← Bring down the 6, then divide 8 into 76.
        72
         4
```

Example 9   Find 746 ÷ 32.

```
              2
STEP 1:   32)746
              64
              10
```
Trial divide 30 into 70 or 3 into 7, giving 2 in the tens position. Note that 10 is less than 32.

```
            23 R10
STEP 2:   32)746
              64
             106
              96
              10
```
Trial divide 30 into 100 or 3 into 10, giving 3 in the units position.

CHECK:

```
    23        736
   ×32       + 10
    46        746
   69
  736
```

Example 10   Find $9325 \div 45$.

STEP 1:
$$\begin{array}{r} 2 \\ 45\overline{)9325} \\ 90 \\ \hline 3 \end{array}$$

Trial divide 40 into 90 or 4 into 9, giving 2 in the hundreds position.

STEP 2:
$$\begin{array}{r} 20 \\ 45\overline{)9325} \\ 90 \\ \hline 32 \\ 0 \end{array}$$

45 will not divide into 32, so write 0 in the tens column and multiply $0 \cdot 45 = 0$.

STEP 3:
$$\begin{array}{r} 208 \\ 45\overline{)9325} \\ 90 \\ \hline 32 \\ 0 \\ \hline 325 \\ 340 \end{array}$$

Trial divide 45 into 325 or 4 into 32. The trial quotient is too large, since $8 \cdot 45 = 340$ and 340 is larger than 325.

$$\begin{array}{r} 207 \text{ R}10 \\ 45\overline{)9325} \\ 90 \\ \hline 32 \\ 0 \\ \hline 325 \\ 315 \\ \hline 10 \end{array}$$

Now the trial divisor is 7. Since $7 \cdot 45 = 315$ and 315 is smaller than 325, 7 is the desired number.

CHECK:

$$\begin{array}{r} 207 \\ \times 45 \\ \hline 1035 \\ 828 \\ \hline 9315 \end{array} \qquad \begin{array}{r} 9315 \\ + 10 \\ \hline 9325 \end{array}$$

Special Note: In Step 2 of Example 10 we wrote 0 in the quotient because 45 did not divide into 32. Be sure to write 0 in the quotient whenever the divisor does not divide into one of the partial remainders.

Completion Example 11   Find the quotient and remainder.

$$
\begin{array}{r}
8 \\
4\overline{)334} \\
32 \\
\hline
1
\end{array}
$$

———— remainder

Completion Example 12   Find the quotient and remainder.

$$
\begin{array}{r}
30 \\
21\overline{)6451} \\
63 \\
\hline
15
\end{array}
$$

————

———— remainder

☐ **NOW WORK EXERCISES 3–5 IN THE MARGIN.**

> If the remainder is 0, then
>
> 1. both the divisor and quotient are factors of the dividend;
> 2. we say that both factors divide **exactly** into the dividend;
> 3. both factors are called divisors of the dividend.

Divide using the division algorithm, then check your answer.

3. $325 \div 7$

4. $16\overline{)324}$

5. $41\overline{)24,682}$

Completion Example Answers

11.
$$
\begin{array}{r}
83 \quad \text{quotient} \\
4\overline{)334} \\
32 \\
\hline
14 \\
12 \\
\hline
2 \quad \text{remainder}
\end{array}
$$

12.
$$
\begin{array}{r}
307 \quad \text{quotient} \\
21\overline{)6451} \\
63 \\
\hline
15 \\
0 \\
\hline
151 \\
147 \\
\hline
4 \quad \text{remainder}
\end{array}
$$

Estimating Quotients _____

By rounding off both divisor and dividend, then dividing, we can estimate the quotient. As with any estimation, the purpose is to ensure that the actual value calculated is reasonable and does not contain a major error.

The idea is to find an approximate quotient using rounded-off numbers.

> To estimate a quotient:
>
> 1. Round off both the divisor and dividend to the place of the last digit on the left.
> 2. Divide using the rounded-off numbers.
>
> This process is very similar to the trial dividing step in the division algorithm.

Example 13 Estimate the quotient $325 \div 42$ using rounded-off values; then find the quotient.

Solution

(a) Estimation: $325 \div 42 \rightarrow 300 \div 40$

$$
\begin{array}{r}
7 \quad \text{estimated quotient} \\
40\overline{)300} \\
\underline{280} \\
20
\end{array}
$$

(b) The quotient should be near 7.

$$
\begin{array}{r}
7 \quad \text{quotient} \\
42\overline{)325} \\
\underline{294} \\
31 \quad \text{remainder}
\end{array}
$$

In this case, the quotient is the same as the estimated value. The true remainder is different.

Example 14 First estimate the quotient $48{,}062 \div 26$; then find the quotient.

Solution

(a) Estimation: $48{,}062 \div 26 \rightarrow 50{,}000 \div 30$

```
         1,666    approximate quotient
   30)50,000
         30
         20 0
         18 0
            2 00
            1 80
               200
               180
                20
```

(b) The quotient should be near 1600 or 1700.

```
            1,848
   26)48,062
         26
         22 0
         20 8
            1 26
            1 04
               222
               208
                14   remainder
```

We close this section with two rules about division involving 0.

DIVISION INVOLVING 0

1. If a is any nonzero whole number, then

$$0 \div a = 0.$$

2. If a is any whole number,

$$a \div 0 \text{ is undefined.}$$

These rules will be discussed in detail in Chapter 3.

Answers to Marginal Exercises

```
                      548
         94            8
          4           40
         90           500
1. 9)849      2. 11)6038
        810          5500
         39           538
         36           440
          3  R         98
                       88
                       10  R
```

3. 46 R3   **4.** 20 R4   **5.** 602 R0

NAME _____ SECTION _____ DATE _____

EXERCISES **1.7** _____

Find the following quotients. (You should be able to do these mentally since they are related to the multiplication tables.)

1. $12 \div 3$

2. $6 \div 2$

3. $18 \div 2$

4. $49 \div 7$

5. $64 \div 8$

6. $20 \div 5$

7. $20 \div 4$

8. $25 \div 5$

9. $30 \div 6$

10. $35 \div 7$

11. $40 \div 8$

12. $42 \div 7$

13. $56 \div 7$

14. $30 \div 5$

15. $72 \div 8$

16. $63 \div 9$

17. $63 \div 7$

18. $15 \div 3$

19. $24 \div 3$

20. $54 \div 6$

21. $0 \div 8$

22. $0 \div 5$

23. $6 \div 6$

24. $2 \div 2$

Answers

1. _____

2. _____

3. _____

4. _____

5. _____

6. _____

7. _____

8. _____

9. _____

10. _____

11. _____

12. _____

13. _____

14. _____

15. _____

16. _____

17. _____

18. _____

19. _____

20. _____

21. _____

22. _____

23. _____

24. _____

Answers

25. _____

26. _____

27. _____

28. _____

29. _____

30. _____

31. _____

32. _____

33. _____

34. _____

35. _____

36. _____

37. _____

38. _____

39. _____

40. _____

In Exercises 25–37, find the quotient and remainder using the method of repeated subtraction. State whether or not the divisor and quotient are factors of the dividend.

25. $210 \div 7$ **26.** $168 \div 8$ **27.** $132 \div 11$

28. $75 \div 15$ **29.** $51 \div 3$ **30.** $600 \div 25$

31. $413 \div 20$ **32.** $161 \div 15$ **33.** $182 \div 13$

34. $3\overline{)98}$ **35.** $14\overline{)52}$ **36.** $12\overline{)108}$

37. $11\overline{)424}$

In Exercises 38–55, divide using the division algorithm after first estimating the quotient using rounded-off numbers.

38. $16\overline{)128}$ **39.** $20\overline{)305}$ **40.** $18\overline{)206}$

NAME _____   SECTION _____   DATE _____

41. $10\overline{)423}$ **42.** $15\overline{)750}$ **43.** $13\overline{)260}$

44. $12\overline{)360}$ **45.** $19\overline{)7603}$ **46.** $16\overline{)4813}$

47. $13\overline{)3917}$ **48.** $73\overline{)148}$ **49.** $68\overline{)207}$

50. $50\overline{)3065}$ **51.** $40\overline{)2163}$ **52.** $105\overline{)210}$

53. $213\overline{)4760}$ **54.** $716\overline{)3056}$ **55.** $630\overline{)4768}$

Answers

41. _____

42. _____

43. _____

44. _____

45. _____

46. _____

47. _____

48. _____

49. _____

50. _____

51. _____

52. _____

53. _____

54. _____

55. _____

Answers

56. _____

57. _____

58. _____

59. _____

60. _____

56. Show that 27 and 30 are both factors of 810 by using division.

57. Show that 35 and 46 are both factors of 1610 by using division.

58. What number multiplied by 73 gives a product of 1606?

59. What number multiplied by 18 gives a product by 3654?

60. A purchasing agent bought 19 new cars for his company for a total price of $190,665. What price did he pay per car?

1.8

APPLICATIONS USING WHOLE NUMBERS _____

In this section, the applications are varied and can relate to various combinations of the four operations of addition, subtraction, multiplication, and division. The decisions on what operations are to be used are based generally on experience and practice.

> **RECOMMENDED APPROACH TO WORD PROBLEMS**
>
> 1. Read each problem carefully two or three times.
> 2. Identify what is being asked for.
> 3. Decide what operations are needed: addition, subtraction, multiplication, or division.
> 4. Perform these operations.
> 5. Mentally check to see if your answer seems reasonable.

Example 1 *CONSUMER ITEMS* Sara went to the pharmacy and bought 3 prescriptions for $12 each, a tube of toothpaste for $2, skin lotion for $5, 2 pairs of hosiery for $4 a pair, and 3 bottles of nail polish for $3 a bottle. How much did she spend? (Ignore tax for now.)

Solution

Find the total expense by multiplying and adding.

$$
\begin{array}{ccccc}
\$\ 12 & \$\ 2 & \$\ 5 & \$\ 4 & \$\ 3 \\
\times\ 3 & \times 1 & \times 1 & \times 2 & \times 3 \\
\hline
36 & 2 & 5 & 8 & 9
\end{array}
$$

$$
\begin{array}{r}
\$\ 36 \\
2 \\
5 \\
8 \\
+\ 9 \\
\hline
\$\ 60 \quad \text{total expenses}
\end{array}
$$

She spent a total of $60.

Example 2 *CONSUMER ITEMS* Bill bought a car for $9000. The salesman added $540 for taxes and $150 for license fees. If Bill made a down payment of $3500 and financed the rest with his credit union, how much did he finance?

Solution

Find the total cost and subtract the down payment.

$ 9000	$ 9690 total cost
540	−3500 down payment
+ 150	$ 6190 to be financed
$ 9690 total cost	

Bill financed $6190.

(Mental checking is difficult here. But you might think that if he paid about $10,000 and put down $3,500, an answer around $6500 is reasonable.)

Example 3 *CHECKING* In July, Ms. Smith opened a checking account and deposited $1528. She wrote checks for $132, $425, $196, and $350. What was her balance at the end of July?

Solution

To find the balance, find the sum of the checks and subtract the sum from $1528.

21	
$ 132	$ 1528
425	−1103
196	$ 425 balance
+ 350	
$ 1103 total checks	

The balance was $425 at the end of July.

(Mentally round off each check to hundreds to see if the answer is reasonable.)

Example 4 *GEOMETRY* A triangle has three sides: a base of 3 feet, a height of 4 feet, and a third side of 5 feet. (See diagram below.) This triangle is called a right triangle because one angle is 90°. Find the perimeter (distance around) of the triangle. Find the area of the triangle in square feet. (To find area, multiply base times height, then divide by 2.)

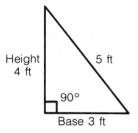

Height
4 ft

5 ft

90°

Base 3 ft

Solution   *PERIMETER* (in feet)   *AREA* (in square feet)

$$3$$
$$4$$
$$\underline{+\ 5}$$
$$12 \quad \text{feet (perimeter)}$$

$$\begin{array}{r} 3 \\ \times\ 4 \\ \hline 12 \end{array} \qquad \begin{array}{r} 6 \quad \text{square feet (area)} \\ 2\overline{)12} \\ \underline{12} \\ 0 \end{array}$$

The perimeter is 12 feet and the area is 6 square feet.

Example 5 *STUDENT EXPENSES* Mrs. James decided to return to college. She found that for the first semester her tuition would be $750, books would be $85, food would be $300, and student body fees would be $45. What is the total of these expenses for her first semester?

Solution

Add all the amounts.

$$\begin{array}{r} \$\quad 750 \\ 85 \\ 300 \\ \underline{+\quad 45} \\ \$\ 1180 \quad \text{total} \end{array}$$

The total of these expenses is $1180.

NAME _____ SECTION _____ DATE _____

EXERCISES **1.8** _____

Answers

Numbers

1. Find the sum of 846, 950, and 783. Then subtract 579. What is the quotient if the difference is divided by 125?

1. _____

2. The difference between 8000 and 1750 is added to 975. If this sum is multiplied by 600, what is the product?

2. _____

3. If the product of 307 and 69 is divided by 23, what is the quotient?

3. _____

Consumer Items

4. To purchase a new refrigerator for $1200 including tax, Mr. Kline paid $240 down and the remainder in six equal monthly payments. What were his monthly payments?

4. _____

5. Mike decided to go shopping for school clothes before college started in the fall. How much did he spend if he bought four pairs of pants for $21 a pair, five shirts for $18 each, three pairs of socks for $4 a pair, and two pairs of shoes for $38 a pair?

5. _____

6. To purchase a new dining room set for $1200, Mrs. Steel had to pay $72 in sales tax. If she made a deposit of $486, how much did she still owe?

6. _____

Answers

7. _____

8. _____

9. _____

10. _____

11. _____

7. Alan wanted to buy a new car. He could buy a red one for $8500 plus $510 in sales tax and $135 in fees; or he could buy a blue one for $8700 plus $522 in sales tax and $140 in fees. If the manufacturer was giving a $250 rebate on the blue model, which car would be cheaper for Alan? How much cheaper?

8. Lynn decided to take up surfing. She bought a new surfboard for $675, a wet suit for $130, a beach towel for $12, and a new swim suit for $57. How much money did she spend? (Sales tax was included in the prices.)

9. Pat needed art supplies for a new course at the local community college. She bought a portfolio for $32, a zinc plate for $44, etching ink for $12, and three sheets of rag paper totaling $6. She received a student discount of $9. What did she spend on art supplies?

Geometry

10. A rectangle is a four-sided figure with opposite sides equal and all four angles equal. (Each angle is 90°). **(a)** Find the perimeter of a rectangle that has a width of 15 meters and a length of 37 meters. **(b)** Also find its area (multiply length times width) in square meters.

15 m
37 m

11. A regular hexagon is a six-sided figure with all six sides equal and all six angles equal. Find the perimeter of a regular hexagon with one side of 19 centimeters.

19 cm

NAME _____ SECTION _____ DATE _____

12. An isosceles triangle (two sides equal) is placed on top of a square to form a window as shown in the figure below. If each of the two equal sides of the triangle is 18 inches long and the square is 28 inches on a side, what is the perimeter of the window?

12. _____

18 in.

28 in.

13. A rectangular picture is mounted in a rectangular frame with a border (called a mat). If the picture is 12 inches by 18 inches and the frame is 16 inches by 24 inches, what is the area of the mat? (Area of a rectangle is length times width.)

13. _____

12 in.

18 in.

24 in.

16 in.

Student Expenses

14. You bought the following texts and supplies: special paper for $30, pencils for $2, erasers for $2, three art brushes for $5 each, two texts for $21 each, and one text for $28. How much did you spend?

14. _____

15. For a class in statistics, Ann bought a new calculator for $35, special graph paper for $10, disks for use on the computer for $6, a text for $22, and a workbook for $17. What did Ann spend for her statistics class?

15. _____

16. For his physical education class, Bill bought a pair of shoes for $27, two pairs of socks for $4 a pair, shorts for $18, and two shirts for $11 each. What were his expenses for this class?

Checking

17. If you opened a checking account with $875, then wrote checks for $20, $35, $115, $8, and $212, what would be your balance?

18. Your friend has a checking balance of $1250 and writes checks for $375, $52, $83, and $246. What is her balance?

19. On August 1, Matt had a balance of $250. During August, he made deposits of $200, $350, and $236. He wrote checks for $487, $25, $33, and $175. What was his balance on September 1?

20. Steve deposited $500, $2470, $800, $3562, and $2875 in his checking account during a five-month period. He wrote checks totaling $6742. If his beginning balance was $1400, what was his balance at the end of the five months?

NAME _____ SECTION _____ DATE _____

CHAPTER **1** TEST _____

Answers

1. Write 8952 in expanded notation and in its English word equivalent.

1. _____

2. The number 1 is called the multiplicative _____.

2. _____

3. Give an example that illustrates the commutative property of multiplication.

3. _____

4. _____

4. Round off 997 to the nearest thousand.

5. _____

5. Round off 135,721 to the nearest ten-thousand.

6. _____

7. _____

8. _____

Add.

9. _____

6.	9586	7.	37	8.	1,480,900
	345		486		2,576,850
	+2078		493		5,200,635
			162		+4,523,276
			+557		

10. _____

11. _____

Subtract.

9.	850	10.	5097	11.	6000
	−362		−3868		− 293

Answers

12. _____

13. _____

14. _____

15. _____

16. _____

17. _____

18. _____

19. _____

20. _____

Multiply.

12.   34
       ×76

13.   2593
      × 85

14.   793
      ×266

Divide.

15. $25\overline{)10{,}075}$

16. $462\overline{)79{,}852}$

17. $603\overline{)1{,}209{,}015}$

18. Find the sum of the numbers 82, 96, 49, and 69; then divide this sum by 4.

19. If the quotient of 51 and 17 is subtracted from the product of 19 and 3, what is the difference?

20. Robert and his brother were saving money to buy a new TV set for their parents. If Robert saved $23 a week and his brother saved $28 a week, how much did they save in six weeks? If the set they wanted to buy cost $530 including tax, how much did they still need after the six weeks?

Prime Numbers

2

2.1
EXPONENTS

Repeated addition is shortened by using multiplication:

$$3 + 3 + 3 + 3 = 4 \cdot 3 = 12$$

Repeated multiplication can be shortened by using exponents.

> **Base:** A number used as a repeated factor
>
> **Exponent:** A number that tells how many times its base is used as a factor
>
> **Power:** The product of a repeated factor

If 2 is used as a factor three times, we can write

$$2 \cdot 2 \cdot 2 = 2^3 = 8$$

with *exponent* pointing to the 3, *base* pointing to the 2, and *power* pointing to the 8.

Example 1

REPEATED MULTIPLICATION	USING EXPONENTS
$7 \cdot 7 = 49$	$7^2 = 49$
$3 \cdot 3 = 9$	$3^2 = 9$
$2 \cdot 2 \cdot 2 \cdot 2 = 16$	$2^4 = 16$
$10 \cdot 10 \cdot 10 = 1000$	$10^3 = 1000$

In each expression (a) name the exponent; (b) name the base; (c) find the value of the expression.

1. 8^2

2. 6^3

3. 9^2

4. 2^6

5. 1^4

Note: Do not multiply by an exponent. Many beginning students make the mistake of thinking $6^3 = 6 \cdot 3 = 18$. This is **WRONG.** In fact,

$$6^3 = 6 \cdot 6 \cdot 6 = 216$$

Example 2   Expressions with exponent 2 are read "squared."

(a) $5^2 = 25$ is read "five squared is equal to twenty-five"

$$(5^2 = 5 \cdot 5)$$

(b) $9^2 = 81$ is read "nine squared is equal to eighty-one"

$$(9^2 = 9 \cdot 9)$$

Example 3   Expressions with exponent 3 are read "cubed."

(a) $4^3 = 64$ is read "four cubed is equal to sixty-four"

$$(4^3 = 4 \cdot 4 \cdot 4)$$

(b) $5^3 = 125$ is read "five cubed is equal to one hundred twenty-five"

$$(5^3 = 5 \cdot 5 \cdot 5)$$

Example 4   Expressions with other exponents are read "to the _____ power."

(a) $2^5 = 32$ is read "two to the fifth power is equal to thirty-two"

$$(2^5 = 2 \cdot 2 \cdot 2 \cdot 2 \cdot 2)$$

(b) $3^4 = 81$ is read "three to the fourth power is equal to eighty-one"

$$(3^4 = 3 \cdot 3 \cdot 3 \cdot 3)$$

Note: A power is not an exponent. A power is the product indicated by an exponent. Thus the phrase "two to the fifth power" is to be thought of in its entirety and the corresponding power is the product, 32.

☐ **NOW WORK EXERCISES 1–5 IN THE MARGIN.**

If there is no exponent, the exponent is understood to be 1. That is, $a = a^1$.

Example 5   The exponent 1 is understood.

(a) $8 = 8^1$

(b) $6 = 6^1$

(c) $942 = 942^1$

| For any nonzero whole number a, $\mathbf{a^0 = 1}$.

When the exponent 0 is used for any base except 0, the value of the power is 1. To help in understanding the use of 0 as an exponent, consider the following rule for exponents (which will be studied in algebra).

Subtract the exponents when dividing numbers with the same base. For example,

$$\frac{2^6}{2^2} = \frac{\cancel{2} \cdot \cancel{2} \cdot 2 \cdot 2 \cdot 2 \cdot 2}{\cancel{2} \cdot \cancel{2}} = 2 \cdot 2 \cdot 2 \cdot 2 = 2^4 \qquad \text{or} \qquad \frac{2^6}{2^2} = 2^{6-2} = 2^4$$

$$\frac{5^4}{5^3} = \frac{\cancel{5} \cdot \cancel{5} \cdot \cancel{5} \cdot 5}{\cancel{5} \cdot \cancel{5} \cdot \cancel{5}} = 5 \qquad \text{or} \qquad \frac{5^4}{5^3} = 5^{4-3} = 5^1$$

So,

$$\frac{3^4}{3^4} = 3^{4-4} = 3^0, \qquad \text{but} \qquad \frac{3^4}{3^4} = \frac{81}{81} = 1$$

$$\frac{5^2}{5^2} = 5^{2-2} = 5^0, \qquad \text{but} \qquad \frac{5^2}{5^2} = \frac{25}{25} = 1$$

Thus,

$$3^0 = 1 \qquad \text{and} \qquad 5^0 = 1.$$

Example 6

(a) $2^0 = 1$

(b) $46^0 = 1$

(c) $10^0 = 1$

Rewrite each product using exponents.

6. $8 \cdot 8 \cdot 8$

7. $2 \cdot 2 \cdot 3 \cdot 5 \cdot 5$

8. $3 \cdot 3 \cdot 4 \cdot 4 \cdot 4$

Find the following squares.

9. 16^2

10. 13^2

To improve your speed in factoring (Section 2.5) and working with fractions (Chapter 3), you should try to memorize all the squares of the whole numbers from 1 to 20. The following table lists these squares.

Number (n)	1	2	3	4	5	6	7	8	9	10
Square (n^2)	1	4	9	16	25	36	49	64	81	100

Number (n)	11	12	13	14	15	16	17	18	19	20
Square (n^2)	121	144	169	196	225	256	289	324	361	400

☐ **NOW WORK EXERCISES 6–10 IN THE MARGIN.**

Answers to Marginal Exercises

	(a) exponent	(b) base	(c) value
1.	2	8	64
2.	3	6	216
3.	2	9	81
4.	6	2	64
5.	4	1	1

6. 8^3　　**7.** $2^2 \cdot 3 \cdot 5^2$　　**8.** $3^2 \cdot 4^3$　　**9.** 256　　**10.** 169

NAME _____   SECTION _____   DATE _____

EXERCISES **2.1** _____

In each of the following expressions, (a) name the exponent; (b) name the base; (c) find the value of each expression.

1. 2^3

2. 2^5

3. 5^2

4. 6^2

5. 7^0

6. 11^2

7. 1^4

8. 4^3

9. 4^0

10. 3^6

11. 3^2

12. 2^4

13. 5^0

14. 1^{50}

15. 62^1

16. 12^2

17. 10^2

18. 10^3

19. 10^4

20. 10^5

Write a base-and-exponent form for each of the following numbers without using the exponent 1.

21. 4

22. 25

23. 16

24. 27

25. 32

26. 121

27. 49

28. 8

Answers

1. (a) _____ (b) _____ (c) _____
2. (a) _____ (b) _____ (c) _____
3. (a) _____ (b) _____ (c) _____
4. (a) _____ (b) _____ (c) _____
5. (a) _____ (b) _____ (c) _____
6. (a) _____ (b) _____ (c) _____
7. (a) _____ (b) _____ (c) _____
8. (a) _____ (b) _____ (c) _____
9. (a) _____ (b) _____ (c) _____
10. (a) _____ (b) _____ (c) _____
11. (a) _____ (b) _____ (c) _____
12. (a) _____ (b) _____ (c) _____
13. (a) _____ (b) _____ (c) _____
14. (a) _____ (b) _____ (c) _____
15. (a) _____ (b) _____ (c) _____
16. (a) _____ (b) _____ (c) _____
17. (a) _____ (b) _____ (c) _____
18. (a) _____ (b) _____ (c) _____
19. (a) _____ (b) _____ (c) _____
20. (a) _____ (b) _____ (c) _____
21. _____
22. _____
23. _____
24. _____
25. _____
26. _____
27. _____
28. _____

Answers

29. _____	**29.** 9 **30.** 36 **31.** 125 **32.** 81
30. _____	
31. _____	**33.** 64 **34.** 100 **35.** 1000 **36.** 10,000
32. _____	
33. _____	
34. _____	Rewrite the following products using exponents.
35. _____	
36. _____	**37.** $6 \cdot 6 \cdot 6 \cdot 6 \cdot 6$ **38.** $7 \cdot 7 \cdot 7 \cdot 7$ **39.** $2 \cdot 2 \cdot 7 \cdot 7$
37. _____	
38. _____	
39. _____	**40.** $5 \cdot 5 \cdot 9 \cdot 9 \cdot 9$ **41.** $2 \cdot 2 \cdot 3 \cdot 3 \cdot 3$ **42.** $2 \cdot 2 \cdot 2 \cdot 3 \cdot 3$
40. _____	
41. _____	
42. _____	
43. _____	**43.** $3 \cdot 3 \cdot 5 \cdot 5 \cdot 5$ **44.** $7 \cdot 7 \cdot 13$ **45.** $11 \cdot 11 \cdot 11$
44. _____	
45. _____	
46. _____	**46.** $2 \cdot 3 \cdot 3 \cdot 7 \cdot 7$ **47.** $3 \cdot 5 \cdot 5 \cdot 5$ **48.** $5 \cdot 5 \cdot 5 \cdot 13 \cdot 13$
47. _____	
48. _____	
49. _____	
50. _____	Find the value of each of the following squares. Write as many of them as you can from memory.
51. _____	
52. _____	**49.** 8^2 **50.** 20^2 **51.** 12^2 **52.** 11^2
53. _____	
54. _____	
55. _____	**53.** 15^2 **54.** 14^2 **55.** 18^2 **56.** 9^2
56. _____	

2.2
ORDER OF OPERATIONS _PEMDAS_

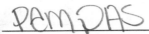

To evaluate an expression that has several operations involves a set of rules. The rules are very explicit. Read them carefully.

RULES FOR ORDER OF OPERATIONS

1. First, simplify within grouping symbols, such as parentheses ( ), brackets [ ], or braces { }. Start with the innermost grouping.

2. Second, find any powers indicated by exponents.

3. Third, moving from **left to right,** perform any multiplications or divisions in the order in which they appear.

4. Fourth, moving from **left to right,** perform any additions or subtractions in the order in which they appear.

Note that in Rule 3, neither multiplication nor division has priority over the other. Whichever of these operations occurs first, moving left to right, is done first. In Rule 4, addition and subtraction are handled in the same way.

Example 1   Evaluate the expression $5 \cdot 2 + 14 \div 2$.

WRONG WAY

$$5 \cdot 2 + 14 \div 2 = 10 + 14 \div 2$$
$$= 24 \div 2$$
$$= 12$$

RIGHT WAY   **Multiply and divide before adding.**

$$5 \cdot 2 + 14 \div 2 = 10 + 7$$
$$= 17$$

The correct answer is 17.

Example 2   Evaluate $14 \div 7 + 3 \cdot 2 - 5$.

$$14 \div 7 + 3 \cdot 2 - 5$$
$$= \quad 2 \quad + \quad 6 \quad - 5 = 8 - 5 = 3$$

Example 3   Evaluate $3 \cdot 6 \div 9 - 1 + 4 \cdot 7$.

$$3 \cdot 6 \div 9 - 1 + 4 \cdot 7 \quad \text{Multiply.}$$
$$= 18 \div 9 - 1 + 4 \cdot 7 \quad \text{Divide.}$$
$$= \quad 2 \quad - 1 + 4 \cdot 7 \quad \text{Multiply.}$$
$$= \quad 2 \quad - 1 + 28 \quad \text{Subtract.}$$
$$= \quad \quad 1 \quad + 28 \quad \text{Add.}$$
$$= \quad \quad 29$$

Example 4   Evaluate $(6 + 2) + (8 + 1) \div 9$.

$$(6 + 2) + (8 + 1) \div 9 \quad \text{Do the operations in parentheses first.}$$
$$= \quad 8 \quad + \quad 9 \quad \div 9 \quad \text{Divide.}$$
$$= \quad 8 \quad + \quad \quad 1 \quad \quad \text{Add.}$$
$$= \quad \quad 9$$

Example 5   Evaluate $30 \div 3 \cdot 2 + 3(6 - 21 \div 7)$.

$$30 \div 3 \cdot 2 + 3(6 - 21 \div 7) \quad \text{Do the operations in parentheses.}$$
$$= 30 \div 3 \cdot 2 + 3(6 - \quad 3)$$
$$= 30 \div 3 \cdot 2 + 3(\quad 3 \quad) \quad \text{Divide.}$$
$$= \quad 10 \quad \cdot 2 + 3(3) \quad \text{Multiply.}$$
$$= \quad \quad 20 \quad + \quad 9 \quad \text{Add.}$$
$$= \quad \quad \quad 29$$

Parts of expressions separated by $+$ or $-$ signs can be simplified separately for easier evaluation.

$$30 \div 3 \cdot 2 + 3(6 - 21 \div 7)$$
$$= \quad 10 \quad \cdot 2 + 3(6 - \quad 3)$$
$$= \quad \quad 20 \quad + 3(\quad 3 \quad)$$
$$= \quad \quad 20 \quad + \quad \quad 9$$
$$= \quad \quad \quad 29$$

Example 6   Evaluate $[(5 + 2^2) \div 3 + 8](14 - 10)$.

$$[(5 + 2^2) \div 3 + 8](14 - 10)$$ Find the power and operate in parentheses.

$$= [(5 + 4) \div 3 + 8](4)$$ Operate in parentheses.

$$= [9 \div 3 + 8](4)$$ Divide.

$$= [3 + 8](4)$$ Operate in brackets.

$$= [11](4) = 44$$ Multiply.

Completion Example 7   Evaluate $2 \cdot 3^2 + 18 \div 3^2$.

$$2 \cdot 3^2 + 18 \div 3^2$$

$$= 2 \cdot \underline{} + 18 \div \underline{}$$

$$= \underline{} + \underline{}$$

$$= \underline{}$$

Completion Example 8   Evaluate $3(2 + 2^2) - 6 - 3 \cdot 2^2$.

$$3(2 + 2^2) - 6 - 3 \cdot 2^2$$

$$= 3(2 + \underline{}) - 6 - 3 \cdot \underline{}$$

$$= 3(\underline{}) - 6 - \underline{}$$

$$= \underline{} - 6 - \underline{}$$

$$= \underline{}$$

☐ **NOW WORK EXERCISES 1–5 IN THE MARGIN.**

Completion Example Answers

7. $= 2 \cdot 9 + 18 \div 9$

$= 18 + 2$

$= 20$

8. $= 3(2 + 4) - 6 - 3 \cdot 4$

$= 3(6) - 6 - 12$

$= 18 - 6 - 12$

$= 0$

Answers to Marginal Exercises

1. 23   **2.** 13   **3.** 5   **4.** 9   **5.** 12

Find the value of each expression using the rules for order of operations.

1. $15 \div 5 + 10 \cdot 2$

2. $4 \div 2^2 + 3 \cdot 2^2$

3. $(5 + 7) \div 3 + 1$

4. $19 - 5(3 - 1)$

5. $3 \cdot 2^4 - 18 - 2 \cdot 3^2$

NAME *Liliana Brooks*   SECTION _____   DATE _____

EXERCISES **2.2** _____

Find the value of each expression using the rules for order of operations.

1. $4 \div 2 + 7 - 3 \cdot 2$

2. $8 \cdot 3 \div 12 + 13$

3. $6 + 3 \cdot 2 - 10 \div 2$

4. $14 \cdot 3 \div 7 \div 2 + 6$

5. $6 \div 2 \cdot 3 - 1 + 2 \cdot 7$

6. $5 \cdot 1 \cdot 3 - 4 \div 2 + 6 \cdot 3$

7. $72 \div 4 \div 9 - 2 + 3$

8. $14 + 63 \div 3 - 35$

9. $(2 + 3 \cdot 4) \div 7 + 3$

10. $(2 + 3) \cdot 4 \div 5 + 3 \cdot 2$

11. $(7 - 3) + (2 + 5) \div 7$

12. $16(2 + 4) - 90 - 3 \cdot 2$

Answers

1. _____

2. _____

3. _____

4. _____

5. _____

6. _____

7. _____

8. _____

9. _____

10. _____ 10 _____

11. _____ 7 _____

12. _____ 0 _____

Answers

13. $35 \div (6 - 1) - 5 + 6 \div 2$

14. $22 - 11 \cdot 2 + 15 - 5 \cdot 3$

13. _____

14. _____

15. $(42 - 2 \div 2 \cdot 3) \div 13$

16. $18 + 18 \div 2 \div 3 - 3 \cdot 1$

15. _____

16. _____

17. $4(7 - 2) \div 10 + 5$

18. $(33 - 2 \cdot 6) \div 7 + 3 - 6$

17. _____

18. _____

19. $72 \div 8 + 3 \cdot 4 - 105 \div 5$

20. $6(14 - 6 \div 2 - 11)$

19. _____

20. _____

21. $48 \div 12 \div 4 - 1 + 6$

22. $5 - 1 \cdot 2 + 4(6 - 18 \div 3)$

21. _____

22. _____

23. $8 - 1 \cdot 5 + 6(13 - 39 \div 3)$

24. $(21 \div 7 - 3)42 + 6$

23. _____

24. _____

25. _____

25. $16 - 16 \div 2 - 2 + 7 \cdot 3$

26. $(135 \div 3 + 21 \div 7) \div 12 - 4$

26. _____

27. _____

27. $(13 - 5) \div 4 + 12 \cdot 4 \div 3 - 72 \div 18 \cdot 2 + 16$

NAME _____ SECTION _____ DATE _____

28. $15 \div 3 + 2 - 6 + (3)(2)(18)(0)(5)$

29. $100 \div 10 \div 10 + 1000 \div 10 \div 10 \div 10 - 2$

30. $[(85 + 5) \div 3 \cdot 2 + 15] \div 15$ **31.** $2 \cdot 5^2 - 4 \div 2 + 3 \cdot 7$

32. $16 \div 2^4 - 9 \div 3^2$ **33.** $(4^2 - 7) \cdot 2^3 - 8 \cdot 5 \div 10$

34. $4^2 - 2^4 + 5 \cdot 6^2 - 10^2$ **35.** $(2^5 + 1) \div 11 - 3 + 7(3^3 - 7)$

36. $(6 + 8^2 - 10 \div 2) \div 5$ **37.** $(5 + 7) \div 4 + 2$

38. $(2^3 + 2) \div 5 + 5$ **39.** $(5^2 + 7) \div 8 - (14 \div 7 \cdot 2)$

40. $(3 \cdot 2^2 - 5 \cdot 2 + 2) - (1 \cdot 2^2 + 5 \cdot 2 - 10)$

Answers

28. _____

29. _____

30. _____

31. _____

32. _____

33. _____

34. _____

35. _____

36. _____

37. _____

38. _____

39. _____

40. _____

Answers

41. _____

42. _____

43. _____

44. _____

45. _____

46. _____

47. _____

48. _____

49. _____

50. _____

41. $2^3 \cdot 3^2 \div 24 - 3 + 6^2 \div 4$

42. $2 \cdot 3^2 + 5 \cdot 9 + 15^2 - (21 \cdot 3^2 + 6)$

43. $3 \cdot 8 - 2^2 + 4 \cdot 2 - 2^4$

44. $2 \cdot 5^2 - 4(21 \div 3 - 7) + 10^3 - 1000$

45. $(4 + 3)(4 + 3) - (2 + 3)(2 + 3)$

46. $(5 + 1)(5 + 1) + (3 + 1)(3 + 1)$

47. $40 \div 2 \cdot 5 + 1 \cdot 9 \cdot 2$ **48.** $20 - 2(3 - 1) + 6^2 \div 2 \cdot 3$

49. $(2^4 - 16)[13 - (5^2 - 20)]$ **50.** $100 + 2[(7^2 - 9)(5 + 1)^2]$

2.3
TESTS FOR DIVISIBILITY (2, 3, 4, 5, 9, AND 10) _____

In our work with factoring (Section 2.5) and fractions (Chapter 3), we will need to be able to divide quickly by small numbers. We will want to know if a number is **exactly divisible** (remainder 0) by some number **before** we divide. In this section, we will discuss a few simple tests that can be used to determine whether a number is divisible by 2, 3, 4, 5, 9, or 10 **without actually dividing.**

> If a number can be divided exactly by another number (the remainder is 0), then we say
>
> 1. the first number is **divisible** by the second, or
> 2. the second number **divides** the first.

Example 1   Is 360 divisible by 4?

$$
\begin{array}{r}
90 \\
4\overline{)360} \\
\underline{36} \\
00 \\
\underline{0} \\
0 \quad \text{remainder}
\end{array}
$$

Since the remainder is 0, 360 is divisible by 4.

Example 2   Is 360 divisible by 10?

$$
\begin{array}{r}
36 \\
10\overline{)360} \\
\underline{30} \\
60 \\
\underline{60} \\
0 \quad \text{remainder}
\end{array}
$$

Since the remainder is 0, 360 is divisible by 10.

Example 3   Is 104 divisible by 5?

$$
\begin{array}{r}
20 \\
5\overline{)104} \\
\underline{10} \\
04 \\
\underline{0} \\
4 \ \ \text{remainder}
\end{array}
$$

Since the remainder is not 0, 104 is not divisible by 5.

Example 4   Does 9 divide 1,809,333?

We could divide by 9 to find the answer. Remember, we do not want to find the quotient. We just want to know if the remainder is 0. We can determine this without dividing.

TESTS FOR DIVISIBILITY BY 2, 3, 4, 5, 9, AND 10

For 2:   If the last digit (units digit) of a whole number is 0, 2, 4, 6, or 8, then the whole number is divisible by 2. In other words, even whole numbers are divisible by 2; odd whole numbers are not divisible by 2.

For 3:   If the sum of the digits of a whole number is divisible by 3, then the number is divisible by 3.

For 4:   If the last two digits of a whole number form a number that is divisible by 4, then the number is divisible by 4. (00 is considered to be divisible by 4.)

For 5:   If the last digit of a whole number is 0 or 5, then the number is divisible by 5.

For 9:   If the sum of the digits of a whole number is divisible by 9, then the number is divisible by 9.

For 10: If the last digit (units digit) of a whole number is 0, then the number is divisible by 10.

Example 5   356 is divisible by 2 since the last digit is 6, an even digit.

Example 6   6801 is divisible by 3 since $6 + 8 + 0 + 1 = 15$ and 15 is divisible by 3.

Example 7   9036 is divisible by 4 since 36 (last 2 digits) is divisible by 4. (9036 is also divisible by 2.)

Example 8   1365 is divisible by 5 since 5 is the last digit.

Example 9   9657 is divisible by 9 since $9 + 6 + 5 + 7 = 27$ and 27 is divisible by 9.

Example 10   3590 is divisible by 10 since 0 is the last digit.

Example 11   9 divides 1,809,333 because $1 + 8 + 0 + 9 + 3 + 3 + 3 = 27$ and 27 is divisible by 9.

☐ **NOW WORK EXERCISES 1–5 IN THE MARGIN.**

Use all six tests to determine which of the numbers 2, 3, 4, 5, 9, and 10 will divide into each number in Examples 12 and 13.

Example 12   2530

    (a) divisible by 2 (last digit is 0, an even digit)
    (b) not divisible by 3 ($2 + 5 + 3 + 0 = 10$ and 10 is not divisible by 3)
    (c) not divisible by 4 (30 is not divisible by 4)
    (d) divisible by 5 (last digit is 0)
    (e) not divisible by 9 ($2 + 5 + 3 + 0 = 10$ and 10 is not divisible by 9)
    (f) divisible by 10 (last digit is 0)

Example 13   5712

    (a) divisible by 2 (last digit is 2, an even digit)
    (b) divisible by 3 ($5 + 7 + 1 + 2 = 15$ and 15 is divisible by 3)
    (c) divisible by 4 (12 is divisible by 4)
    (d) not divisible by 5 (last digit is not 0 or 5)
    (e) not divisible by 9 ($5 + 7 + 1 + 2 = 15$ and 15 is not divisible by 9)
    (f) not divisible by 10 (last digit is not 0)

---

1. Does 4 divide 9044?

2. Does 3 divide 106?

3. Is 306 divisible by 3? By 9?

4. Is 3165 divisible by 5? By 10?

5. Is 463 divisible by 2? By 4? By 5?

Determine which of the numbers 2, 3, 4, 5, 9, and 10 divides into each number.

6. 842

7. 9030

8. 4031

Completion Example 14

(a) 250 is divisible by 10 because

_____ .

(b) 250 is not divisible by 3 because

_____ .

(c) 250 is not divisible by 4 because

_____ .

Completion Example 15

(a) 512 is divisible by 4 because

_____ .

(b) 512 is not divisible by 9 because

_____ .

(c) 512 is not divisible by 5 because

_____ .

☐ **NOW WORK EXERCISES 6–8 IN THE MARGIN.**

Completion Example Answers

14. (a) The last digit is 0.
(b) $2 + 5 + 0 = 7$ and 7 is not divisible by 3.
(c) 50 is not divisible by 4.

15. (a) 12 is divisible by 4.
(b) $5 + 1 + 2 = 8$ and 8 is not divisible by 9.
(c) The last digit is not 0 or 5.

> Numbers divisible by 9 are also divisible by 3, but numbers divisible by 3 might not be divisible by 9.

Example 16   9675

$$9 + 6 + 7 + 5 = 27$$

27 is divisible by 9 and by 3, so 9675 is divisible by both 9 and 3.

Example 17   6108

$$6 + 1 + 0 + 8 = 15$$

15 is divisible by 3 but not by 9, so 6108 is divisible by 3 but not by 9.

> Numbers divisible by 10 are also divisible by 5, but numbers divisible by 5 might not be divisible by 10.

Example 18   2580

The last digit is 0, so 2580 is divisible by 10 and by 5.

Example 19   5365

The last digit is 5, so 5365 is divisible by 5 but not by 10.

Answers to Marginal Exercises

1. Yes; 44 is divisible by 4.   **2.** No; $1 + 0 + 6 = 7$ and 7 is not divisible by 3.
3. Yes; $3 + 0 + 6 = 9$ and 9 is divisible by 3. Yes; $3 + 0 + 6 = 9$ and 9 is divisible by 9.
4. Yes; last digit is 5. No; last digit is not 0.   **5.** No; last digit is not even. No; 63 is not
divisible by 4. No; last digit is not 5.   **6.** 2   **7.** 2, 3, 5, 10   **8.** None

---

NAME _____ SECTION _____ DATE _____

EXERCISES **2.3** _____

Using the tests for divisibility, determine which of the numbers 2, 3, 4, 5, 9, and 10 will divide exactly into each of the following numbers.

1. 72	**2.** 81	**3.** 105	**4.** 333
5. 150	**6.** 471	**7.** 664	**8.** 154
9. 372	**10.** 375	**11.** 443	**12.** 173
13. 567	**14.** 480	**15.** 331	**16.** 370
17. 571	**18.** 466	**19.** 897	**20.** 695
21. 795	**22.** 777	**23.** 45,000	**24.** 885
25. 4422	**26.** 1234	**27.** 4321	**28.** 8765
29. 5678	**30.** 402	**31.** 705	**32.** 732

Answers

1. _____   2. _____

3. _____   4. _____

5. _____   6. _____

7. _____   8. _____

9. _____   10. _____

11. _____   12. _____

13. _____   14. _____

15. _____   16. _____

17. _____   18. _____

19. _____   20. _____

21. _____   22. _____

23. _____   24. _____

25. _____   26. _____

27. _____   28. _____

29. _____   30. _____

31. _____   32. _____

Answers

33. _____ 34. _____

35. _____ 36. _____

37. _____ 38. _____

39. _____ 40. _____

41. _____ 42. _____

43. _____ 44. _____

45. _____ 46. _____

47. _____ 48. _____

49. _____ 50. _____

51. _____ 52. _____

53. _____ 54. _____

55. _____ 56. _____

57. _____ 58. _____

59. _____ 60. _____

61. _____

62. _____

33. 441   34. 555   35. 666   36. 9000

37. 10,000   38. 576   39. 549   40. 792

41. 5700   42. 4391   43. 5476   44. 6930

45. 4380   46. 510   47. 8805   48. 7155

49. 8377   50. 2222   51. 35,622   52. 75,495

53. 12,324   54. 55,555   55. 632,448   56. 578,400

57. 9,737,001   58. 17,158,514   59. 36,762,252   60. 20,498,105

61. If a number is divisible by both 2 and 9, will it be divisible by 18? Give five examples to support your answer.

62. If a number is divisible by both 3 and 9, will it be divisible by 27? Give five examples to support your answer.

2.4
PRIME NUMBERS AND COMPOSITE NUMBERS ____

Counting Numbers (or Natural Numbers): Nonzero whole numbers:

$$1, 2, 3, 4, 5, 6, 7, 8, 9, 10, 11, \ldots$$

Example 1 Every counting number, except 1, has at least two factors.

COUNTING NUMBER	FACTORS
18	1, 2, 3, 6, 9, 18
26	1, 2, 13, 26
12	1, 2, 3, 4, 6, 12
17	1, 17
3	1, 3
29	1, 29

17, 3, and 29 have exactly two different factors.

Prime Number: A counting number with exactly two different factors (or divisors)

Composite Number: A counting number with more than two different factors (or divisors)

Example 2 Is 1 a prime number or a composite number?

$1 = 1 \cdot 1$ and 1 is the only factor of 1.

1 does not have two **different** factors and it does not have more than two different factors.

1 is **neither** a prime number nor a composite number.

Example 3 Some prime numbers:

2   2 has exactly two different factors, 1 and 2.

7   7 has exactly two different factors, 1 and 7.

11   11 has exactly two different factors, 1 and 11.

23   23 has exactly two different factors, 1 and 23.

Determine whether each of the numbers is prime or composite.

1. 17

2. 28

3. 25

4. 2

5. 31

Example 4   Some composite numbers:

$$14 \quad 14 = 1 \cdot 14$$
$$14 = 2 \cdot 7$$

1, 2, 7, 14 are all factors of 14.

$$15 \quad 15 = 1 \cdot 15$$
$$15 = 3 \cdot 5$$

1, 3, 5, 15 are all factors of 15.

$$36 \quad 36 = 1 \cdot 36$$
$$36 = 2 \cdot 18$$
$$36 = 3 \cdot 12$$
$$36 = 4 \cdot 9$$
$$36 = 6 \cdot 6$$

1, 2, 3, 4, 6, 9, 12, 18, 36 are all factors of 36.

☐ **NOW WORK EXERCISES 1–5 IN THE MARGIN.**

> To find the **multiples** of any counting number, multiply each of the counting numbers by that number.

Example 5

counting numbers	1, 2, 3, 4, 5, 6, 7, 8, . . .
multiples of 8	8, 16, 24, 32, 40, 48, 56, 64, . . .
multiples of 2	2, 4, 6, 8, 10, 12, 14, 16, . . .
multiples of 3	3, 6, 9, 12, 15, 18, 21, 24, . . .
multiples of 10	10, 20, 30, 40, 50, 60, 70, 80, . . .

None of the multiples of a number, except possibly the number itself, can be prime since each has that number as a factor. A famous Greek mathematician, Eratosthenes, developed a technique called the **Sieve of Eratosthenes** to help find prime numbers. This technique is discussed in the following steps.

STEP 1: To find the prime numbers from 1 to 50, list all the counting numbers from 1 to 50 in rows of ten.

1	2	3	4	5	6	7	8	9	10
11	12	13	14	15	16	17	18	19	20
21	22	23	24	25	26	27	28	29	30
31	32	33	34	35	36	37	38	39	40
41	42	43	44	45	46	47	48	49	50

STEP 2: Start by crossing out 1 (since 1 is not a prime number). Next, circle 2 and cross out all the other multiples of 2; that is, cross out every second number.

1̶	②	3	4̶	5	6̶	7	8̶	9	1̶0̶
11	1̶2̶	13	1̶4̶	15	1̶6̶	17	1̶8̶	19	2̶0̶
21	2̶2̶	23	2̶4̶	25	2̶6̶	27	2̶8̶	29	3̶0̶
31	3̶2̶	33	3̶4̶	35	3̶6̶	37	3̶8̶	39	4̶0̶
41	4̶2̶	43	4̶4̶	45	4̶6̶	47	4̶8̶	49	5̶0̶

STEP 3: The first number after 2 not crossed out is 3. Circle 3 and cross out all multiples of 3 that are not already crossed out; that is, after 3, every third number should be crossed out.

1̶	②	③	4̶	5	6̶	7	8̶	9̶	10
11	1̶2̶	13	1̶4̶	1̶5̶	1̶6̶	17	1̶8̶	19	2̶0̶
2̶1̶	2̶2̶	23	2̶4̶	25	2̶6̶	2̶7̶	2̶8̶	29	3̶0̶
31	3̶2̶	3̶3̶	3̶4̶	35	3̶6̶	37	3̶8̶	3̶9̶	4̶0̶
41	4̶2̶	43	4̶4̶	4̶5̶	4̶6̶	47	4̶8̶	49	5̶0̶

STEP 4: The next number not crossed out is 5. Circle 5 and cross out all multiples of 5 that are not already crossed out. If we proceed this way, we will have the prime numbers circled and the composite numbers crossed out. The final table is as follows.

1̶	②	③	4̶	⑤	6̶	⑦	8̶	9̶	1̶0̶
⑪	1̶2̶	⑬	1̶4̶	1̶5̶	1̶6̶	⑰	1̶8̶	⑲	2̶0̶
2̶1̶	2̶2̶	㉓	2̶4̶	2̶5̶	2̶6̶	2̶7̶	2̶8̶	㉙	3̶0̶
㉛	3̶2̶	3̶3̶	3̶4̶	3̶5̶	3̶6̶	㉗	3̶8̶	3̶9̶	4̶0̶
㊶	4̶2̶	㊸	4̶4̶	4̶5̶	4̶6̶	㊼	4̶8̶	4̶9̶	5̶0̶

> The prime numbers less than 50 are:
>
> $$2, 3, 5, 7, 11, 13, 17, 19, 23, 29, 31, 37, 41, 43, 47$$

> To decide whether a number (greater than 50) is prime,
>
> 1. use the tests for divisibility for 2, 3, and 5,
> 2. continue to divide by progressively larger **prime** numbers until
>   (a) you find a 0 remainder (meaning the number is composite).
>   (b) you find a quotient less than the divisor (meaning the number is prime).

Example 6   Is 103 prime?

Tests for 2, 3, and 5 fail. (103 is not even; $1 + 0 + 3 = 4$ and 4 is not divisible by 3; the last digit is not 0 or 5.)

Divide by 7:

$$
\begin{array}{r}
14 \\
7\overline{)103} \\
7 \\
\hline
33 \\
28 \\
\hline
5
\end{array}
$$

14   quotient greater than divisor

5   remainder not 0

Divide by 11:

$$
\begin{array}{r}
9 \\
11\overline{)103} \\
99 \\
\hline
4
\end{array}
$$

9   quotient is less than divisor

103 is prime.

Example 7   Is 221 prime?

Tests for 2, 3, and 5 fail.

Divide by 7:

$$
\begin{array}{r}
31 \\
7\overline{)221} \\
21 \\
\hline
11 \\
7 \\
\hline
4
\end{array}
$$

31   quotient greater than divisor

4   remainder not 0

Divide by 11:
$$
\begin{array}{r}
20 \\
11\overline{)221} \\
22 \\
\hline
01 \\
0 \\
\hline
1
\end{array}
$$
    quotient greater than divisor

    remainder not 0

Divide by 13:
$$
\begin{array}{r}
17 \\
13\overline{)221} \\
13 \\
\hline
91 \\
91 \\
\hline
0
\end{array}
$$

    remainder is 0

221 is composite and not prime.

Note: $221 = 13 \cdot 17$. That is, 13 and 17 are factors of 221.

Completion Example 8 Is 105 prime or composite?

105 is divisible by 5, so 105 is _____ .

Two factors of 105 are 5 and _____ .

Completion Example 9 Is 211 prime or composite?

Tests for 2, 3, and 5 all fail.

Divide by 7:	Divide by 11:
$7\overline{)211}$	$11\overline{)211}$

Divide by 13:	Divide by ____:
$13\overline{)211}$	$\overline{)211}$

211 is _____ .

Completion Example Answers

8. composite; 5 and 21 (Note: There are other factors, too.)

9.
$$
\begin{array}{r}
30 \\
7\overline{)211} \\
21 \\
\hline
01 \\
0 \\
\hline
1
\end{array}
\quad
\begin{array}{r}
19 \\
11\overline{)211} \\
11 \\
\hline
101 \\
99 \\
\hline
2
\end{array}
\quad
\begin{array}{r}
16 \\
13\overline{)211} \\
13 \\
\hline
81 \\
78 \\
\hline
3
\end{array}
\quad
\begin{array}{r}
12 \\
17\overline{)211} \\
17 \\
\hline
41 \\
34 \\
\hline
7
\end{array}
$$

211 is prime.

6. Is 187 prime or composite?

☐ **NOW WORK EXERCISES 6 AND 7 IN THE MARGIN.**

Example 10 One interesting application of factors of counting numbers involves finding two factors that add up to be some specific number.

For example, find two factors of 72 such that their product is 72 and their sum is 27.

Solution

72 has several factors, but by experimenting, you will find the numbers are 3 and 24.

$$3 \cdot 24 = 72 \quad \text{and} \quad 3 + 24 = 27$$

7. Is 233 prime or composite?

Answers to Marginal Exercises

1. prime   2. composite   3. composite   4. prime   5. prime   6. composite
7. prime

NAME _____ SECTION _____ DATE _____

EXERCISES **2.4** _____

List the multiples for each of the following numbers.

1. 5   2. 7   3. 11   4. 13   5. 12

6. 9   7. 20   8. 17   9. 16   10. 25

11. Construct a *Sieve of Eratosthenes* for the numbers from 1 to 100. List the prime numbers from 1 to 100. (See pages 102–103.)

Decide whether each of the following numbers is prime or composite. If the number is composite, find at least two pairs of factors for the number.

12. 17   13. 19   14. 28   15. 32   16. 47

17. 59   18. 16   19. 63   20. 14   21. 51

1. _____
2. _____
3. _____
4. _____
5. _____
6. _____
7. _____
8. _____
9. _____
10. _____
11. _____
12. _____
13. _____
14. _____
15. _____
16. _____
17. _____
18. _____
19. _____
20. _____
21. _____

Answers

22. _____

23. _____

24. _____

25. _____

26. _____

27. _____

28. _____

29. _____

30. _____

31. _____

32. _____

33. _____

34. _____

35. _____

36. _____

37. _____

38. _____

39. _____

40. _____

41. _____

42. _____

43. _____

44. _____

45. _____

46. _____

47. _____

48. _____

49. _____

50. _____

51. _____

52. _____

22. 67   **23.** 89   **24.** 73   **25.** 61   **26.** 52

27. 57   **28.** 98   **29.** 86   **30.** 53   **31.** 37

Two numbers are given. Find two factors of the first number such that their product is the first number and their sum is the second number.

Example: 12, 8 Two factors of 12 whose product is 12 and whose sum
                is 8 are 6 and 2 since $6 \cdot 2 = 12$ and $6 + 2 = 8$.

32. 24, 10   **33.** 12, 7   **34.** 16, 10   **35.** 12, 13

36. 14, 9   **37.** 50, 27   **38.** 20, 9   **39.** 24, 11

40. 48, 19   **41.** 36, 15   **42.** 7, 8   **43.** 63, 24

44. 51, 20   **45.** 25, 10   **46.** 16, 8   **47.** 60, 17

48. 52, 17   **49.** 27, 12   **50.** 72, 22

51. List the set of prime numbers less than 100 that **are not** odd.

52. List the set of prime numbers less than 100 that **are** odd.

2.5
PRIME FACTORIZATION

To find common denominators for fractions, which will be discussed in Chapter 3, finding **all** the prime factors of composite numbers will be very useful. For example, $28 = 4 \cdot 7$ and 7 is a prime number, but 4 is not prime. We can write $28 = 4 \cdot 7 = 2 \cdot 2 \cdot 7$ and this product $(2 \cdot 2 \cdot 7)$ contains all prime factors. It is called the **prime factorization** of 28.

> To find the **prime factorization** of a composite number:
>
> 1. Factor the composite number into **any** two factors.
> 2. If either or both factors are not prime, factor each of these.
> 3. Continue this process until all factors are prime.

Example 1 Find the prime factorization for 60.

$$60 = \quad 6 \cdot 10 \qquad \text{Since the last digit is 0, we know 10 is a factor.}$$
$$= 2 \cdot 3 \cdot 2 \cdot 5 \qquad \text{6 and 10 can both be factored so that each factor is a prime number. This is the prime factorization of 60.}$$

or,

$$60 = \quad 3 \cdot 20 \qquad \text{3 is prime, but 20 is not.}$$
$$= 3 \cdot 4 \cdot 5 \qquad \text{4 is not prime.}$$
$$= 3 \cdot 2 \cdot 2 \cdot 5 \qquad \text{All factors are prime.}$$

Since multiplication is commutative, the order of the factors is not important. What is important is that **all the factors must be prime numbers.**

Writing the factors in order, the prime factorization of 60 is $2 \cdot 2 \cdot 3 \cdot 5$ or, using exponents, $2^2 \cdot 3 \cdot 5$.

Example 2 Find the prime factorization of 70.

$$70 = 7 \cdot 10 \qquad \text{10 is a factor since the last digit is 0.}$$
$$= 7 \cdot 2 \cdot 5$$
$$= 2 \cdot 5 \cdot 7 \qquad \text{Writing the factors in order is not necessary, but it is convenient for comparing answers.}$$

Example 3   Find the prime factorization of each number.

(a) $85 = 5 \cdot 17$ 5 is a factor since the last digit is 5. Since both
5 and 17 are prime, $5 \cdot 17$ is the prime factorization.

(b) $72 = 8 \cdot 9$ or $72 = 2 \cdot 36$

$\quad\quad = 2 \cdot 4 \cdot 3 \cdot 3$ $\quad\quad\quad = 2 \cdot 6 \cdot 6$

$\quad\quad = 2 \cdot 2 \cdot 2 \cdot 3 \cdot 3$ $\quad\quad\quad = 2 \cdot 2 \cdot 3 \cdot 2 \cdot 3$

$\quad\quad = 2^3 \cdot 3^2$ $\quad\quad\quad = 2^3 \cdot 3^2$ using exponents

(c) $245 = 5 \cdot 49$

$\quad\quad\quad = 5 \cdot 7 \cdot 7$

$\quad\quad\quad = 5 \cdot 7^2$

(d) $264 = 2 \cdot 132$ or $264 = 4 \cdot 66$

$\quad\quad\quad = 2 \cdot 2 \cdot 66$ $\quad\quad\quad = 2 \cdot 2 \cdot 6 \cdot 11$

$\quad\quad\quad = 2 \cdot 2 \cdot 2 \cdot 33$ $\quad\quad\quad = 2 \cdot 2 \cdot 2 \cdot 3 \cdot 11$

$\quad\quad\quad = 2 \cdot 2 \cdot 2 \cdot 3 \cdot 11$ $\quad\quad\quad = 2^3 \cdot 3 \cdot 11$

$\quad\quad\quad = 2^3 \cdot 3 \cdot 11$

Regardless of your choices for the first two factors, there is only one prime factorization for any composite number.

Completion Example 4   Find the prime factorization of 90.

$$90 = 9 \cdot \underline{\quad\quad}$$

$$= 3 \cdot 3 \cdot \underline{\quad\quad} \cdot \underline{\quad\quad}$$

$$= \underline{\quad\quad\quad\quad}   \text{using exponents}$$

Completion Example 5   Find the prime factorization of 925.

$$925 = 5 \cdot \underline{\quad\quad}$$

$$= 5 \cdot \underline{\quad\quad} \cdot \underline{\quad\quad}$$

$$= \underline{\quad\quad\quad\quad}   \text{using exponents}$$

Completion Example 6   Find the prime factorization of 196.

$$196 = 2 \cdot 98$$

$$= 2 \cdot \underline{\quad\quad} \cdot \underline{\quad\quad}$$

$$= 2 \cdot \underline{\quad\quad} \cdot \underline{\quad\quad} \cdot \underline{\quad\quad}$$

$$= \underline{\quad\quad\quad\quad}   \text{using exponents}$$

Completion Example Answers

4. $90 = 9 \cdot 10$
$\quad\quad = 3 \cdot 3 \cdot 2 \cdot 5$
$\quad\quad = 2 \cdot 3^2 \cdot 5$

5. $925 = 5 \cdot 185$
$\quad\quad = 5 \cdot 5 \cdot 37$
$\quad\quad = 5^2 \cdot 37$

6. $196 = 2 \cdot 98$
$\quad\quad = 2 \cdot 2 \cdot 49$
$\quad\quad = 2 \cdot 2 \cdot 7 \cdot 7$
$\quad\quad = 2^2 \cdot 7^2$

☐ **NOW WORK EXERCISES 1–4 IN THE MARGIN.**

The following important theorem guarantees us that, no matter how we proceed to find prime factorizations, the result will be the same.

> **THE FUNDAMENTAL THEOREM OF ARITHMETIC**
>
> Every composite number has exactly one prime factorization

Once a prime factorization of a composite number is known, all the factors (or divisors) of that number can be found using all the various combinations of these factors. That is, for a number to be a factor of a composite number, it must be either 1 or involve some of the prime factors of that composite number.

> The only factors (or divisors) of a composite number are
>
> 1. 1 and the number itself,
> 2. each prime factor, and
> 3. products of various combinations of prime factors.

Example 7   Find all factors of 30.

Solution

Since $30 = 2 \cdot 3 \cdot 5$, the factors are

(a) 1 and the number itself: 1 and 30.
(b) Each prime factor: 2, 3, 5.
(c) Products of all combinations of the prime factors:

$$2 \cdot 3 = 6, \quad 2 \cdot 5 = 10, \quad 3 \cdot 5 = 15$$

The factors are 1, 30, 2, 3, 5, 6, 10, 15. These are the only factors of 30.

Example 8   Find all the factors of 140.

Solution
$$140 = 14 \cdot 10$$
$$= 2 \cdot 7 \cdot 2 \cdot 5$$
$$= 2 \cdot 2 \cdot 5 \cdot 7$$

Find the prime factorization of each number.

1. 42

2. 56

3. 230

4. 165

5. Find all the factors of 18.

The factors are

(a) 1 and the number itself: 1 and 140.

(b) Each prime factor: 2, 5, 7.

(c) Products of all combinations of the prime factors:
$2 \cdot 2 = 4$, $2 \cdot 5 = 10$, $2 \cdot 7 = 14$, $5 \cdot 7 = 35$, $2 \cdot 2 \cdot 5 = 20$,
$2 \cdot 2 \cdot 7 = 28$, $2 \cdot 5 \cdot 7 = 70$

6. Find all the factors of 63.

The factors are

1, 140, 2, 5, 7, 4, 10, 14, 35, 20, 28, 70.

There are no other factors (or divisors) of 140.

☐ **NOW WORK EXERCISES 5 AND 6 IN THE MARGIN.**

Answers to Marginal Exercises

1. $2 \cdot 3 \cdot 7$ **2.** $2 \cdot 2 \cdot 2 \cdot 7 = 2^3 \cdot 7$ **3.** $2 \cdot 5 \cdot 23$ **4.** $3 \cdot 5 \cdot 11$ **5.** 1, 18, 2, 3, 6, 9
6. 1, 63, 3, 7, 9, 21

NAME _____ SECTION _____ DATE _____

EXERCISES **2.5** _____

Find the prime factorization for each of the following numbers. Use the tests for divisibility given in Section 2.3 whenever you need them to help you get started.

1. 24   **2.** 28   **3.** 27   **4.** 16

5. 36   **6.** 60   **7.** 72   **8.** 90

9. 81   **10.** 105   **11.** 125   **12.** 160

13. 75   **14.** 150   **15.** 210   **16.** 40

17. 250   **18.** 93   **19.** 168   **20.** 360

21. 126   **22.** 48   **23.** 17   **24.** 47

25. 51   **26.** 144   **27.** 121   **28.** 169

Answers

1. _____
2. _____
3. _____
4. _____
5. _____
6. _____
7. _____
8. _____
9. _____
10. _____
11. _____
12. _____
13. _____
14. _____
15. _____
16. _____
17. _____
18. _____
19. _____
20. _____
21. _____
22. _____
23. _____
24. _____
25. _____
26. _____
27. _____
28. _____

Answers

29. _____

30. _____

31. _____

32. _____

33. _____

34. _____

35. _____

36. _____

37. _____

38. _____

39. _____

40. _____

41. _____

42. _____

43. _____

44. _____

45. _____

46. _____

47. _____

48. _____

49. _____

50. _____

29. 225   **30.** 52   **31.** 32   **32.** 98

33. 108   **34.** 103   **35.** 101   **36.** 202

37. 78   **38.** 500   **39.** 10,000   **40.** 100,000

Using the prime factorization of each number, find all the factors (or divisors) of each number.

41. 12   **42.** 18   **43.** 28   **44.** 98

45. 121   **46.** 45   **47.** 105   **48.** 54

49. 97   **50.** 144

2.6
LEAST COMMON MULTIPLE (LCM) _____

The techniques discussed in this section are used throughout Chapter 3 with fractions. Study these ideas thoroughly because they will make your work with fractions much easier.

> **Least Common Multiple (LCM):** The least common multiple of a set of counting numbers is the smallest number that is divisible by each number in the set.

Remember that the multiples of a number are the products of that number with the counting numbers. The first multiple of a number is that number, and all other multiples are larger than the number.

counting numbers	1, 2, 3, 4, 5, 6, 7, 8, 9, 10, . . .
multiples of 6	6, 12, 18, 24, 30, 36, 42, 48, 54, 60, . . .
multiples of 10	10, 20, 30, 40, 50, 60, 70, 80, 90, 100, . . .

The common multiples of 6 and 10 are 30, 60, 90, 120, ... The **Least Common Multiple** is 30.

Listing all the multiples as we just did for 6 and 10, and then choosing the least common multiple (LCM) is not very efficient. The following technique involving prime factorization is much easier to use.

> To find the LCM of a set of counting numbers:
>
> 1. Find the prime factorization of each number.
> 2. Find the prime factors that appear in **any one** of the prime factorizations.
> 3. Form the product of these primes using each prime the most number of times it appears in **any one** of the prime factorizations.

Example 1 Find the LCM of 6 and 10.

$6 \cdot 10 = 60$ and 60 is divisible by both 6 and 10, but 60 is not the smallest number divisible by both 6 and 10.

(a) prime factorizations:
$$6 = 2 \cdot 3$$
$$10 = 2 \cdot 5$$

(b) the prime factors: 2, 3, 5

1. Find the LCM of 28 and 70.

(c) 2 appears once in 6 and once in 10, so the most number of times 2 appears in any one prime factorization is once. The same is true for both 3 and 5.

$$LCM = 2 \cdot 3 \cdot 5 = 30$$

Each factor appears only once in the LCM and 30 is the smallest number divisible by both 6 and 10.

Example 2 Find the LCM of 18, 30, and 45.

(a) prime factorizations:

$$18 = 2 \cdot 9 = 2 \cdot 3 \cdot 3 \quad \text{one 2, two 3's}$$

$$30 = 6 \cdot 5 = 2 \cdot 3 \cdot 5 \quad \text{one 2, one 3, one 5}$$

$$45 = 9 \cdot 5 = 3 \cdot 3 \cdot 5 \quad \text{two 3's, one 5}$$

(b) 2, 3, and 5 are the only prime factors.

(c) most of each factor in any one factorization:

one 2   (in 18 and in 30)
two 3's   (in 18 and in 45)
one 5   (in 30 and in 45)

$$LCM = 2 \cdot 3 \cdot 3 \cdot 5 = 2 \cdot 3^2 \cdot 5 = 90$$

90 is the smallest number divisible by all the numbers 18, 30, and 45.

Completion Example 3 Find the LCM of 36, 24, and 48.

(a) prime factorizations:

$$36 = \underline{\hspace{3cm}}$$

$$24 = \underline{\hspace{3cm}}$$

$$48 = \underline{\hspace{3cm}}$$

(b) $\underline{\hspace{1.5cm}}$ and $\underline{\hspace{1.5cm}}$ are the only prime factors.

(c) most of each factor in any one factorization:

$$\underline{\hspace{2.5cm}} \quad \text{(in 48)}$$

$$\underline{\hspace{2.5cm}} \quad \text{(in 36)}$$

$$LCM = \underline{\hspace{3cm}} = \underline{\hspace{2cm}} = 144$$

$\underline{\hspace{2.5cm}}$ is the smallest number divisible by all the numbers 36, 24, and 48.

☐ **NOW WORK EXERCISE 1 IN THE MARGIN.**

Completion Example Answer

3. (a) $36 = 2 \cdot 2 \cdot 3 \cdot 3$
$24 = 2 \cdot 2 \cdot 2 \cdot 3$
$48 = 2 \cdot 2 \cdot 2 \cdot 2 \cdot 3$

(b) 2 and 3 are the only prime factors.

(c) four 2's (in 48)
two 3's (in 36)

$LCM = 2 \cdot 2 \cdot 2 \cdot 2 \cdot 3 \cdot 3 = 2^4 \cdot 3^2 = 144$

144 is the smallest number divisible by all the numbers 36, 24, and 48.

In our work with fractions, we will want to know how many times each number in a set divides into the LCM. We could perform long division:

$$
\begin{array}{r}
6 \\
24\overline{)144} \\
144 \\
\hline
0
\end{array}
$$

Thus, $24 \cdot 6 = 144$. However, once we have the prime factorization of 144, as we do from the Completion Example 3, we can group the factors as follows to find how many times 24 divides into 144 without performing long division.

$$
\begin{aligned}
144 &= 2 \cdot 2 \cdot 2 \cdot 2 \cdot 3 \cdot 3 \\
&= \underbrace{2 \cdot 2 \cdot 2 \cdot 3}_{} \cdot \underbrace{2 \cdot 3}_{} \\
&= \quad 24 \quad \cdot \quad 6
\end{aligned}
$$

> To find how many times a number in a set of counting numbers divides into the LCM of that set of numbers:
>
> 1. Find the LCM.
> 2. In the LCM, group together all the prime factors that make up the number.
>
> The product of the remaining factors in the LCM tells how many times that number divides into the LCM.

Example 4   (a) Find the LCM of 12, 18, and 20. (b) State the number of times each number divides into the LCM.

Solution

(a) $\left.\begin{array}{l} 12 = 2 \cdot 2 \cdot 3 \\ 18 = 2 \cdot 3 \cdot 3 \\ 20 = 2 \cdot 2 \cdot 5 \end{array}\right\}$ $\begin{aligned} \text{LCM} &= 2 \cdot 2 \cdot 3 \cdot 3 \cdot 5 \\ &= 2^2 \cdot 3^2 \cdot 5 = 180 \end{aligned}$

(b) Group the factors of 12:

$$
180 = 2 \cdot 2 \cdot 3 \cdot 3 \cdot 5 = \underbrace{(2 \cdot 2 \cdot 3)}_{} \cdot \underbrace{(3 \cdot 5)}_{}
$$
$$
= \quad 12 \quad \cdot \quad 15
$$

Group the factors of 18:

$$
180 = 2 \cdot 2 \cdot 3 \cdot 3 \cdot 5 = \underbrace{(2 \cdot 3 \cdot 3)}_{} \cdot \underbrace{(2 \cdot 5)}_{}
$$
$$
= \quad 18 \quad \cdot \quad 10
$$

(a) Find the LCM of each set of numbers. (b) State the number of times each number in the set divides into the LCM.

2. 30, 50

Group the factors of 20:

$$180 = 2 \cdot 2 \cdot 3 \cdot 3 \cdot 5 = \underbrace{(2 \cdot 2 \cdot 5)} \cdot \underbrace{(3 \cdot 3)}$$
$$= \quad 20 \quad \cdot \quad 9$$

Thus,

12 divides 15 times into 180.

18 divides 10 times into 180.

20 divides 9 times into 180.

Completion Example 5   (a) Find the LCM of 27, 30, and 42. (b) State the number of times each number divides into the LCM.

(a) 27 = ⎯⎯⎯⎯⎯⎯

30 = ⎯⎯⎯⎯⎯⎯ } LCM = ⎯⎯⎯⎯⎯⎯

42 = ⎯⎯⎯⎯⎯   = ⎯⎯⎯⎯⎯ = 1890

(b) 1890 = ⎯⎯⎯⎯⎯ = $(3 \cdot 3 \cdot 3) \cdot$ (⎯⎯⎯) = $27 \cdot$ ⎯⎯⎯

1890 = ⎯⎯⎯⎯⎯ = $(2 \cdot 3 \cdot 5) \cdot$ (⎯⎯⎯) = $30 \cdot$ ⎯⎯⎯

1890 = ⎯⎯⎯⎯⎯ = $(2 \cdot 3 \cdot 7) \cdot$ (⎯⎯⎯) = $42 \cdot$ ⎯⎯⎯

☐ **NOW WORK EXERCISES 2 AND 3 IN THE MARGIN.**

3. 18, 20, 25

Completion Example Answer

5. (a) $27 = 3 \cdot 3 \cdot 3$
     $30 = 2 \cdot 3 \cdot 5$
     $42 = 2 \cdot 3 \cdot 7$
   LCM $= 2 \cdot 3 \cdot 3 \cdot 3 \cdot 5 \cdot 7$
       $= 2 \cdot 3^3 \cdot 5 \cdot 7 = 1890$

(b) $1890 = 2 \cdot 3 \cdot 3 \cdot 3 \cdot 5 \cdot 7 = (3 \cdot 3 \cdot 3) \cdot (2 \cdot 5 \cdot 7) = 27 \cdot 70$
   $1890 = 2 \cdot 3 \cdot 3 \cdot 3 \cdot 5 \cdot 7 = (2 \cdot 3 \cdot 5) \cdot (3 \cdot 3 \cdot 7) = 30 \cdot 63$
   $1890 = 2 \cdot 3 \cdot 3 \cdot 3 \cdot 5 \cdot 7 = (2 \cdot 3 \cdot 7) \cdot (3 \cdot 3 \cdot 5) = 42 \cdot 45$

Example 6 Suppose three weather satellites A, B, and C orbit the earth in different times: A takes 18 hours, B takes 14 hours, and C takes 10 hours. If they are directly above each other now (as shown in the figure), in how many hours will they again be directly above each other in exactly the same position as shown?

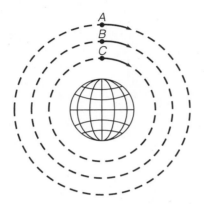

Solution

Note that, when the three satellites are in this position again, each one will have made some number of complete orbits. Since satellite A takes 18 hours to make one complete orbit of the earth, our solution must be a multiple of 18. Similarly, our solution must also be a multiple of 14 and a multiple of 10 to account for the complete orbits of satellites B and C.

The LCM of 18, 14, and 10 will tell us the next time satellites A, B, and C will all complete an orbit at the same time.

$$\left.\begin{array}{l} 18 = 2 \cdot 3^2 \\ 14 = 2 \cdot 7 \\ 10 = 2 \cdot 5 \end{array}\right\} \quad \text{LCM} = 2 \cdot 3^2 \cdot 5 \cdot 7 = 630$$

Thus, the satellites will not align again for 630 hours (or 26 days and 6 hours).

Note: Satellite A will have made 35 orbits, since $630 = 18 \cdot 35$.
　　　Satellite B will have made 45 orbits, since $630 = 14 \cdot 45$.
　　　Satellite C will have made 63 orbits, since $630 = 10 \cdot 63$.

Answers to Marginal Exercises

1. LCM = 140　　**2.** (a) LCM = 150　(b) $150 = 30 \cdot 5$;　$150 = 50 \cdot 3$
3. (a) LCM = 900　(b) $900 = 18 \cdot 50$;　$900 = 20 \cdot 45$;　$900 = 25 \cdot 36$

NAME _____ SECTION _____ DATE _____

EXERCISES **2.6** _____

Find the LCM of each of the following sets of numbers.

1. 8, 12   **2.** 3, 5, 7   **3.** 4, 6, 9

4. 3, 5, 9   **5.** 2, 5, 11   **6.** 4, 14, 18

7. 6, 15, 12   **8.** 6, 8, 27   **9.** 25, 40

10. 40, 75   **11.** 28, 98   **12.** 30, 75

13. 30, 80   **14.** 16, 28   **15.** 25, 100

16. 20, 50   **17.** 35, 100   **18.** 144, 216

Answers

1. _____

2. _____

3. _____

4. _____

5. _____

6. _____

7. _____

8. _____

9. _____

10. _____

11. _____

12. _____

13. _____

14. _____

15. _____

16. _____

17. _____

18. _____

Answers

19. _____

20. _____

21. _____

22. _____

23. _____

24. _____

25. _____

26. _____

27. _____

28. _____

29. _____

30. _____

31. _____

32. _____

33. _____

34. _____

35. _____

36. _____

19. 36, 42

20. 40, 100

21. 2, 4, 8

22. 10, 15, 35

23. 8, 13, 15

24. 25, 35, 49

25. 6, 12, 15

26. 8, 10, 120

27. 6, 15, 80

28. 13, 26, 169

29. 45, 125, 150

30. 34, 51, 54

31. 33, 66, 121

32. 36, 54, 72

33. 45, 145, 290

34. 54, 81, 108

35. 45, 75, 135

36. 35, 40, 72

37. 10, 20, 30, 40   **38.** 15, 25, 30, 40   **39.** 24, 40, 48, 56

40. 169, 637, 845

(a) Find the LCM. (b) State the number of times each number divides into the LCM.

41. 8, 10, 15   **42.** 6, 15, 30   **43.** 10, 15, 24

44. 8, 10, 120   **45.** 6, 18, 27, 45   **46.** 12, 95, 228

47. 45, 63, 98   **48.** 40, 56, 196   **49.** 99, 143, 363

50. 125, 135, 225

51. Two long-distance joggers are running on the same course in the same direction. They meet at the water fountain and say "Hi." One jogger goes around the course in 10 minutes and the other goes around the course in 14 minutes. They continue to jog until they meet again at the water fountain. (a) How many minutes elapsed between meetings at the fountain? (b) How many times will each have jogged around the course?

Answers

37. _____

38. _____

39. _____

40. _____

41. _____

42. _____

43. _____

44. _____

45. _____

46. _____

47. _____

48. _____

49. _____

50. _____

51. _____

52. _____

53. _____

54. _____

55. _____

52. Three night watchmen walk around inspecting buildings at a shopping center. The watchmen take 9, 12, and 14 minutes, respectively, for the inspection trip. (a) If they start at the same time, in how many minutes will they meet? (b) How many inspection trips will each watchman have made?

53. Two astronauts miss connections at their first rendezvous in space. (a) If one astronaut circles the earth every 12 hours and the other every 16 hours, in how many hours will they rendezvous again? (b) How many orbits will each astronaut have made between the first and second rendezvouses?

54. Three truck drivers eat lunch together whenever all three are at the routing station at the same time. The first driver's route takes 5 days, the second driver's takes 15 days, and the third driver's takes 6 days. (a) How often do the three drivers each lunch together? (b) If the first driver's route was changed to take 6 days, how often would they eat lunch together?

55. Four book salespersons leave the home office the same day. They take 10 days, 12 days, 15 days, and 18 days, respectively, to travel their own sales regions. (a) In how many days will they all meet again at the home office? (b) How many sales trips will each have made?

NAME _____ SECTION _____ DATE _____

CHAPTER **2** TEST _____

In Problems 1–4, find the value of each expression by using the rules for order of operations.

1. $12 + 9 \div 3 - 2$ **2.** $60 \div 4(4 - 1) + 3$

3. $2 \cdot 3^2 - (2^2 \cdot 3 \div 2) - 3(5 - 1)$ **4.** $12 \cdot 2^3 \div 4 \cdot 3$

In Problems 5–8, use the tests for divisibility to determine if the given number can be divided exactly by 2, 3, 4, 5, 9, and 10.

5. 90   **6.** 221

7. 324   **8.** 1700

9. List the multiples of 13 that are less than 100.

10. The only counting number that is neither prime nor composite is

_____ .

Answers

1. _____

2. _____

3. _____

4. _____

5. _____

6. _____

7. _____

8. _____

9. _____

10. _____

Answers

11. _____

12. _____

13. _____

14. _____

15. _____

16. _____

17. _____

18. _____

19. _____

20. _____

11. Circle all of the prime numbers in the following list:

1, 2, 4, 7, 16, 18, 21, 23, 33, 39, 42, 47, 51

In Problems 12–14, find the prime factorization of each number.

12. 70

13. 124

14. 650

15. List all the factors of 60.

In Problems 16–19, find the LCM of each set of numbers.

16. 24 and 54   **17.** 4, 14, and 21

18. 6, 15, and 60   **19.** 4, 7, and 15

20. Three salesmen have lunch together each time they are in the home office on the same day. If it takes Salesman A 6 days to cover his territory, Salesman B 9 days, and Salesman C 12 days, how often do they have lunch together? How many times does each salesman cover his territory between lunches with his friends?

Fractions

3.1
BASIC MULTIPLICATION

Fraction:* A number that can be written in the form $\frac{a}{b}$, where a is a whole number and b is a nonzero whole number

$$\frac{a}{b} \begin{array}{l} \leftarrow \text{numerator} \\ \leftarrow \text{denominator} \end{array}$$

Fractions can be used to indicate

1. division,
2. equal parts of a whole.

Example 1 $\frac{1}{2}$ 1 of 2 equal parts

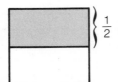

Example 2 $\frac{3}{4}$ 3 of 4 equal parts

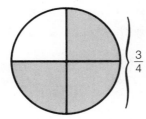

---

*The term **rational number** applies to this type of fraction. Other types of fractions are discussed in later courses in mathematics.

Example 3 $\frac{2}{3}$ 2 of 3 equal parts

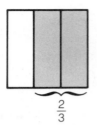

$\underbrace{\hspace{1.2cm}}$
$\frac{2}{3}$

> **Improper fraction:** Numerator is larger than denominator
>
> **Note:** There is nothing "improper" about such fractions. Improper fractions are used in algebra and other mathematics courses.

Example 4 $\frac{5}{3}$ 5 of 3 equal parts (more than a whole)

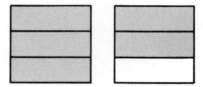

Example 5 $\frac{13}{6}$ 13 of 6 equal parts

To multiply fractions,

1. multiply the numerators;
2. multiply the denominators.

$$\frac{a}{b} \cdot \frac{c}{d} = \frac{a \cdot c}{b \cdot d}$$

Finding the product of two fractions can be thought of as finding one fractional part **of** another fraction. Thus, when we multiply $\frac{1}{3} \cdot \frac{1}{2}$, we are finding $\frac{1}{3}$ of $\frac{1}{2}$.

Example 6   Find the product of $\frac{2}{3}$ and $\frac{4}{5}$.

$$\frac{2}{3} \cdot \frac{4}{5} = \frac{2 \cdot 4}{3 \cdot 5} = \frac{8}{15}$$

This product can be illustrated graphically. We can think of the multiplication as finding $\frac{2}{3}$ **of** $\frac{4}{5}$.

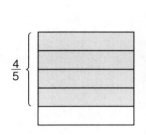 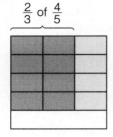 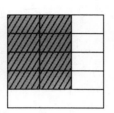

(a) Shade $\frac{4}{5}$.   (b) Shade $\frac{2}{3}$ of $\frac{4}{5}$.   (c) Shaded region is

$$\frac{2}{3} \text{ of } \frac{4}{5} \text{ or } \frac{2}{3} \cdot \frac{4}{5} = \frac{8}{15} \text{ .}$$

Example 7   Find the product of $\frac{1}{3}$ and $\frac{2}{5}$ and illustrate the product as in Example 6 by shading parts of a square.

$$\frac{1}{3} \cdot \frac{2}{5} = \frac{1 \cdot 2}{3 \cdot 5} = \frac{2}{15}$$

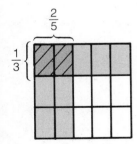

The square is separated into thirds in one direction and into fifths in the other direction. There are 15 equal regions. The overlapping shaded region represents

$$\frac{1}{3} \text{ of } \frac{2}{5} \text{ or } \frac{1}{3} \cdot \frac{2}{5} = \frac{2}{15}$$

☐ **NOW WORK EXERCISE 1 IN THE MARGIN.**

---

1. Find the product of $\frac{3}{4}$ and $\frac{3}{5}$ and illustrate the product with appropriate shading in a square.

Find the following products.

2. $\dfrac{1}{7} \cdot \dfrac{3}{5} =$

3. $\dfrac{0}{4} \cdot \dfrac{5}{8} =$

4. $\dfrac{3}{14} \cdot 5 =$

5. $\dfrac{1}{2} \cdot \dfrac{3}{4} \cdot \dfrac{7}{5} =$

6. $\dfrac{7}{8} \cdot \dfrac{3}{4} \cdot 1 =$

Completion Example 8   Find the following products. Remember that for any whole number a, $a = \dfrac{a}{1}$. $\left(\text{For example, } 6 = \dfrac{6}{1}.\right)$

(a) $\dfrac{2}{3} \cdot \dfrac{5}{7} = \dfrac{2 \cdot 5}{3 \cdot 7} = $ _____

(b) $\dfrac{1}{4} \cdot \dfrac{3}{5} = \dfrac{1 \cdot 3}{4 \cdot 5} = $ _____

(c) $\dfrac{7}{5} \cdot 2 = \dfrac{7}{5} \cdot \dfrac{2}{1} = $ _____

(d) $\dfrac{1}{4} \cdot \dfrac{3}{5} \cdot \dfrac{7}{2} = \dfrac{1 \cdot 3 \cdot 7}{\rule{1.5cm}{0.4pt}} = $ _____

(e) $\dfrac{2}{3} \cdot \dfrac{11}{15} \cdot \dfrac{1}{7} = $ _____

☐ **NOW WORK EXERCISES 2–6 IN THE MARGIN.**

> **COMMUTATIVE PROPERTY OF MULTIPLICATION**
>
> If $\dfrac{a}{b}$ and $\dfrac{c}{d}$ are fractions, then
>
> $$\dfrac{a}{b} \cdot \dfrac{c}{d} = \dfrac{c}{d} \cdot \dfrac{a}{b}$$

> **ASSOCIATIVE PROPERTY OF MULTIPLICATION**
>
> If $\dfrac{a}{b}, \dfrac{c}{d}$, and $\dfrac{e}{f}$ are fractions, then
>
> $$\dfrac{a}{b} \cdot \dfrac{c}{d} \cdot \dfrac{e}{f} = \left(\dfrac{a}{b} \cdot \dfrac{c}{d}\right) \cdot \dfrac{e}{f} = \dfrac{a}{b} \cdot \left(\dfrac{c}{d} \cdot \dfrac{e}{f}\right)$$

Completion Example Answer

8. (a) $\dfrac{10}{21}$

   (b) $\dfrac{3}{20}$

   (c) $\dfrac{14}{5}$

   (d) $\dfrac{1 \cdot 3 \cdot 7}{4 \cdot 5 \cdot 2} = \dfrac{21}{40}$

   (e) $\dfrac{2 \cdot 11 \cdot 1}{3 \cdot 15 \cdot 7} = \dfrac{22}{315}$

Example 9 $\dfrac{3}{4} \cdot \dfrac{1}{7} = \dfrac{1}{7} \cdot \dfrac{3}{4}$ Illustrates the commutative property of multiplication.

$$\dfrac{3}{4} \cdot \dfrac{1}{7} = \dfrac{3 \cdot 1}{4 \cdot 7} = \dfrac{3}{28} \quad \text{and} \quad \dfrac{1}{7} \cdot \dfrac{3}{4} = \dfrac{1 \cdot 3}{7 \cdot 4} = \dfrac{3}{28}$$

Example 10 $\left(\dfrac{5}{8} \cdot \dfrac{3}{2}\right) \cdot \dfrac{3}{11} = \dfrac{5}{8} \cdot \left(\dfrac{3}{2} \cdot \dfrac{3}{11}\right)$ Illustrates the associative property of multiplication.

$$\left(\dfrac{5}{8} \cdot \dfrac{3}{2}\right) \cdot \dfrac{3}{11} = \left(\dfrac{5 \cdot 3}{8 \cdot 2}\right) \cdot \dfrac{3}{11} = \dfrac{15}{16} \cdot \dfrac{3}{11} = \dfrac{15 \cdot 3}{16 \cdot 11} = \dfrac{45}{176}$$

$$\text{and} \quad \dfrac{5}{8} \cdot \left(\dfrac{3}{2} \cdot \dfrac{3}{11}\right) = \dfrac{5}{8} \cdot \left(\dfrac{3 \cdot 3}{2 \cdot 11}\right) = \dfrac{5}{8} \cdot \dfrac{9}{22} = \dfrac{5 \cdot 9}{8 \cdot 22} = \dfrac{45}{176}$$

☐ **NOW WORK EXERCISES 7–10 IN THE MARGIN.**

We stated earlier that in a fraction $\dfrac{a}{b}$, b is nonzero. That is, $b \neq 0$ ($\neq$ means is not equal to). Under the division meaning of fractions, $\dfrac{15}{5}$ is the same as $15 \div 5$. By the meaning of division related to multiplication, $\dfrac{15}{5} = 3$ because $15 = 5 \cdot 3$. Similarly, $\dfrac{0}{4} = 0$ because $0 = 4 \cdot 0$.

In general,

$$\dfrac{0}{b} = 0 \quad \text{if } b \neq 0.$$

The following discussion explains why division by 0 is undefined.

> **DIVISION BY 0 IS UNDEFINED. NO DENOMINATOR CAN BE 0.**
>
> 1. Consider $\dfrac{5}{0} = \boxed{?}$. Then, we must have $5 = 0 \cdot \boxed{?}$. But this is impossible because $0 \cdot \boxed{?} = 0$ and $5 \neq 0$. Therefore, $\dfrac{5}{0}$ is undefined.
>
> 2. Next, consider $\dfrac{0}{0} = \boxed{?}$. Then, we must have $0 = 0 \cdot \boxed{?}$. But $0 = 0 \cdot \boxed{?}$ is true for any value of $\boxed{?}$, and in arithmetic, an operation such as division cannot give more than one answer. Thus, we agree that $\dfrac{0}{0}$ is undefined.
>
> **Thus, $\dfrac{a}{0}$ is undefined for any value of a.**

State which property of multiplication is illustrated in each problem.

7. $\dfrac{1}{2} \cdot \dfrac{5}{8} = \dfrac{5}{8} \cdot \dfrac{1}{2}$

8. $\dfrac{3}{4} \cdot 6 = 6 \cdot \dfrac{3}{4}$

9. $\dfrac{1}{3} \cdot \left(\dfrac{1}{2} \cdot \dfrac{1}{4}\right) = \left(\dfrac{1}{3} \cdot \dfrac{1}{2}\right) \cdot \dfrac{1}{4}$

10. $\dfrac{5}{6} \cdot \left(\dfrac{7}{8} \cdot \dfrac{11}{13}\right) = \left(\dfrac{5}{6} \cdot \dfrac{7}{8}\right) \cdot \dfrac{11}{13}$

The following examples illustrate uses of 0 in fractions.

Example 11 $\frac{13}{0}$ is undefined.

Example 12 $\frac{0}{20} = 0$

Example 13 $\frac{0}{3} \cdot \frac{5}{8} = \frac{0}{24} = 0$

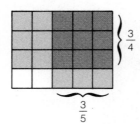

NAME _____ SECTION _____ DATE _____

EXERCISES **3.1** _____

1. Each of the following products has the same value. What is that value?

   (a) $\dfrac{0}{4} \cdot \dfrac{5}{6}$ (b) $\dfrac{3}{10} \cdot \dfrac{0}{8}$ (c) $\dfrac{1}{10} \cdot \dfrac{0}{10}$ (d) $\dfrac{0}{6} \cdot \dfrac{0}{52}$

2. What is the value, if any, of each of the following expressions?

   (a) $\dfrac{7}{0}$ (b) $\dfrac{16}{0}$ (c) $\dfrac{75}{0} \cdot \dfrac{2}{0}$ (d) $\dfrac{1}{0} \cdot \dfrac{0}{2}$

Write a rational number that represents the shaded parts in each of the following diagrams.

3.

4.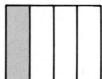

5.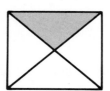

6.

7.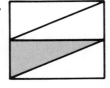

8.

Answers

1. (a) _____

   (b) _____

   (c) _____

   (d) _____

2. (a) _____

   (b) _____

   (c) _____

   (d) _____

3. _____

4. _____

5. _____

6. _____

7. _____

8. _____

Answers

9. _____

10. _____

11. _____

12. _____

13. _____

14. _____

15. _____

16. _____

17. _____

18. _____

19. _____

20. _____

21. _____

22. _____

23. _____

24. _____

25. _____

In Exercises 9–13, illustrate each product with appropriate shading in a square (similar to Examples 6 and 7 in the text).

9. $\dfrac{1}{5} \cdot \dfrac{3}{4}$

10. $\dfrac{5}{6} \cdot \dfrac{5}{6}$

11. $\dfrac{7}{8} \cdot \dfrac{1}{4}$

12. $\dfrac{2}{3} \cdot \dfrac{4}{7}$

13. $\dfrac{2}{9} \cdot \dfrac{2}{3}$

In Exercises 14–19, find the fractional parts as indicated.

14. Find $\dfrac{1}{5}$ of $\dfrac{3}{5}$.

15. Find $\dfrac{1}{4}$ of $\dfrac{1}{4}$.

16. Find $\dfrac{1}{3}$ of $\dfrac{2}{3}$.

17. Find $\dfrac{6}{7}$ of $\dfrac{2}{5}$.

18. Find $\dfrac{4}{7}$ of $\dfrac{3}{5}$.

19. Find $\dfrac{1}{3}$ of $\dfrac{1}{3}$.

Find the following products.

20. $\dfrac{1}{2} \cdot \dfrac{1}{2}$

21. $\dfrac{3}{16} \cdot \dfrac{1}{2}$

22. $\dfrac{2}{5} \cdot \dfrac{2}{5}$

23. $\dfrac{3}{7} \cdot \dfrac{3}{7}$

24. $\dfrac{1}{2} \cdot \dfrac{3}{4}$

25. $\dfrac{5}{8} \cdot \dfrac{3}{4}$

26. $\dfrac{1}{9} \cdot \dfrac{4}{9}$

27. $\dfrac{0}{3} \cdot \dfrac{5}{7}$

28. $\dfrac{0}{4} \cdot \dfrac{7}{6}$

29. $\dfrac{7}{6} \cdot \dfrac{5}{2}$

30. $\dfrac{4}{1} \cdot \dfrac{3}{1}$

31. $\dfrac{2}{1} \cdot \dfrac{5}{1}$

32. $\dfrac{14}{1} \cdot \dfrac{0}{2}$

33. $\dfrac{15}{1} \cdot \dfrac{3}{2}$

34. $\dfrac{6}{5} \cdot \dfrac{7}{1}$

35. $\dfrac{8}{5} \cdot \dfrac{4}{3}$

36. $\dfrac{5}{6} \cdot \dfrac{11}{3}$

37. $\dfrac{9}{4} \cdot \dfrac{11}{5}$

38. $\dfrac{1}{5} \cdot \dfrac{2}{7} \cdot \dfrac{3}{11}$

39. $\dfrac{4}{13} \cdot \dfrac{2}{5} \cdot \dfrac{6}{7}$

40. $\dfrac{7}{8} \cdot \dfrac{7}{9} \cdot \dfrac{7}{3}$

41. $\dfrac{1}{6} \cdot \dfrac{1}{10} \cdot \dfrac{1}{6}$

42. $\dfrac{9}{100} \cdot 1 \cdot 3$

43. $\dfrac{5}{1} \cdot \dfrac{12}{1} \cdot \dfrac{14}{1}$

Answers

26. _____

27. _____

28. _____

29. _____

30. _____

31. _____

32. _____

33. _____

34. _____

35. _____

36. _____

37. _____

38. _____

39. _____

40. _____

41. _____

42. _____

43. _____

Answers

44. _____

45. _____

46. _____

47. _____

48. _____

49. _____

50. _____

51. _____

52. _____

53. _____

54. _____

44. $\dfrac{1}{3} \cdot \dfrac{1}{10} \cdot \dfrac{7}{3}$

45. $\dfrac{3}{11} \cdot \dfrac{0}{8} \cdot \dfrac{6}{7}$

46. $\dfrac{0}{4} \cdot \dfrac{3}{8} \cdot \dfrac{1}{5}$

47. $\dfrac{5}{6} \cdot \dfrac{5}{11} \cdot \dfrac{5}{7} \cdot \dfrac{0}{3}$

48. $\left(\dfrac{3}{4}\right)\left(\dfrac{5}{8}\right)\left(\dfrac{11}{13}\right)$

49. $\left(\dfrac{7}{5}\right)\left(\dfrac{8}{3}\right)\left(\dfrac{13}{3}\right)$

50. $\left(\dfrac{1}{10}\right)\left(\dfrac{3}{5}\right)\left(\dfrac{3}{10}\right)$

Tell which property of multiplication is illustrated in each problem.

51. $\dfrac{5}{9} \cdot \dfrac{11}{14} = \dfrac{11}{14} \cdot \dfrac{5}{9}$

52. $\dfrac{3}{5} \cdot \left(\dfrac{7}{8} \cdot \dfrac{1}{2}\right) = \left(\dfrac{3}{5} \cdot \dfrac{7}{8}\right) \cdot \dfrac{1}{2}$

53. $\left(\dfrac{6}{11} \cdot \dfrac{13}{5}\right) \cdot \dfrac{7}{5} = \dfrac{6}{11} \cdot \left(\dfrac{13}{5} \cdot \dfrac{7}{5}\right)$

54. $\dfrac{2}{3} \cdot 4 = 4 \cdot \dfrac{2}{3}$

3.2
HIGHER TERMS AND REDUCING

> **Equal fractions:** Fractions that represent the same amount of the whole (Equal fractions are also called **equivalent fractions**).

Example 1   The fractions $\frac{1}{2}, \frac{2}{4}, \frac{3}{6}, \frac{4}{8}$, and $\frac{5}{10}$ are all equal (or equivalent).

Each represents the same part of a whole.   Thus, all the fractions represent the same number, even though they have different representations.

$$\frac{1}{2} = \frac{2}{4} = \frac{3}{6} = \frac{4}{8} = \frac{5}{10}$$

Example 2 $\frac{1}{2} = \frac{2}{4} = \frac{3}{6} = \frac{4}{8} = \frac{5}{10} = \frac{6}{12} = \frac{7}{14}$ and so on.

$\frac{2}{3} = \frac{4}{6} = \frac{6}{9} = \frac{8}{12} = \frac{10}{15}$ and so on.

With whole numbers,

$$0 = \frac{0}{1} = \frac{0}{2} = \frac{0}{3} = \frac{0}{4} = \frac{0}{5}$$ and so on.

$$1 = \frac{1}{1} = \frac{2}{2} = \frac{3}{3} = \frac{4}{4} = \frac{5}{5}$$ and so on.   In fact, $1 = \frac{k}{k}$ where k is not zero.

$$2 = \frac{2}{1} = \frac{4}{2} = \frac{6}{3} = \frac{8}{4} = \frac{10}{5}$$ and so on.

---

1 IS THE MULTIPLICATIVE IDENTITY

1. If $\dfrac{a}{b}$ is a fraction, then $\dfrac{a}{b} \cdot 1 = \dfrac{a}{b}$.

2. $\dfrac{a}{b} = \dfrac{a}{b} \cdot 1 = \dfrac{a}{b} \cdot \dfrac{k}{k} = \dfrac{a \cdot k}{b \cdot k}$ where $k \neq 0 \left(1 = \dfrac{k}{k} \right)$.

In effect, the value of a fraction $\dfrac{a}{b}$ is unchanged if both the numerator a and the denominator b are multiplied by the same nonzero number k.

We can use the multiplicative identity to

1. raise to higher terms (find an equivalent fraction with a larger denominator).

2. reduce (find an equivalent fraction with a smaller denominator).

In Examples 3–6, raise to higher terms as indicated using nonzero k.

Example 3 $\dfrac{3}{4} = \dfrac{?}{28}$

We know $4 \cdot 7 = 28$, so use $1 = \dfrac{k}{k} = \dfrac{7}{7}$.

$$\frac{3}{4} = \frac{3}{4} \cdot 1 = \frac{3}{4} \cdot \frac{7}{7} = \frac{21}{28}$$

Example 4 $\dfrac{9}{10} = \dfrac{?}{30}$

Since $10 \cdot 3 = 30$, use $1 = \dfrac{k}{k} = \dfrac{3}{3}$.

$$\frac{9}{10} = \frac{9}{10} \cdot 1 = \frac{9}{10} \cdot \frac{3}{3} = \frac{27}{30}$$

Example 5 $\dfrac{9}{10} = \dfrac{?}{40}$

Since $10 \cdot 4 = 40$, use $1 = \dfrac{k}{k} = \dfrac{4}{4}$.

$$\frac{9}{10} = \frac{9}{10} \cdot 1 = \frac{9}{10} \cdot \frac{4}{4} = \frac{36}{40}$$

Example 6 $\dfrac{11}{8} = \dfrac{?}{40}$

Since $8 \cdot 5 = 40$, use $1 = \dfrac{5}{5}$.

$$\frac{11}{8} = \frac{11}{8} \cdot \frac{5}{5} = \frac{55}{40}$$

Completion Example 7 Find a fraction with denominator 35 equal to $\dfrac{3}{5}$.

$$\frac{3}{5} = \frac{3}{5} \cdot 1 = \frac{3}{5} \cdot \underline{\quad} = \frac{\quad}{35}$$

Completion Example 8 Find a fraction with numerator 28 equal to $\dfrac{2}{3}$.

$$\frac{2}{3} = \frac{2}{3} \cdot 1 = \frac{2}{3} \cdot \underline{\quad} = \frac{28}{\quad}$$

☐ **NOW WORK EXERCISES 1–4 IN THE MARGIN.**

> A fraction is in **lowest terms** if the numerator and denominator have no common factor other than 1. To reduce to lowest terms,
>
> 1. factor the numerator and denominator into prime factors;
>
> 2. use the fact that $\dfrac{k}{k} = 1$.

Reduce each fraction in Examples 9–13 to its lowest terms.

Example 9 $\dfrac{14}{21} = \dfrac{2 \cdot 7}{3 \cdot 7} = \dfrac{2}{3} \cdot \dfrac{7}{7} = \dfrac{2}{3} \cdot 1 = \dfrac{2}{3}$

Example 10 $\dfrac{15}{20} = \dfrac{3 \cdot 5}{2 \cdot 2 \cdot 5} = \dfrac{3}{4} \cdot \dfrac{5}{5} = \dfrac{3}{4} \cdot 1 = \dfrac{3}{4}$

Find the missing numerator or denominator that will make the fractions equal.

1. $\dfrac{3}{5} = \dfrac{}{20}$

2. $\dfrac{1}{9} = \dfrac{}{72}$

3. $\dfrac{5}{8} = \dfrac{25}{}$

4. $\dfrac{3}{11} = \dfrac{12}{}$

Completion Example Answers

7. $\dfrac{3}{5} = \dfrac{3}{5} \cdot 1 = \dfrac{3}{5} \cdot \dfrac{7}{7} = \dfrac{21}{35}$

8. $\dfrac{2}{3} = \dfrac{2}{3} \cdot 1 = \dfrac{2}{3} \cdot \dfrac{14}{14} = \dfrac{28}{42}$

Example 11 $\dfrac{44}{20}$

We can just cross out common factors (prime or not) with the understanding that $\dfrac{k}{k} = 1$.

(a) $\dfrac{44}{20} = \dfrac{\cancel{4} \cdot 11}{\cancel{4} \cdot 5} = \dfrac{11}{5}$

Or, using prime factors,

(b) $\dfrac{44}{20} = \dfrac{\cancel{2} \cdot \cancel{2} \cdot 11}{\cancel{2} \cdot \cancel{2} \cdot 5} = \dfrac{11}{5}$

Example 12 $\dfrac{8}{72}$

Remember that 1 is a factor of any whole number. So, if all the factors in the numerator or denominator are crossed out, 1 must be used as a factor.

(a) Using prime factors,

$$\dfrac{8}{72} = \dfrac{\cancel{2} \cdot \cancel{2} \cdot \cancel{2} \cdot 1}{\cancel{2} \cdot \cancel{2} \cdot \cancel{2} \cdot 3 \cdot 3} = \dfrac{1}{9} \qquad \text{1 is used as a factor in the numerator.}$$

(b) Or, if you see that 8 is a common factor, cross it out. But remember that 1 is a factor.

$$\dfrac{8}{72} = \dfrac{\cancel{8} \cdot 1}{\cancel{8} \cdot 9} = \dfrac{1}{9}$$

Example 13 $\dfrac{36}{27}$

(a) Using prime factors,

$$\dfrac{36}{27} = \dfrac{2 \cdot 2 \cdot \cancel{3} \cdot \cancel{3}}{3 \cdot \cancel{3} \cdot \cancel{3}} = \dfrac{4}{3}$$

(b) Or, using 3 as a factor, we get an incomplete answer.

$$\dfrac{36}{27} = \dfrac{\cancel{3} \cdot 12}{\cancel{3} \cdot 9} = \dfrac{12}{9} \qquad \text{not in lowest terms}$$

By using prime factors, you can be certain that the fraction is reduced to lowest terms. You may use larger numbers but be sure you have the largest common factor.

$$\dfrac{36}{27} = \dfrac{4 \cdot \cancel{9}}{\cancel{9} \cdot 3} = \dfrac{4}{3}$$

Completion Example 14 Reduce $\dfrac{12}{18}$ to lowest terms.

(a) $\dfrac{12}{18} = \dfrac{2 \cdot 2 \cdot 3}{2 \cdot 3 \cdot 3} =$ _____

Or,

(b) $\dfrac{12}{18} = \dfrac{2 \cdot ?}{3 \cdot ?} =$ _____

Completion Example 15 Reduce $\dfrac{52}{65}$ to lowest terms.

Finding a common factor could be difficult here. Prime factoring helps.

$$\frac{52}{65} = \frac{2 \cdot 2 \cdot ?}{5 \cdot ?} = \underline{\hspace{2cm}}$$

☐ **NOW WORK EXERCISES 5–10 IN THE MARGIN.**

Reduce each fraction to lowest terms.

5. $\dfrac{8}{36} =$

6. $\dfrac{35}{40} =$

7. $\dfrac{66}{44} =$

8. $\dfrac{12}{25} =$

9. $\dfrac{10}{60} =$

10. $\dfrac{78}{104} =$

Completion Example Answers

14. (a) $\dfrac{12}{18} = \dfrac{2 \cdot \cancel{2} \cdot \cancel{3}}{\cancel{2} \cdot \cancel{3} \cdot 3} = \dfrac{2}{3}$

(b) $\dfrac{12}{18} = \dfrac{2 \cdot \cancel{6}}{3 \cdot \cancel{6}} = \dfrac{2}{3}$

15. $\dfrac{52}{65} = \dfrac{2 \cdot 2 \cdot \cancel{13}}{5 \cdot \cancel{13}} = \dfrac{4}{5}$

Answers to Marginal Exercises

1. $\dfrac{3}{5} = \dfrac{12}{20}$ **2.** $\dfrac{1}{9} = \dfrac{8}{72}$ **3.** $\dfrac{5}{8} = \dfrac{25}{40}$ **4.** $\dfrac{3}{11} = \dfrac{12}{44}$ **5.** $\dfrac{2}{9}$ **6.** $\dfrac{7}{8}$ **7.** $\dfrac{3}{2}$

8. $\dfrac{12}{25}$ (already reduced)   **9.** $\dfrac{1}{6}$ **10.** $\dfrac{3}{4}$

NAME _____ SECTION _____ DATE _____

EXERCISES **3.2** _____

Find the missing numerator or denominator.

1. $\dfrac{7}{8} = \dfrac{}{24}$

2. $\dfrac{1}{16} = \dfrac{}{64}$

3. $\dfrac{2}{5} = \dfrac{}{25}$

4. $\dfrac{6}{7} = \dfrac{}{49}$

5. $\dfrac{1}{9} = \dfrac{5}{}$

6. $\dfrac{3}{4} = \dfrac{15}{}$

7. $\dfrac{5}{8} = \dfrac{10}{}$

8. $\dfrac{6}{5} = \dfrac{}{45}$

9. $\dfrac{14}{3} = \dfrac{}{9}$

10. $\dfrac{5}{8} = \dfrac{}{96}$

11. $\dfrac{9}{16} = \dfrac{}{96}$

12. $\dfrac{7}{2} = \dfrac{}{20}$

13. $\dfrac{10}{11} = \dfrac{}{44}$

14. $\dfrac{3}{16} = \dfrac{}{80}$

15. $\dfrac{11}{12} = \dfrac{}{48}$

Answers

1. _____

2. _____

3. _____

4. _____

5. _____

6. _____

7. _____

8. _____

9. _____

10. _____

11. _____

12. _____

13. _____

14. _____

15. _____

Answers

16. _____

17. _____

18. _____

19. _____

20. _____

21. _____

22. _____

23. _____

24. _____

25. _____

26. _____

27. _____

28. _____

29. _____

30. _____

31. _____

32. _____

16. $\dfrac{3}{7} = \dfrac{}{105}$ 17. $\dfrac{5}{21} = \dfrac{}{42}$ 18. $\dfrac{2}{3} = \dfrac{}{48}$

19. $\dfrac{5}{12} = \dfrac{}{108}$ 20. $\dfrac{1}{13} = \dfrac{}{39}$ 21. $\dfrac{12}{14} = \dfrac{}{7}$

22. $\dfrac{18}{40} = \dfrac{}{20}$ 23. $\dfrac{90}{100} = \dfrac{}{10}$ 24. $\dfrac{80}{100} = \dfrac{}{10}$

25. $\dfrac{60}{100} = \dfrac{}{10}$ 26. $\dfrac{42}{70} = \dfrac{}{10}$ 27. $\dfrac{10}{100} = \dfrac{}{10}$

28. $\dfrac{66}{60} = \dfrac{}{10}$ 29. $\dfrac{88}{80} = \dfrac{}{10}$ 30. $\dfrac{64}{48} = \dfrac{}{6}$

31. $\dfrac{3}{100} = \dfrac{}{1000}$ 32. $\dfrac{9}{10} = \dfrac{}{1000}$

NAME _____ SECTION _____ DATE _____

Reduce to lowest terms. Rewrite the fraction if it is already reduced.

33. $\dfrac{3}{9}$

34. $\dfrac{16}{24}$

35. $\dfrac{9}{12}$

36. $\dfrac{6}{20}$

37. $\dfrac{16}{40}$

38. $\dfrac{24}{30}$

39. $\dfrac{14}{36}$

40. $\dfrac{5}{11}$

41. $\dfrac{0}{25}$

42. $\dfrac{75}{100}$

43. $\dfrac{22}{55}$

44. $\dfrac{60}{75}$

45. $\dfrac{30}{36}$

46. $\dfrac{7}{28}$

47. $\dfrac{26}{39}$

48. $\dfrac{27}{56}$

Answers

33. _____

34. _____

35. _____

36. _____

37. _____

38. _____

39. _____

40. _____

41. _____

42. _____

43. _____

44. _____

45. _____

46. _____

47. _____

48. _____

Answers

49. _____

50. _____

51. _____

52. _____

53. _____

54. _____

55. _____

56. _____

57. _____

58. _____

59. _____

60. _____

61. _____

62. _____

63. _____

64. _____

65. _____

49. $\dfrac{34}{51}$ **50.** $\dfrac{36}{48}$ **51.** $\dfrac{24}{100}$ **52.** $\dfrac{16}{32}$

53. $\dfrac{30}{45}$ **54.** $\dfrac{28}{42}$ **55.** $\dfrac{12}{35}$ **56.** $\dfrac{66}{84}$

57. $\dfrac{14}{63}$ **58.** $\dfrac{30}{70}$ **59.** $\dfrac{25}{76}$ **60.** $\dfrac{70}{84}$

61. $\dfrac{50}{100}$ **62.** $\dfrac{48}{12}$ **63.** $\dfrac{54}{9}$ **64.** $\dfrac{51}{6}$

65. $\dfrac{6}{51}$

3.3
MULTIPLICATION AND REDUCING

> **MULTIPLYING TWO OR MORE FRACTIONS**
>
> 1. Poor method
>   (a) Multiply all numerators and all denominators.
>   (b) Factor the numerator and denominator and reduce.
> 2. Good method
>   (a) Factor numerators and denominators into prime factors and reduce (one-step process).

Example 1 $\dfrac{15}{28} \cdot \dfrac{7}{9}$

POOR METHOD

$$\frac{15}{28} \cdot \frac{7}{9} = \frac{15 \cdot 7}{28 \cdot 9} = \frac{105}{252}$$

Now factor and reduce.

$$\frac{105}{252} = \frac{5 \cdot 21}{4 \cdot 63} = \frac{5 \cdot \cancel{3} \cdot \cancel{7}}{2 \cdot 2 \cdot 3 \cdot \cancel{3} \cdot \cancel{7}} = \frac{5}{12}$$

GOOD METHOD

$$\frac{15}{28} \cdot \frac{7}{9} = \frac{15 \cdot 7}{28 \cdot 9} = \frac{\cancel{3} \cdot 5 \cdot \cancel{7}}{2 \cdot 2 \cdot \cancel{7} \cdot 3 \cdot \cancel{3}} = \frac{5}{2 \cdot 2 \cdot 3} = \frac{5}{12}$$

Only the good method is used to find the products in Examples 2 and 3.

Example 2 $\dfrac{9}{10} \cdot \dfrac{25}{32} \cdot \dfrac{44}{33}$

$$\frac{9}{10} \cdot \frac{25}{32} \cdot \frac{44}{33} = \frac{9 \cdot 25 \cdot 44}{10 \cdot 32 \cdot 33}$$

$$= \frac{\cancel{3} \cdot 3 \cdot \cancel{5} \cdot 5 \cdot \cancel{2} \cdot \cancel{2} \cdot \cancel{11}}{2 \cdot \cancel{5} \cdot 2 \cdot 2 \cdot 2 \cdot \cancel{2} \cdot \cancel{2} \cdot \cancel{3} \cdot \cancel{11}}$$

$$= \frac{3 \cdot 5}{2 \cdot 2 \cdot 2 \cdot 2}$$

$$= \frac{15}{16}$$

Find each product in lowest terms.

1. $\dfrac{2}{5} \cdot \dfrac{8}{5}$

2. $\dfrac{10}{9} \cdot \dfrac{3}{5}$

3. $\dfrac{5}{6} \cdot \dfrac{8}{7} \cdot \dfrac{14}{10}$

4. $6 \cdot \dfrac{7}{12} \cdot \dfrac{3}{28} \cdot \dfrac{1}{9}$

    $\left(\text{Remember } 6 = \dfrac{6}{1}.\right)$

Example 3 $\dfrac{36}{49} \cdot \dfrac{14}{75} \cdot \dfrac{15}{18}$

$$\dfrac{36}{49} \cdot \dfrac{14}{75} \cdot \dfrac{15}{18} = \dfrac{36 \cdot 14 \cdot 15}{49 \cdot 75 \cdot 18}$$

$$= \dfrac{2 \cdot 2 \cdot \cancel{3} \cdot \cancel{3} \cdot 2 \cdot \cancel{7} \cdot \cancel{3} \cdot \cancel{5}}{\cancel{7} \cdot 7 \cdot \cancel{3} \cdot 5 \cdot \cancel{5} \cdot \cancel{2} \cdot \cancel{3} \cdot \cancel{3}}$$

$$= \dfrac{2 \cdot 2}{7 \cdot 5}$$

$$= \dfrac{4}{35}$$

Completion Example 4 $\dfrac{55}{26} \cdot \dfrac{8}{44} \cdot \dfrac{91}{35}$

$$\dfrac{55}{26} \cdot \dfrac{8}{44} \cdot \dfrac{91}{35} = \dfrac{55 \cdot 8 \cdot 91}{26 \cdot 44 \cdot 35}$$

$$= \dfrac{?}{?}$$

$$= \dfrac{?}{?}$$

$$= \underline{\quad ? \quad}$$

☐ **NOW WORK EXERCISES 1–4 IN THE MARGIN.**

> **ANOTHER METHOD**
>
> Common factors (ones that you can spot easily) are divided into both a numerator and a denominator. (Be careful and be organized.)
>
> **Special note:** This method can be used only with multiplication and *not* with addition or subtraction.

Examples 1, 2, 3, and 4 are shown again using this new method.

Example 5 $\dfrac{\overset{5}{\cancel{15}}}{\underset{4}{\cancel{28}}} \cdot \dfrac{\overset{1}{\cancel{7}}}{\underset{3}{\cancel{9}}} = \dfrac{5}{12}$ 3 is divided into both 15 and 9.
7 is divided into both 7 and 28.

Completion Example Answer

4. $\dfrac{5 \cdot 11 \cdot 2 \cdot 2 \cdot 2 \cdot 7 \cdot 13}{2 \cdot 13 \cdot 2 \cdot 2 \cdot 11 \cdot 7 \cdot 5} = \dfrac{1}{1} = 1$

Example 6

$$\frac{\overset{3}{\cancel{9}}}{\underset{2}{\cancel{10}}} \cdot \frac{\overset{5}{\cancel{25}}}{\underset{8}{\cancel{32}}} \cdot \frac{\overset{1}{\cancel{44}}}{\underset{\cancel{3}}{\cancel{33}}} = \frac{15}{16}$$

11 is divided into both 44 and 33.
5 is divided into both 25 and 10.
4 is divided into both 4 and 32.
3 is divided into both 3 and 9.

Example 7

$$\frac{\overset{2}{\cancel{36}}}{\underset{7}{\cancel{49}}} \cdot \frac{\overset{2}{\cancel{14}}}{\underset{5}{\cancel{75}}} \cdot \frac{\overset{1}{\cancel{15}}}{\underset{1}{\cancel{18}}} = \frac{4}{35}$$

18 is divided into both 18 and 36.
7 is divided into both 14 and 49.
15 is divided into both 15 and 75.

Example 8

$$\frac{\overset{1}{\underset{2}{\overset{\cancel{5}}{\cancel{55}}}}}{\underset{1}{\cancel{26}}} \cdot \frac{\overset{1}{\underset{\cancel{4}}{\overset{\cancel{8}}{\cancel{8}}}}}{\underset{1}{\cancel{44}}} \cdot \frac{\overset{1}{\underset{\cancel{7}}{\overset{\cancel{7}}{\cancel{91}}}}}{\underset{1}{\cancel{35}}} = \frac{1}{1}$$

11 is divided into both 55 and 44.
13 is divided into both 26 and 91.
5 is divided into 5 and 35.
7 is divided into both 7 and 7.
2 is divided into both 2 and 8.
4 is divided into both 4 and 4.

☐ **NOW WORK EXERCISES 5–7 IN THE MARGIN.**

Multiply and reduce. Use the method that best suits your own needs and abilities.

5. $\dfrac{16}{27} \cdot \dfrac{9}{4}$

6. $\dfrac{14}{15} \cdot \dfrac{25}{21} \cdot \dfrac{3}{10}$

7. $\dfrac{17}{100} \cdot \dfrac{27}{34} \cdot \dfrac{25}{9} \cdot 6$

Answers to Marginal Exercises

1. $\dfrac{16}{25}$ 2. $\dfrac{2}{3}$ 3. $\dfrac{4}{3}$ 4. $\dfrac{1}{24}$ 5. $\dfrac{4}{3}$ 6. $\dfrac{1}{3}$ 7. $\dfrac{9}{4}$

EXERCISES **3.3** _____

Reduce to lowest terms.

1. $\dfrac{27}{72}$ 2. $\dfrac{18}{40}$ 3. $\dfrac{144}{156}$ 4. $\dfrac{150}{135}$ 5. $\dfrac{121}{165}$

6. $\dfrac{140}{112}$ 7. $\dfrac{96}{108}$ 8. $\dfrac{72}{36}$ 9. $\dfrac{84}{42}$ 10. $\dfrac{51}{85}$

Find each product in lowest terms.

11. $\dfrac{2}{3} \cdot \dfrac{4}{3}$ 12. $\dfrac{1}{5} \cdot \dfrac{4}{7}$ 13. $\dfrac{3}{7} \cdot \dfrac{5}{3}$ 14. $\dfrac{2}{11} \cdot \dfrac{3}{2}$

15. $\dfrac{5}{16} \cdot \dfrac{16}{15}$ 16. $\dfrac{7}{18} \cdot \dfrac{9}{14}$ 17. $\dfrac{10}{18} \cdot \dfrac{9}{5}$ 18. $\dfrac{11}{22} \cdot \dfrac{6}{8}$

19. $\dfrac{15}{27} \cdot \dfrac{9}{30}$ 20. $\dfrac{35}{20} \cdot \dfrac{36}{14}$ 21. $\dfrac{25}{9} \cdot \dfrac{3}{100}$ 22. $\dfrac{30}{42} \cdot \dfrac{7}{100}$

1. _____
2. _____
3. _____
4. _____
5. _____
6. _____
7. _____
8. _____
9. _____
10. _____
11. _____
12. _____
13. _____
14. _____
15. _____
16. _____
17. _____
18. _____
19. _____
20. _____
21. _____
22. _____

Answers

23. _____

24. _____

25. _____

26. _____

27. _____

28. _____

29. _____

30. _____

31. _____

32. _____

33. _____

34. _____

35. _____

36. _____

37. _____

38. _____

39. _____

40. _____

23. $\dfrac{18}{42} \cdot \dfrac{14}{75}$

24. $\dfrac{42}{70} \cdot \dfrac{20}{12}$

25. $8 \cdot \dfrac{5}{12}$

26. $9 \cdot \dfrac{7}{24}$

27. $\dfrac{6}{85} \cdot \dfrac{34}{9}$

28. $\dfrac{13}{91} \cdot \dfrac{34}{65}$

Find each product. **Hint:** Factor before multiplying.

29. $\dfrac{23}{36} \cdot \dfrac{20}{46}$

30. $\dfrac{7}{8} \cdot \dfrac{4}{21}$

31. $\dfrac{5}{15} \cdot \dfrac{18}{24}$

32. $\dfrac{20}{32} \cdot \dfrac{9}{13} \cdot \dfrac{26}{7}$

33. $\dfrac{69}{15} \cdot \dfrac{30}{8} \cdot \dfrac{14}{46}$

34. $\dfrac{42}{52} \cdot \dfrac{27}{22} \cdot \dfrac{33}{9}$

35. $\dfrac{3}{4} \cdot 18 \cdot \dfrac{7}{2} \cdot \dfrac{22}{54}$

36. $\dfrac{9}{10} \cdot \dfrac{35}{40} \cdot \dfrac{65}{15}$

37. $\dfrac{66}{84} \cdot \dfrac{12}{5} \cdot \dfrac{28}{33}$

38. $\dfrac{24}{100} \cdot \dfrac{36}{48} \cdot \dfrac{15}{9}$

39. $\dfrac{17}{10} \cdot \dfrac{5}{42} \cdot \dfrac{18}{51} \cdot 4$

40. $\dfrac{75}{8} \cdot \dfrac{16}{36} \cdot 9 \cdot \dfrac{7}{25}$

3.4
DIVISION

> **Reciprocals:** The reciprocal of $\dfrac{a}{b}$ is $\dfrac{b}{a}$ ($a \neq 0$ and $b \neq 0$).
>
> The product of a nonzero number and its reciprocal is always 1.
>
> $$\frac{a}{b} \cdot \frac{b}{a} = 1$$

Note: There is no reciprocal for 0.

Example 1   The reciprocal of $\dfrac{2}{3}$ is $\boxed{\dfrac{3}{2}}$.

$$\frac{2}{3} \cdot \frac{3}{2} = \frac{2 \cdot 3}{3 \cdot 2} = 1$$

Example 2   The reciprocal of $\dfrac{5}{8}$ is $\boxed{\dfrac{8}{5}}$.

$$\frac{5}{8} \cdot \frac{8}{5} = \frac{5 \cdot 8}{8 \cdot 5} = 1$$

Example 3   The reciprocal of 10 is $\boxed{\dfrac{1}{10}}$.

$$10 \cdot \frac{1}{10} = \frac{10}{1} \cdot \frac{1}{10} = \frac{10 \cdot 1}{1 \cdot 10} = 1$$

☐ **NOW WORK EXERCISES 1–3 IN THE MARGIN.**

> To divide by any number (except 0), multiply by its reciprocal.
>
> $$\frac{a}{b} \div \frac{c}{d} = \frac{a}{b} \cdot \frac{d}{c} \quad \text{where } b, c, d \neq 0$$

State the reciprocal of each number.

1. $\dfrac{7}{8}$

2. $\dfrac{1}{10}$

3. 16

As an explanation of why division is accomplished by multiplying by the reciprocal of the divisor, consider treating the division in fraction form. Thus,

$$\frac{a}{b} \div \frac{c}{d} = \frac{\dfrac{a}{b}}{\dfrac{c}{d}} = \frac{\dfrac{a}{b}}{\dfrac{c}{d}} \cdot 1$$

$$= \frac{\dfrac{a}{b}}{\dfrac{c}{d}} \cdot \frac{\dfrac{d}{c}}{\dfrac{d}{c}} = \frac{\dfrac{a}{b} \cdot \dfrac{d}{c}}{\dfrac{c}{d} \cdot \dfrac{d}{c}}$$

$$= \frac{\dfrac{a}{b} \cdot \dfrac{d}{c}}{1} = \frac{a}{b} \cdot \frac{d}{c}$$

Divide as indicated in Examples 4–10.

Example 4 $\dfrac{5}{6} \div \dfrac{1}{6}$ How many $\dfrac{1}{6}$'s are there in $\dfrac{5}{6}$?

$$\frac{5}{6} \div \frac{1}{6} = \frac{5}{6} \cdot \frac{6}{1} \qquad \frac{6}{1} \text{ is the reciprocal of } \frac{1}{6}.$$

$$= 5$$

Thus, there are five $\dfrac{1}{6}$'s in $\dfrac{5}{6}$.

Example 5 $\dfrac{2}{3} \div \dfrac{3}{4}$ The divisor is $\dfrac{3}{4}$. Its reciprocal is $\dfrac{4}{3}$.

$$\frac{2}{3} \div \frac{3}{4} = \frac{2}{3} \cdot \boxed{\frac{4}{3}} = \frac{8}{9}$$

Example 6 $\dfrac{7}{16} \div 7$ The divisor is 7. Its reciprocal is $\dfrac{1}{7}$.

$$\frac{7}{16} \div 7 = \frac{7}{16} \cdot \boxed{\frac{1}{7}} = \frac{\cancel{7} \cdot 1}{16 \cdot \cancel{7}} = \frac{1}{16}$$

Example 7 $\dfrac{16}{27} \div \dfrac{4}{9}$ The divisor is $\dfrac{4}{9}$. Its reciprocal is $\dfrac{9}{4}$.

$$\frac{16}{27} \div \frac{4}{9} = \frac{16}{27} \cdot \boxed{\frac{9}{4}} = \frac{4 \cdot \cancel{4} \cdot \cancel{9}}{\cancel{9} \cdot 3 \cdot \cancel{4}} = \frac{4}{3}$$

Example 8 $\dfrac{\frac{9}{4}}{\frac{9}{2}}$ The divisor is $\dfrac{9}{2}$. Its reciprocal is $\dfrac{2}{9}$.

$$\frac{\frac{9}{4}}{\frac{9}{2}} = \frac{9}{4} \cdot \boxed{\frac{2}{9}} = \frac{\cancel{9} \cdot \cancel{2} \cdot 1}{\cancel{2} \cdot 2 \cdot \cancel{9}} = \frac{1}{2}$$

Completion Example 9 $\dfrac{13}{4} \div \dfrac{39}{5}$

$$\frac{13}{4} \div \frac{39}{5} = \frac{13}{4} \cdot \underline{\hspace{1.5cm}} = \frac{13 \cdot \underline{\hspace{0.5cm}}}{4 \cdot \underline{\hspace{0.5cm}}} = \underline{\hspace{2cm}}$$

Completion Example 10 $\dfrac{\frac{4}{9}}{\frac{4}{9}}$

$$\frac{\frac{4}{9}}{\frac{4}{9}} = \frac{4}{9} \cdot \underline{\hspace{1.5cm}} = \frac{4 \cdot \underline{\hspace{0.5cm}}}{9 \cdot \underline{\hspace{0.5cm}}} = \underline{\hspace{2cm}}$$

☐ **NOW WORK EXERCISES 4–7 IN THE MARGIN.**

Completion Example Answers

9. $\dfrac{13}{4} \div \dfrac{39}{5} = \dfrac{13}{4} \cdot \dfrac{5}{39} = \dfrac{\cancel{13} \cdot 5}{4 \cdot \cancel{13} \cdot 3} = \dfrac{5}{12}$

10. $\dfrac{\frac{4}{9}}{\frac{4}{9}} = \dfrac{4}{9} \cdot \dfrac{9}{4} = \dfrac{4 \cdot 9}{9 \cdot 4} = 1$

Divide as indicated.

4. $\dfrac{3}{4} \div \dfrac{5}{8}$

5. $\dfrac{1}{2} \div 2$

6. $\dfrac{1}{2} \div \dfrac{1}{2}$

7. $\dfrac{\frac{9}{10}}{\frac{3}{4}}$

Example 11 The result of multiplying two numbers is $\frac{7}{16}$. If one of the numbers is $\frac{3}{4}$, what is the other number? **Hint:** Think in terms of whole numbers to convince yourself that this is a division problem. If the product of two numbers is 24 and one of the numbers is 6, what is the other number? You would divide 24 by 6.

$$\begin{array}{r} 4 \\ 6\overline{)24} \\ \underline{24} \\ 0 \end{array}$$ The other number is 4.

Solution

$\frac{7}{16}$ is the result of multiplying two numbers. Divide $\frac{7}{16}$ by $\frac{3}{4}$ to find the second number.

$$\frac{7}{16} \div \frac{3}{4} = \frac{7}{16} \cdot \frac{4}{3} = \frac{7 \cdot \cancel{4}}{\cancel{4} \cdot 4 \cdot 3} = \frac{7}{12}$$

The other number is $\frac{7}{12}$.

Check by multiplying: $\frac{3}{4} \cdot \frac{7}{12} = \frac{\cancel{3} \cdot 7}{4 \cdot 4 \cdot \cancel{3}} = \frac{7}{16}$

Example 12 If the product of $\frac{3}{2}$ with another number is $\frac{5}{18}$, what is the other number?

Solution

As in Example 11, we know the product of two numbers. So divide the product by the given number to find the other number.

$$\frac{5}{18} \div \frac{3}{2} = \frac{5}{18} \cdot \frac{2}{3} = \frac{5 \cdot \cancel{2}}{\cancel{2} \cdot 9 \cdot 3} = \frac{5}{27}$$

$\frac{5}{27}$ is the other number.

Check by multiplying: $\frac{3}{2} \cdot \frac{5}{27} = \frac{\cancel{3} \cdot 5}{2 \cdot \cancel{3} \cdot 9} = \frac{5}{18}$

Answers to Marginal Exercises

1. $\frac{8}{7}$ 2. 10   3. $\frac{1}{16}$ 4. $\frac{6}{5}$ 5. $\frac{1}{4}$ 6. 1   7. $\frac{6}{5}$

NAME _____ SECTION _____ DATE _____

EXERCISES **3.4** _____

Find each quotient.

1. $\dfrac{2}{3} \div \dfrac{3}{4}$ 2. $\dfrac{1}{5} \div \dfrac{3}{4}$ 3. $\dfrac{3}{7} \div \dfrac{3}{5}$ 4. $\dfrac{2}{11} \div \dfrac{2}{3}$

1. _____

2. _____

3. _____

4. _____

5. _____

5. $\dfrac{3}{5} \div \dfrac{3}{7}$ 6. $\dfrac{2}{3} \div \dfrac{2}{11}$ 7. $\dfrac{5}{16} \div \dfrac{15}{16}$ 8. $\dfrac{7}{18} \div \dfrac{3}{9}$

6. _____

7. _____

8. _____

9. _____

9. $\dfrac{3}{14} \div \dfrac{2}{7}$ 10. $\dfrac{13}{40} \div \dfrac{26}{35}$ 11. $\dfrac{5}{12} \div \dfrac{15}{16}$ 12. $\dfrac{12}{27} \div \dfrac{10}{18}$

10. _____

11. _____

12. _____

13. _____

13. $\dfrac{17}{48} \div \dfrac{51}{90}$ 14. $\dfrac{3}{5} \div \dfrac{7}{8}$ 15. $\dfrac{13}{16} \div \dfrac{2}{3}$ 16. $\dfrac{5}{6} \div \dfrac{3}{4}$

14. _____

15. _____

16. _____

Answers

17. _____

18. _____

19. _____

20. _____

21. _____

22. _____

23. _____

24. _____

25. _____

26. _____

27. _____

28. _____

29. _____

30. _____

31. _____

32. _____

17. $\dfrac{3}{4} \div \dfrac{5}{6}$ **18.** $\dfrac{14}{15} \div \dfrac{21}{25}$ **19.** $\dfrac{3}{7} \div \dfrac{3}{7}$ **20.** $\dfrac{6}{13} \div \dfrac{6}{13}$

21. $\dfrac{16}{27} \div \dfrac{7}{18}$ **22.** $\dfrac{20}{21} \div \dfrac{15}{42}$ **23.** $\dfrac{25}{36} \div \dfrac{5}{24}$ **24.** $\dfrac{17}{20} \div \dfrac{3}{14}$

25. $\dfrac{26}{35} \div \dfrac{39}{40}$ **26.** $\dfrac{5}{6} \div \dfrac{13}{4}$ **27.** $\dfrac{7}{8} \div \dfrac{15}{2}$ **28.** $\dfrac{29}{50} \div \dfrac{31}{10}$

29. $\dfrac{21}{5} \div \dfrac{10}{3}$ **30.** $\dfrac{35}{17} \div \dfrac{5}{4}$ **31.** $\dfrac{21}{5} \div 3$ **32.** $\dfrac{41}{6} \div 2$

NAME _____ SECTION _____ DATE _____

33. $3 \div \dfrac{21}{5}$ **34.** $2 \div \dfrac{41}{6}$ **35.** $5 \div \dfrac{15}{8}$ **36.** $14 \div \dfrac{1}{7}$

37. $\dfrac{15}{8} \div 5$ **38.** $\dfrac{1}{7} \div 14$ **39.** $\dfrac{3}{4} \div \dfrac{1}{2}$ **40.** $\dfrac{3}{4} \div 2$

41. $\dfrac{3}{4} \div 3$ **42.** $\dfrac{3}{4} \div \dfrac{1}{3}$ **43.** $\dfrac{92}{7} \div \dfrac{46}{11}$

44. $\dfrac{33}{32} \div \dfrac{11}{4}$ **45.** $\dfrac{37}{8} \div \dfrac{1}{8}$

Answers

33. _____

34. _____

35. _____

36. _____

37. _____

38. _____

39. _____

40. _____

41. _____

42. _____

43. _____

44. _____

45. _____

46. The product of $\frac{9}{10}$ with another number is $\frac{5}{3}$. What is the other number?

47. The result of multiplying two numbers is $\frac{31}{3}$. If one of the numbers is $\frac{43}{6}$, what is the other one?

48. The result of multiplying two numbers is 150. If one of the numbers is $\frac{5}{7}$, what is the other number?

49. The product of $\frac{4}{5}$ with another number is 180. What is the other number?

50. The product of $\frac{7}{8}$ and $\frac{5}{10}$ is divided by $\frac{21}{20}$. Find the quotient.

3.5
ADDITION

> To add fractions that have the same denominator,
>
> 1. add the numerators;
> 2. keep the common denominator.
>
> $$\frac{a}{b} + \frac{c}{b} = \frac{a+c}{b} \qquad (b \neq 0)$$

Example 1 $\frac{3}{7} + \frac{1}{7} = \frac{3+1}{7} = \frac{4}{7}$

Think of a candy bar broken into 7 equal pieces.

Candy bar in 7 pieces

If you eat 3 pieces now and 1 piece later, what part of the candy bar will you have eaten?

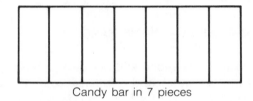

$\frac{3}{7}$ eaten now

(a)

$\frac{1}{7}$ eaten later

(b)

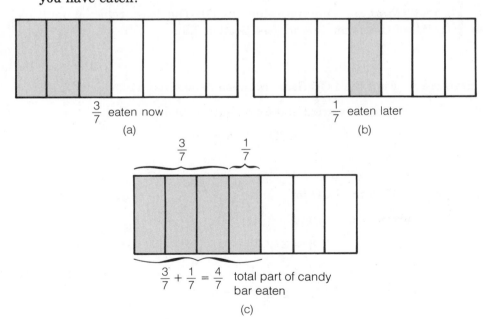

$\frac{3}{7} + \frac{1}{7} = \frac{4}{7}$ total part of candy bar eaten

(c)

Example 2 $\frac{1}{3} + \frac{1}{3} = \frac{1+1}{3} = \frac{2}{3}$

Add and reduce if possible.

1. $\dfrac{1}{7} + \dfrac{4}{7} =$

2. $\dfrac{3}{8} + \dfrac{1}{8} =$

3. $\dfrac{3}{16} + \dfrac{3}{16} + \dfrac{5}{16} =$

4. $\dfrac{7}{12} + \dfrac{8}{12} =$

Example 3 $\dfrac{4}{5} + \dfrac{3}{5} = \dfrac{4+3}{5} = \dfrac{7}{5}$

Example 4   You can add any number of fractions with the same denominator.

$$\dfrac{2}{7} + \dfrac{3}{7} + \dfrac{1}{7} + \dfrac{6}{7} = \dfrac{2+3+1+6}{7} = \dfrac{12}{7}$$

Example 5   You may be able to reduce after adding.

$$\dfrac{4}{15} + \dfrac{6}{15} = \dfrac{4+6}{15} = \dfrac{10}{15} = \dfrac{2 \cdot \cancel{5}}{3 \cdot \cancel{5}} = \dfrac{2}{3}$$

Example 6 $\dfrac{1}{4} + \dfrac{2}{4} + \dfrac{3}{4} = \dfrac{1+2+3}{4} = \dfrac{6}{4} = \dfrac{\cancel{2} \cdot 3}{\cancel{2} \cdot 2} = \dfrac{3}{2}$

☐ **NOW WORK EXERCISES 1–4 IN THE MARGIN.**

Of course, fractions to be added will not always have the same denominator. In these cases, we must find the least common denominator and then change each fraction so that all the fractions have the same denominator.

> **Least Common Denominator (LCD):** The least common multiple (LCM) of the denominators (see Section 2.6).

Example 7   Find the LCD (least common denominator) for $\dfrac{1}{2}$ and $\dfrac{3}{5}$.

The denominators are 2 and 5 and each is prime.

$$LCD = 2 \cdot 5 = 10$$

Example 8   Find the LCD for $\dfrac{5}{8}$ and $\dfrac{5}{12}$.

Using prime factorization,

$$8 = 2 \cdot 2 \cdot 2$$
$$12 = 2 \cdot 2 \cdot 3$$
$$LCD = 2 \cdot 2 \cdot 2 \cdot 3 = 2^3 \cdot 3 = 24$$

Example 9 Find the LCD for $\frac{5}{21}$ and $\frac{9}{28}$.

$$21 = 3 \cdot 7$$

$$28 = 2 \cdot 2 \cdot 7$$

$$\text{LCD} = 2 \cdot 2 \cdot 3 \cdot 7 = 2^2 \cdot 3 \cdot 7 = 84$$

☐ **NOW WORK EXERCISES 5 AND 6 IN THE MARGIN.**

> To add fractions with different denominators,
> 1. find the LCD of the denominators;
> 2. change each fraction to an equal fraction with that denominator;
> 3. add the new fractions;
> 4. reduce if possible.

Example 10 $\frac{1}{2} + \frac{3}{5}$

(a) Find the LCD.
$$\text{LCD} = 2 \cdot 5 = 10$$

(b) Find equal fractions with denominator 10.

$$\frac{1}{2} = \frac{1}{2} \cdot 1 = \frac{1}{2} \cdot \frac{5}{5} = \frac{5}{10}$$

$$\frac{3}{5} = \frac{3}{5} \cdot 1 = \frac{3}{5} \cdot \frac{2}{2} = \frac{6}{10}$$

(c) Add.

$$\frac{1}{2} + \frac{3}{5} = \frac{5}{10} + \frac{6}{10} = \frac{5+6}{10} = \frac{11}{10}$$

Example 11 $\frac{3}{8} + \frac{11}{12}$

(a) Find the LCD.

$$\left.\begin{array}{l} 8 = 2 \cdot 2 \cdot 2 \\ 12 = 2 \cdot 2 \cdot 3 \end{array}\right\} \quad \text{LCD} = 2 \cdot 2 \cdot 2 \cdot 3 = 2^3 \cdot 3 = 24$$

(b) Find equal fractions with denominator 24.

$$\frac{3}{8} = \frac{3}{8} \cdot 1 = \frac{3}{8} \cdot \frac{3}{3} = \frac{9}{24} \qquad \text{Multiply by } \frac{3}{3} \text{ since } 8 \cdot 3 = 24.$$

$$\frac{11}{12} = \frac{11}{12} \cdot 1 = \frac{11}{12} \cdot \frac{2}{2} = \frac{22}{24} \qquad \text{Multiply by } \frac{2}{2} \text{ since } 12 \cdot 2 = 24.$$

(c) Add.

$$\frac{3}{8} + \frac{11}{12} = \frac{9}{24} + \frac{22}{24} = \frac{9+22}{24} = \frac{31}{24}$$

5. Find the LCD for $\frac{1}{4}, \frac{3}{8}$, and $\frac{7}{10}$.

6. Find the LCD for $\frac{1}{8}, \frac{1}{9}$, and $\frac{1}{12}$.

Add and reduce if possible.

7. $\dfrac{1}{6} + \dfrac{3}{10} =$

8. $\dfrac{1}{4} + \dfrac{3}{8} + \dfrac{7}{10} =$

Example 12 $\dfrac{5}{21} + \dfrac{5}{28}$

(a) Find the LCD.

$$\left.\begin{array}{l} 21 = 3 \cdot 7 \\ 28 = 2 \cdot 2 \cdot 7 \end{array}\right\} \quad \text{LCD} = 2 \cdot 2 \cdot 3 \cdot 7 = 84$$

(b) Find equal fractions with denominator 84.

$$\frac{5}{21} = \frac{5}{21} \cdot \frac{4}{4} = \frac{20}{84}$$

$$\frac{5}{28} = \frac{5}{28} \cdot \frac{3}{3} = \frac{15}{84}$$

(c) $\dfrac{5}{21} + \dfrac{5}{28} = \dfrac{20}{84} + \dfrac{15}{84} = \dfrac{20 + 15}{84} = \dfrac{35}{84}$

(d) Now reduce. (Remember: The prime factorization of 84 is in the LCD.)

$$\frac{35}{84} = \frac{\cancel{7} \cdot 5}{2 \cdot 2 \cdot 3 \cdot \cancel{7}} = \frac{5}{12}$$

☐ **NOW WORK EXERCISES 7 AND 8 IN THE MARGIN.**

Steps (b), (c), and (d) can all be written together as one process.

Example 13 $\dfrac{2}{3} + \dfrac{1}{6} + \dfrac{5}{12}$

(a) LCD = 12   You can simply observe this or use prime factorizations.

(b) $\dfrac{2}{3} + \dfrac{1}{6} + \dfrac{5}{12} = \dfrac{2}{3} \cdot \dfrac{4}{4} + \dfrac{1}{6} \cdot \dfrac{2}{2} + \dfrac{5}{12}$

$$= \frac{8}{12} + \frac{2}{12} + \frac{5}{12}$$

$$= \frac{15}{12} = \frac{\cancel{3} \cdot 5}{2 \cdot 2 \cdot \cancel{3}} = \frac{5}{4}$$

Completion Example 14 $\dfrac{2}{3} + \dfrac{5}{8} + \dfrac{1}{6}$

(a) $\left.\begin{array}{l} 3 = 3 \\ 8 = 2 \cdot 2 \cdot 2 \\ 6 = 2 \cdot 3 \end{array}\right\}$ LCD $= 2 \cdot 2 \cdot 2 \cdot 3 = 24$

(b) $\dfrac{2}{3} + \dfrac{5}{8} + \dfrac{1}{6} = \dfrac{2}{3} \cdot \underline{\hspace{1cm}} + \dfrac{5}{8} \cdot \underline{\hspace{1cm}} + \dfrac{1}{6} \cdot \underline{\hspace{1cm}}$

$= \underline{\hspace{1cm}} + \underline{\hspace{1cm}} + \underline{\hspace{1cm}}$

$= \underline{\hspace{1cm}}$

Completion Example 15 $\dfrac{7}{10} + \dfrac{1}{100} + \dfrac{5}{1000}$

(a) LCD $= \underline{\hspace{2cm}}$

(b) $\dfrac{7}{10} + \dfrac{1}{100} + \dfrac{5}{1000} = \dfrac{7}{10} \cdot \underline{\hspace{1cm}} + \dfrac{1}{100} \cdot \underline{\hspace{1cm}} + \dfrac{5}{1000}$

$= \underline{\hspace{1.5cm}} + \underline{\hspace{1.5cm}} + \underline{\hspace{1.5cm}}$

$= \underline{\hspace{1.5cm}} = \underline{\hspace{1.5cm}} = \underline{\hspace{1.5cm}}$

Completion Example Answers

14. (b) $\dfrac{2}{3} + \dfrac{5}{8} + \dfrac{1}{6} = \dfrac{2}{3} \cdot \dfrac{8}{8} + \dfrac{5}{8} \cdot \dfrac{3}{3} + \dfrac{1}{6} \cdot \dfrac{4}{4}$

$= \dfrac{16}{24} + \dfrac{15}{24} + \dfrac{4}{24} = \dfrac{35}{24}$

15. (a) LCD $= 1000$

(b) $\dfrac{7}{10} + \dfrac{1}{100} + \dfrac{5}{1000} = \dfrac{7}{10} \cdot \dfrac{100}{100} + \dfrac{1}{100} \cdot \dfrac{10}{10} + \dfrac{5}{1000}$

$= \dfrac{700}{1000} + \dfrac{10}{1000} + \dfrac{5}{1000}$

$= \dfrac{715}{1000} = \dfrac{\cancel{5} \cdot 143}{\cancel{5} \cdot 200} = \dfrac{143}{200}$

COMMON ERROR

The following common error must be avoided.

Find the sum $\dfrac{3}{2} + \dfrac{1}{6}$.

WRONG SOLUTION

$$\dfrac{\overset{1}{\cancel{3}}}{2} + \dfrac{1}{\underset{2}{\cancel{6}}} = \dfrac{1}{2} + \dfrac{1}{2} = 1 \qquad \textbf{WRONG}$$

You cannot cancel across the + sign.

CORRECT SOLUTION using LCD = 6

$$\dfrac{3}{2} + \dfrac{1}{6} = \dfrac{3}{2} \cdot \dfrac{3}{3} + \dfrac{1}{6} = \dfrac{9}{6} + \dfrac{1}{6} = \dfrac{10}{6}$$

NOW reduce:

$$\dfrac{10}{6} = \dfrac{5 \cdot \cancel{2}}{3 \cdot \cancel{2}} = \dfrac{5}{3} \qquad \text{2 is a factor in both the numerator and the denominator.}$$

Answers to Marginal Exercises

1. $\dfrac{5}{7}$ 2. $\dfrac{1}{2}$ 3. $\dfrac{11}{16}$ 4. $\dfrac{5}{4}$ 5. LCD = 40   6. LCD = 72   7. $\dfrac{7}{15}$ 8. $\dfrac{53}{40}$

NAME _____ SECTION _____ DATE _____

EXERCISES **3.5** _____

Add and reduce if possible.

1. $\dfrac{6}{10} + \dfrac{4}{10}$ 2. $\dfrac{3}{14} + \dfrac{2}{14}$ 3. $\dfrac{1}{20} + \dfrac{3}{20}$

4. $\dfrac{3}{4} + \dfrac{3}{4}$ 5. $\dfrac{5}{6} + \dfrac{4}{6}$ 6. $\dfrac{7}{5} + \dfrac{3}{5}$

7. $\dfrac{11}{15} + \dfrac{7}{15}$ 8. $\dfrac{7}{9} + \dfrac{8}{9}$ 9. $\dfrac{3}{25} + \dfrac{12}{25}$

10. $\dfrac{7}{90} + \dfrac{37}{90} + \dfrac{21}{90}$ 11. $\dfrac{11}{75} + \dfrac{12}{75} + \dfrac{62}{75}$ 12. $\dfrac{14}{32} + \dfrac{7}{32} + \dfrac{1}{32}$

13. $\dfrac{4}{100} + \dfrac{35}{100} + \dfrac{76}{100}$ 14. $\dfrac{21}{95} + \dfrac{33}{95} + \dfrac{3}{95}$ 15. $\dfrac{1}{200} + \dfrac{17}{200} + \dfrac{25}{200}$

1. _____

2. _____

3. _____

4. _____

5. _____

6. _____

7. _____

8. _____

9. _____

10. _____

11. _____

12. _____

13. _____

14. _____

15. _____

Answers

16. _____

17. _____

18. _____

19. _____

20. _____

21. _____

22. _____

23. _____

24. _____

25. _____

26. _____

27. _____

28. _____

29. _____

30. _____

31. _____

32. _____

33. _____

16. $\dfrac{1}{12} + \dfrac{2}{3} + \dfrac{1}{4}$ **17.** $\dfrac{3}{8} + \dfrac{5}{16}$ **18.** $\dfrac{2}{5} + \dfrac{3}{10} + \dfrac{3}{20}$

19. $\dfrac{3}{4} + \dfrac{1}{16} + \dfrac{6}{32}$ **20.** $\dfrac{2}{7} + \dfrac{4}{21} + \dfrac{1}{3}$ **21.** $\dfrac{1}{6} + \dfrac{1}{4} + \dfrac{1}{3}$

22. $\dfrac{2}{39} + \dfrac{1}{3} + \dfrac{4}{13}$ **23.** $\dfrac{1}{2} + \dfrac{3}{10} + \dfrac{4}{5}$ **24.** $\dfrac{1}{27} + \dfrac{4}{18} + \dfrac{1}{6}$

25. $\dfrac{2}{7} + \dfrac{3}{20} + \dfrac{9}{14}$ **26.** $\dfrac{1}{8} + \dfrac{1}{12} + \dfrac{1}{9}$ **27.** $\dfrac{2}{5} + \dfrac{4}{7} + \dfrac{3}{8}$

28. $\dfrac{2}{3} + \dfrac{3}{4} + \dfrac{5}{6}$ **29.** $\dfrac{1}{5} + \dfrac{7}{30} + \dfrac{1}{6}$ **30.** $\dfrac{1}{5} + \dfrac{2}{15} + \dfrac{1}{6}$

31. $\dfrac{1}{5} + \dfrac{1}{10} + \dfrac{1}{4}$ **32.** $\dfrac{1}{5} + \dfrac{7}{40} + \dfrac{1}{4}$ **33.** $\dfrac{1}{3} + \dfrac{5}{12} + \dfrac{1}{15}$

NAME _____ SECTION _____ DATE _____

34. $\frac{1}{4} + \frac{1}{20} + \frac{8}{15}$ **35.** $\frac{7}{10} + \frac{3}{25} + \frac{3}{4}$ **36.** $\frac{5}{8} + \frac{4}{27} + \frac{1}{48}$

37. $\frac{3}{16} + \frac{5}{48} + \frac{1}{32}$ **38.** $\frac{72}{105} + \frac{2}{45} + \frac{15}{21}$ **39.** $\frac{1}{63} + \frac{2}{27} + \frac{1}{45}$

40. $\frac{4}{45} + \frac{1}{18} + \frac{3}{10}$ **41.** $\frac{5}{6} + \frac{3}{10} + \frac{1}{90}$ **42.** $\frac{3}{10} + \frac{1}{100} + \frac{7}{1000}$

43. $\frac{11}{100} + \frac{15}{10} + \frac{1}{10}$ **44.** $\frac{17}{1000} + \frac{1}{100} + \frac{1}{10,000}$

45. $6 + \frac{1}{100} + \frac{3}{10}$ **46.** $8 + \frac{1}{10} + \frac{7}{100}$ **47.** $\frac{1}{10} + \frac{3}{10} + \frac{9}{100}$

Answers

34. _____

35. _____

36. _____

37. _____

38. _____

39. _____

40. _____

41. _____

42. _____

43. _____

44. _____

45. _____

46. _____

47. _____

Answers

48. _____

49. _____

50. _____

51. _____

52. _____

53. _____

54. _____

55. _____

48. $\dfrac{7}{10} + \dfrac{5}{100} + \dfrac{3}{1000}$ **49.** $\dfrac{1}{2} + \dfrac{3}{4} + \dfrac{1}{100}$ **50.** $\dfrac{1}{4} + \dfrac{1}{8} + \dfrac{7}{100}$

51. $\dfrac{9}{1000} + \dfrac{7}{1000} + \dfrac{21}{10,000}$ **52.** $\dfrac{11}{100} + \dfrac{1}{2} + \dfrac{3}{1000}$

53. $\dfrac{3}{4} + \dfrac{17}{1000} + \dfrac{13}{10,000}$ **54.** $\dfrac{1}{10} + \dfrac{3}{100} + \dfrac{4}{1000}$

55. $\dfrac{13}{10,000} + \dfrac{1}{100,000} + \dfrac{21}{1,000,000}$

3.6
SUBTRACTION

> To subtract fractions that have the same denominator,
>
> 1. subtract the numerators;
> 2. keep the common denominator.
>
> $$\frac{a}{b} - \frac{c}{b} = \frac{a-c}{b}$$

Example 1 $\dfrac{5}{7} - \dfrac{1}{7} = \dfrac{5-1}{7} = \dfrac{4}{7}$

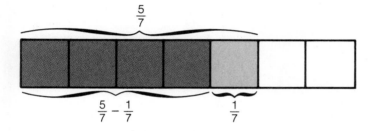

Example 2 $\dfrac{5}{9} - \dfrac{4}{9} = \dfrac{5-4}{9} = \dfrac{1}{9}$

Example 3 You may be able to reduce after subtracting.

$$\frac{7}{8} - \frac{1}{8} = \frac{7-1}{8} = \frac{6}{8} = \frac{\cancel{2} \cdot 3}{\cancel{2} \cdot 4} = \frac{3}{4}$$

Example 4 $\dfrac{19}{10} - \dfrac{11}{10} = \dfrac{19-11}{10} = \dfrac{8}{10} = \dfrac{\cancel{2} \cdot 4}{\cancel{2} \cdot 5} = \dfrac{4}{5}$

☐ **NOW WORK EXERCISES 1–4 IN THE MARGIN.**

Subtract and reduce if possible.

1. $\dfrac{6}{10} - \dfrac{3}{10} =$

2. $\dfrac{27}{30} - \dfrac{5}{30} =$

3. $\dfrac{7}{8} - \dfrac{3}{8} =$

4. $\dfrac{5}{6} - \dfrac{1}{6} =$

Subtract and reduce if possible.

5. $\dfrac{7}{10} - \dfrac{2}{15} =$

6. $\dfrac{15}{16} - \dfrac{5}{12} =$

> To subtract fractions with different denominators,
>
> 1. find the LCD of the denominators;
> 2. change each fraction to an equal fraction with that denominator;
> 3. subtract the new fractions;
> 4. reduce if possible.

Example 5 $\dfrac{9}{10} - \dfrac{2}{15}$

(a) Find the LCD.

$$\left. \begin{array}{l} 10 = 2 \cdot 5 \\ 15 = 3 \cdot 5 \end{array} \right\} \quad LCD = 2 \cdot 3 \cdot 5 = 30$$

(b) Find equal fractions with denominator 30.

$$\dfrac{9}{10} = \dfrac{9}{10} \cdot \dfrac{3}{3} = \dfrac{27}{30}$$

$$\dfrac{2}{15} = \dfrac{2}{15} \cdot \dfrac{2}{2} = \dfrac{4}{30}$$

(c) Subtract.

$$\dfrac{9}{10} - \dfrac{2}{15} = \dfrac{27}{30} - \dfrac{4}{30} = \dfrac{27 - 4}{30} = \dfrac{23}{30}$$

Example 6 $\dfrac{12}{55} - \dfrac{2}{33}$

(a) Find the LCD.

$$\left. \begin{array}{l} 55 = 5 \cdot 11 \\ 33 = 3 \cdot 11 \end{array} \right\} \quad LCD = 3 \cdot 5 \cdot 11 = 165$$

(b) Find equal fractions with denominator 165.

$$\dfrac{12}{55} = \dfrac{12}{55} \cdot \dfrac{3}{3} = \dfrac{36}{165}$$

$$\dfrac{2}{33} = \dfrac{2}{33} \cdot \dfrac{5}{5} = \dfrac{10}{165}$$

(c) Subtract.

$$\dfrac{12}{55} - \dfrac{2}{33} = \dfrac{36}{165} - \dfrac{10}{165} = \dfrac{36 - 10}{165} = \dfrac{26}{165}$$

(d) Reduce if possible.

$$\dfrac{26}{165} = \dfrac{2 \cdot 13}{3 \cdot 5 \cdot 11} = \dfrac{26}{165} \quad \text{cannot be reduced}$$

▢ **NOW WORK EXERCISES 5 AND 6 IN THE MARGIN.**

As with addition, steps (b), (c), and (d) can be written as one process (Examples 7 and 8).

Completion Example 7 $\dfrac{5}{18} - \dfrac{1}{6}$

(a) LCD = 18

(b) $\dfrac{5}{18} - \dfrac{1}{6} = \dfrac{5}{18} - \dfrac{1}{6} \cdot \boxed{\dfrac{3}{3}}$ Use $\dfrac{3}{3}$ since $6 \cdot \boxed{3} = 18$.

$= \underline{\hphantom{xxx}} - \underline{\hphantom{xxx}}$

$= \underline{\hphantom{xxx}} = \underline{\hphantom{xxx}}$

Completion Example 8 $\dfrac{7}{12} - \dfrac{3}{20}$

(a) $\left. \begin{array}{l} 12 = 2 \cdot 2 \cdot 3 \\ 20 = 2 \cdot 2 \cdot 5 \end{array} \right\}$ LCD = \underline{\hphantom{xxxxx}}

(b) $\dfrac{7}{12} - \dfrac{3}{20} = \dfrac{7}{12} \cdot \underline{\hphantom{xx}} - \dfrac{3}{20} \cdot \underline{\hphantom{xx}}$

$= \underline{\hphantom{xxx}} - \underline{\hphantom{xxx}}$

$= \underline{\hphantom{xxx}} = \underline{\hphantom{xxx}}$

Completion Example Answers

7. (b) $\dfrac{5}{18} - \dfrac{1}{6} = \dfrac{5}{18} - \dfrac{1}{6} \cdot \dfrac{3}{3} = \dfrac{5}{18} - \dfrac{3}{18} = \dfrac{2}{18} = \dfrac{1}{9}$

8. (a) LCD = $2 \cdot 2 \cdot 3 \cdot 5 = 60$

(b) $\dfrac{7}{12} - \dfrac{3}{20} = \dfrac{7}{12} \cdot \dfrac{5}{5} - \dfrac{3}{20} \cdot \dfrac{3}{3} = \dfrac{35}{60} - \dfrac{9}{60} = \dfrac{26}{60} = \dfrac{13}{30}$

Answers to Marginal Exercises

1. $\dfrac{3}{10}$ 2. $\dfrac{11}{15}$ 3. $\dfrac{1}{2}$ 4. $\dfrac{2}{3}$ 5. $\dfrac{17}{30}$ 6. $\dfrac{25}{48}$

EXERCISES **3.6** _____

Subtract and reduce if possible.

1. $\dfrac{4}{7} - \dfrac{1}{7}$ 2. $\dfrac{5}{7} - \dfrac{3}{7}$ 3. $\dfrac{9}{10} - \dfrac{3}{10}$ 4. $\dfrac{11}{10} - \dfrac{7}{10}$

5. $\dfrac{5}{8} - \dfrac{1}{8}$ 6. $\dfrac{7}{8} - \dfrac{5}{8}$ 7. $\dfrac{11}{12} - \dfrac{7}{12}$ 8. $\dfrac{7}{12} - \dfrac{3}{12}$

9. $\dfrac{13}{15} - \dfrac{4}{15}$ 10. $\dfrac{21}{15} - \dfrac{11}{15}$ 11. $\dfrac{5}{6} - \dfrac{1}{3}$ 12. $\dfrac{5}{6} - \dfrac{1}{2}$

Answers

1. _____

2. _____

3. _____

4. _____

5. _____

6. _____

7. _____

8. _____

9. _____

10. _____

11. _____

12. _____

Answers

13. _____

14. _____

15. _____

16. _____

17. _____

18. _____

19. _____

20. _____

21. _____

22. _____

23. _____

24. _____

25. _____

26. _____

27. _____

28. _____

13. $\dfrac{11}{15} - \dfrac{3}{10}$ 14. $\dfrac{8}{10} - \dfrac{3}{15}$ 15. $\dfrac{3}{4} - \dfrac{2}{3}$ 16. $\dfrac{2}{3} - \dfrac{1}{4}$

17. $\dfrac{15}{16} - \dfrac{21}{32}$ 18. $\dfrac{3}{8} - \dfrac{1}{16}$ 19. $\dfrac{5}{4} - \dfrac{3}{5}$ 20. $\dfrac{5}{12} - \dfrac{1}{6}$

21. $\dfrac{14}{27} - \dfrac{7}{18}$ 22. $\dfrac{25}{18} - \dfrac{21}{27}$ 23. $\dfrac{8}{45} - \dfrac{11}{72}$ 24. $\dfrac{46}{55} - \dfrac{10}{33}$

25. $\dfrac{5}{36} - \dfrac{1}{45}$ 26. $\dfrac{5}{1} - \dfrac{3}{4}$ 27. $\dfrac{4}{1} - \dfrac{5}{8}$ 28. $2 - \dfrac{9}{16}$

NAME _____ SECTION _____ DATE _____

29. $1 - \dfrac{13}{16}$ **30.** $6 - \dfrac{2}{3}$ **31.** $\dfrac{9}{10} - \dfrac{3}{100}$ **32.** $\dfrac{159}{1000} - \dfrac{1}{10}$

33. $\dfrac{76}{100} - \dfrac{7}{10}$ **34.** $\dfrac{999}{1000} - \dfrac{99}{100}$ **35.** $\dfrac{54}{100} - \dfrac{5}{10}$ **36.** $\dfrac{7}{24} - \dfrac{10}{36}$

37. $\dfrac{31}{40} - \dfrac{5}{8}$ **38.** $\dfrac{14}{35} - \dfrac{12}{30}$ **39.** $\dfrac{20}{35} - \dfrac{24}{42}$ **40.** $\dfrac{3}{10} - \dfrac{298}{1000}$

Answers

29. _____

30. _____

31. _____

32. _____

33. _____

34. _____

35. _____

36. _____

37. _____

38. _____

39. _____

40. _____

Answers

41. _____

42. _____

43. _____

44. _____

41. Find the sum of $\frac{1}{4}$, $\frac{1}{8}$, and $\frac{3}{16}$. Subtract $\frac{1}{3}$ from the sum. What is the difference?

42. Find the difference between $\frac{2}{3}$ and $\frac{5}{9}$. Add $\frac{5}{6}$ to the difference. What is the sum?

43. Find the sum of $\frac{7}{16}$ and $\frac{5}{32}$. Multiply the sum by $\frac{3}{19}$. What is the product?

44. Find the product of $\frac{9}{10}$ and $\frac{9}{10}$. Find the difference between $\frac{3}{4}$ and $\frac{1}{25}$. What is the sum of the product and the difference?

3.7

COMPARISONS AND ORDER OF OPERATIONS

> To compare two fractions (to find which is larger or smaller),
>
> 1. find the LCD of the denominators;
> 2. change each fraction to an equal fraction with that denominator;
> 3. compare the numerators.

Example 1 Which is larger, $\dfrac{5}{6}$ or $\dfrac{7}{8}$? How much larger?

(a) Find the LCD for 6 and 8.

$$\left.\begin{array}{l} 6 = 2\cdot3 \\ 8 = 2\cdot2\cdot2 \end{array}\right\} \quad LCD = 2\cdot2\cdot2\cdot3 = 24$$

(b) Find equal fractions with denominator 24.

$$\frac{5}{6} = \frac{5}{6}\cdot\frac{4}{4} = \frac{20}{24} \quad \text{and} \quad \frac{7}{8} = \frac{7}{8}\cdot\frac{3}{3} = \frac{21}{24}$$

(c) $\dfrac{7}{8}$ is larger than $\dfrac{5}{6}$ since 21 is larger than 20.

(d) $\dfrac{7}{8} - \dfrac{5}{6} = \dfrac{21}{24} - \dfrac{20}{24} = \dfrac{1}{24}$

$\dfrac{7}{8}$ is larger by $\dfrac{1}{24}$.

Example 2 Which is larger, $\dfrac{8}{9}$ or $\dfrac{11}{12}$? How much larger?

(a) $LCD = 2\cdot2\cdot3\cdot3 = 36$

(b) $\dfrac{8}{9} = \dfrac{8}{9}\cdot\dfrac{4}{4} = \dfrac{32}{36}$ and $\dfrac{11}{12} = \dfrac{11}{12}\cdot\dfrac{3}{3} = \dfrac{33}{36}$

(c) $\dfrac{11}{12}$ is larger than $\dfrac{8}{9}$ since 33 is larger than 32.

(d) $\dfrac{11}{12} - \dfrac{8}{9} = \dfrac{33}{36} - \dfrac{32}{36} = \dfrac{1}{36}$

$\dfrac{11}{12}$ is larger by $\dfrac{1}{36}$.

1. Which is larger, $\frac{9}{22}$ or $\frac{10}{33}$? How much larger?

2. Arrange $\frac{7}{12}$, $\frac{5}{9}$, and $\frac{2}{3}$ in order, from smallest to largest.

Example 3　Arrange $\frac{2}{3}$, $\frac{7}{10}$, and $\frac{9}{15}$ in order, from smallest to largest.

(a) LCD = 30

(b) $\frac{2}{3} = \frac{2}{3} \cdot \frac{10}{10} = \frac{20}{30}$;　$\frac{7}{10} = \frac{7}{10} \cdot \frac{3}{3} = \frac{21}{30}$;　$\frac{9}{15} = \frac{9}{15} \cdot \frac{2}{2} = \frac{18}{30}$

(c) Smallest to largest: $\frac{9}{15}, \frac{2}{3}, \frac{7}{10}$

☐ **NOW WORK EXERCISES 1 AND 2 IN THE MARGIN.**

RULES FOR ORDER OF OPERATIONS

1. First, simplify within grouping symbols, such as parentheses ( ), brackets [ ], or braces { }. Start with the innermost grouping.
2. Second, find any powers indicated by exponents.
3. Third, moving from **left to right,** perform any multiplications or divisions in the order that they appear.
4. Fourth, moving from **left to right,** perform any additions or subtractions in the order that they appear.

Example 4　Evaluate the expression $\frac{1}{2} \div \frac{3}{4} + \frac{5}{6} \cdot \frac{1}{5}$.

(a) Divide first.　　　　$\frac{1}{2} \div \frac{3}{4} + \frac{5}{6} \cdot \frac{1}{5}$

$$= \frac{1}{\cancel{2}} \cdot \frac{\cancel{4}^{\,2}}{3} + \frac{5}{6} \cdot \frac{1}{5}$$

(b) Now multiply.　　$= \frac{2}{3} + \frac{\cancel{5}}{6} \cdot \frac{1}{\cancel{5}}$

(c) Now add.　　　　$= \frac{2}{3} + \frac{1}{6}$
　　(LCD = 6)

$$= \frac{2}{3} \cdot \frac{2}{2} + \cdot\frac{1}{6}$$

$$= \frac{4}{6} + \frac{1}{6}$$

$$= \frac{5}{6}$$

Example 5 Evaluate the expression $\left(\dfrac{3}{4} - \dfrac{5}{8}\right) \div \left(\dfrac{15}{16} - \dfrac{1}{2}\right)$.

(a) Work inside the parentheses.

$$\left(\dfrac{3}{4} - \dfrac{5}{8}\right) \div \left(\dfrac{15}{16} - \dfrac{1}{2}\right)$$

$$= \left(\dfrac{6}{8} - \dfrac{5}{8}\right) \div \left(\dfrac{15}{16} - \dfrac{8}{16}\right)$$

(b) Now divide.

$$= \left(\dfrac{1}{8}\right) \div \left(\dfrac{7}{16}\right)$$

$$= \dfrac{1}{\cancel{8}} \cdot \dfrac{\overset{2}{\cancel{16}}}{7}$$

$$= \dfrac{2}{7}$$

Example 6 Evaluate the expression $\dfrac{9}{10} - \left(\dfrac{1}{4}\right)^2 + \dfrac{1}{2}$.

(a) Use the exponent first.

$$\dfrac{9}{10} - \left(\dfrac{1}{4}\right)^2 + \dfrac{1}{2} = \dfrac{9}{10} - \dfrac{1}{16} + \dfrac{1}{2}$$

(b) Now add and subtract.
   (LCD = 80)

$$= \dfrac{9}{10} \cdot \dfrac{8}{8} - \dfrac{1}{16} \cdot \dfrac{5}{5} + \dfrac{1}{2} \cdot \dfrac{40}{40}$$

$$= \dfrac{72}{80} - \dfrac{5}{80} + \dfrac{40}{80}$$

$$= \dfrac{107}{80}$$

Completion Example 7 $\dfrac{1}{2} \cdot \dfrac{5}{6} + \dfrac{7}{15} \div 2$

$$\dfrac{1}{2} \cdot \dfrac{5}{6} + \dfrac{7}{15} \div 2 = \dfrac{5}{12} + \dfrac{7}{15} \div 2$$

$$= \dfrac{5}{12} + \dfrac{7}{15} \cdot \underline{\quad\quad}$$

$$= \dfrac{5}{12} + \underline{\quad\quad}$$

$$(\text{LCD} = \underline{\quad\quad}) = \underline{\quad\quad} + \underline{\quad\quad}$$

$$= \underline{\quad\quad} = \underline{\quad\quad} = \underline{\quad\quad}$$

Completion Example Answer

7. $\dfrac{1}{2} \cdot \dfrac{5}{6} + \dfrac{7}{15} \div 2 = \dfrac{5}{12} + \dfrac{7}{15} \div 2$

$$= \dfrac{5}{12} + \dfrac{7}{15} \cdot \dfrac{1}{2} = \dfrac{5}{12} + \dfrac{7}{30}$$

$$(\text{LCD} = 60) = \dfrac{25}{60} + \dfrac{14}{60} = \dfrac{39}{60} = \dfrac{\cancel{3} \cdot 13}{\cancel{3} \cdot 20} = \dfrac{13}{20}$$

3. $\dfrac{5}{4} \div \left(1 - \dfrac{1}{3}\right) =$

4. $\dfrac{1}{4} + \left(\dfrac{1}{3}\right)^{2} \div \dfrac{7}{3} =$

Completion Example 8 $\left(\dfrac{7}{8} - \dfrac{7}{10}\right) \div \dfrac{7}{2}$

$$\left(\dfrac{7}{8} - \dfrac{7}{10}\right) \div \dfrac{7}{2} = \left(\dfrac{35}{40} - \underline{\quad}\right) \div \dfrac{7}{2}$$

$$= \left(\underline{\quad}\right) \div \dfrac{7}{2}$$

$$= \left(\underline{\quad}\right) \cdot \underline{\quad}$$

$$= \underline{\quad} = \underline{\quad}$$

☐ **NOW WORK EXERCISES 3 AND 4 IN THE MARGIN.**

Completion Example Answer

8. $\left(\dfrac{7}{8} - \dfrac{7}{10}\right) \div \dfrac{7}{2} = \left(\dfrac{35}{40} - \dfrac{28}{40}\right) \div \dfrac{7}{2}$

$$= \left(\dfrac{7}{40}\right) \div \dfrac{7}{2}$$

$$= \left(\dfrac{7}{40}\right) \cdot \dfrac{2}{7} = \dfrac{\cancel{7} \cdot \cancel{2} \cdot 1}{\cancel{2} \cdot 20 \cdot \cancel{7}} = \dfrac{1}{20}$$

Answers to Marginal Exercises

1. $\dfrac{9}{22}$ is larger by $\dfrac{7}{66}$ 2. $\dfrac{5}{9}, \dfrac{7}{12}, \dfrac{2}{3}$ 3. $\dfrac{15}{8}$ 4. $\dfrac{25}{84}$

NAME _____ SECTION _____ DATE _____

EXERCISES **3.7** _____

Find the larger number of each pair and state how much larger it is.

1. $\dfrac{2}{3}, \dfrac{3}{4}$

2. $\dfrac{5}{6}, \dfrac{7}{8}$

3. $\dfrac{4}{5}, \dfrac{17}{20}$

4. $\dfrac{4}{10}, \dfrac{3}{8}$

5. $\dfrac{13}{20}, \dfrac{5}{8}$

6. $\dfrac{13}{16}, \dfrac{21}{25}$

7. $\dfrac{14}{35}, \dfrac{12}{30}$

8. $\dfrac{10}{36}, \dfrac{7}{24}$

9. $\dfrac{17}{80}, \dfrac{11}{48}$

10. $\dfrac{37}{100}, \dfrac{24}{75}$

Arrange the numbers in order, from smallest to largest.

11. $\dfrac{2}{3}, \dfrac{3}{5}, \dfrac{7}{10}$

12. $\dfrac{8}{9}, \dfrac{9}{10}, \dfrac{11}{12}$

13. $\dfrac{7}{6}, \dfrac{11}{12}, \dfrac{19}{20}$

14. $\dfrac{1}{3}, \dfrac{5}{42}, \dfrac{3}{7}$

15. $\dfrac{1}{2}, \dfrac{1}{3}, \dfrac{1}{4}$

16. $\dfrac{2}{3}, \dfrac{3}{4}, \dfrac{5}{8}$

Answers

1. _____

2. _____

3. _____

4. _____

5. _____

6. _____

7. _____

8. _____

9. _____

10. _____

11. _____

12. _____

13. _____

14. _____

15. _____

16. _____

17. $\dfrac{7}{9}, \dfrac{31}{36}, \dfrac{13}{18}$

18. $\dfrac{17}{12}, \dfrac{40}{36}, \dfrac{31}{24}$

19. $\dfrac{1}{100}, \dfrac{3}{1000}, \dfrac{20}{10,000}$

20. $\dfrac{32}{100}, \dfrac{298}{1000}, \dfrac{3}{10}$

Evaluate each expression using the rules of order of operations.

21. $\dfrac{1}{2} \div \dfrac{7}{8} + \dfrac{1}{7} \cdot \dfrac{2}{3}$

22. $\dfrac{3}{5} \cdot \dfrac{1}{6} + \dfrac{1}{5} \div 2$

23. $\dfrac{1}{2} \div \dfrac{1}{2} + \dfrac{2}{3} \cdot \dfrac{2}{3}$

24. $5 - \dfrac{3}{4} \div 3$

25. $6 - \dfrac{5}{8} \div 4$

26. $\dfrac{2}{15} \cdot \dfrac{1}{4} \div \dfrac{3}{5} + \dfrac{1}{25}$

27. $\dfrac{5}{8} \cdot \dfrac{1}{10} \div \dfrac{3}{4} + \dfrac{1}{6}$

28. $\left(\dfrac{7}{15} + \dfrac{8}{21} \right) \div \dfrac{3}{35}$

29. $\left(\dfrac{1}{2} - \dfrac{1}{3} \right) \div \left(\dfrac{5}{8} + \dfrac{3}{16} \right)$

30. $\left(\dfrac{1}{3} + \dfrac{1}{5} \right) \cdot \left(\dfrac{3}{4} - \dfrac{1}{6} \right)$

31. $\left(\dfrac{1}{2} \right)^2 - \left(\dfrac{1}{4} \right)^3$

32. $\dfrac{2}{3} + \dfrac{3}{4} + \left(\dfrac{1}{2} \right)^2$

33. $\left(\dfrac{1}{3} \right)^2 + \left(\dfrac{1}{6} \right)^2 + \dfrac{2}{3}$

34. $\dfrac{1}{2} \div \dfrac{2}{3} + \left(\dfrac{1}{3} \right)^2$

NAME _____ SECTION _____ DATE _____

CHAPTER **3** TEST _____

1. The reciprocal of 9 is _____ .

1. _____

2. A fraction equivalent to $\frac{7}{16}$ with denominator 80 is _____ .

2. _____

3. _____

4. _____

In Problems 3 and 4, reduce each fraction to lowest terms.

3. $\frac{90}{108}$ 4. $\frac{13}{51}$

5. _____

6. _____

5. Find the sum of $\frac{3}{8}$ and $\frac{5}{12}$.

7. _____

6. Find the product of $\frac{2}{7}$ and $\frac{3}{5}$.

8. _____

7. Find the difference between $\frac{5}{6}$ and $\frac{4}{9}$.

8. Find the quotient when $\frac{4}{9}$ is divided by 4.

Answers

9. _____

10. _____

11. _____

12. _____

13. _____

14. _____

15. _____

16. _____

17. _____

18. _____

19. _____

20. _____

21. _____

22. _____

23. _____

24. _____

25. _____

In Problems 9–24, perform the indicated operations and reduce all answers to lowest terms.

9. $\dfrac{3}{7} \cdot \dfrac{14}{27}$ **10.** $2 - \dfrac{21}{19}$ **11.** $10 \div \dfrac{2}{5}$

12. $\dfrac{5}{16} + \left(\dfrac{1}{4}\right)^2$ **13.** $\dfrac{3}{5} \cdot \dfrac{1}{2} \cdot \dfrac{3}{7}$ **14.** $\dfrac{4}{35} + \dfrac{2}{7} + \dfrac{1}{10}$

15. $\dfrac{5}{11} \div \dfrac{2}{5}$ **16.** $\dfrac{51}{16} - 3$ **17.** $\dfrac{2}{7} \cdot \dfrac{3}{4} \cdot \dfrac{7}{9}$

18. $\dfrac{54}{17} - \dfrac{3}{17}$ **19.** $\dfrac{3}{8} \div \dfrac{9}{16}$ **20.** $\dfrac{17}{19} + \dfrac{6}{19} + \dfrac{15}{19}$

21. $\dfrac{4}{5} - \left(\dfrac{1}{2}\right)^2 + \dfrac{1}{10}$ **22.** $\dfrac{1}{3} \cdot \dfrac{5}{6} + \left(\dfrac{1}{3}\right)^2 \div 2$

23. $\left(\dfrac{3}{4} - \dfrac{1}{5}\right) \div \dfrac{3}{2}$ **24.** $\left(\dfrac{1}{2}\right)^3 - \left(\dfrac{1}{4}\right)^2$

25. Arrange the numbers $\dfrac{7}{8}, \dfrac{3}{4},$ and $\dfrac{2}{3}$ in order, from smallest to largest.

Mixed Numbers

4

4.1
CHANGING MIXED NUMBERS

Most people are familiar with mixed numbers and use them daily, as in "I drive $4\frac{1}{2}$ miles to school;" or "IBM stock rose $2\frac{3}{8}$ points today." In this chapter, we will discuss the operations of addition, subtraction, multiplication, and division with mixed numbers, using many of the ideas related to fractions that we discussed in Chapter 3. These ideas can be applied to mixed numbers because a mixed number is really the sum of a whole number and a fraction. Thus, $4\frac{1}{2} = 4 + \frac{1}{2}$ and $2\frac{3}{8} = 2 + \frac{3}{8}$. With this basic fact in mind, you will find that the techniques developed for working with mixed numbers are logically related to those in Chapter 3.

> **Mixed Number:** The sum of a whole number and a fraction, usually written side by side without a plus sign

Example 1 $5 + \frac{2}{3} = 5\frac{2}{3}$ read **five and two-thirds**

Example 2 $3 + \frac{5}{8} = 3\frac{5}{8}$ read **three and five-eighths**

Example 3 $9 + \frac{1}{2} = 9\frac{1}{2}$ read **nine and one-half**

Change each mixed number to an improper fraction.

1. $7 + \dfrac{1}{8}$

2. $11 + \dfrac{3}{4}$

3. $8\dfrac{1}{9}$

4. $5\dfrac{1}{2}$

To change a mixed number to an improper fraction, add the whole number and the fraction. Remember, the whole number can be written with 1 as the denominator.

Example 4 $5\dfrac{2}{3} = 5 + \dfrac{2}{3} = \dfrac{5}{1} \cdot \dfrac{3}{3} + \dfrac{2}{3} = \dfrac{15}{3} + \dfrac{2}{3} = \dfrac{17}{3}$

The LCD = 3 since the denominator of 5 is 1.

Example 5 $3\dfrac{5}{8} = 3 + \dfrac{5}{8} = \dfrac{3}{1} \cdot \dfrac{8}{8} + \dfrac{5}{8} = \dfrac{24}{8} + \dfrac{5}{8} = \dfrac{29}{8}$

Example 6 $9\dfrac{1}{2} = 9 + \dfrac{1}{2} = \dfrac{9}{1} \cdot \dfrac{2}{2} + \dfrac{1}{2} = \dfrac{18}{2} + \dfrac{1}{2} = \dfrac{19}{2}$

SHORTCUT TO CHANGING MIXED NUMBERS TO FRACTION FORM

1. Multiply the denominator by the whole number.
2. Add the numerator to this product.
3. Write this sum over the denominator of the fraction.

Example 7   Change $5\dfrac{2}{3}$ to an improper fraction.

Solution

Multiply $5 \cdot 3 = 15$ and add 2: $15 + 2 = 17$

Write 17 over denominator 3: $5\dfrac{2}{3} = \dfrac{17}{3}$

Example 8   Change $6\dfrac{9}{10}$ to an improper fraction.

Solution

Multiply $10 \cdot 6 = 60$ and add 9: $60 + 9 = 69$

Write 69 over denominator 10: $6\dfrac{9}{10} = \dfrac{69}{10}$

☐ **NOW WORK EXERCISES 1–4 IN THE MARGIN.**

To change an improper fraction to a mixed number

1. divide the numerator by the denominator;
2. write the remainder over the denominator as the fraction part of the mixed number.

Example 9 Change $\dfrac{29}{4}$ to a mixed number.

Solution

Divide 29 by 4:

$$\begin{array}{r} 7 \\ 4\overline{)29} \\ 28 \\ \hline 1 \end{array}$$

$$\frac{29}{4} = 7 + \frac{1}{4} = 7\frac{1}{4}$$

Example 10 Change $\dfrac{59}{3}$ to a mixed number.

Solution

Divide 59 by 3:

$$\begin{array}{r} 19 \\ 3\overline{)59} \\ 3 \\ \hline 29 \\ 27 \\ \hline 2 \end{array}$$

$$\frac{59}{3} = 19 + \frac{2}{3} = 19\frac{2}{3}$$

Example 11 Reduce $\dfrac{24}{10}$.

Solution

$$\frac{24}{10} = \frac{\cancel{2} \cdot 12}{\cancel{2} \cdot 5} = \frac{12}{5} \quad \text{reduced}$$

Example 12 Change $\dfrac{24}{10}$ to a mixed number, then reduce the fraction part.

Solution

Divide 24 by 10:

$$\begin{array}{r} 2 \\ 10\overline{)24} \\ 20 \\ \hline 4 \end{array}$$

$$\frac{24}{10} = 2 + \frac{4}{10} = 2 + \frac{\cancel{4}^2}{\cancel{10}_5} = 2 + \frac{2}{5} = 2\frac{2}{5}$$

Change each improper fraction to a mixed number with the fraction part reduced.

5. $\dfrac{18}{7}$

6. $\dfrac{20}{16}$

7. $\dfrac{25}{2}$

8. $\dfrac{35}{15}$

Example 13   Reduce $\dfrac{51}{34}$, then change it to a mixed number.

Solution

$$\frac{51}{34} = \frac{3 \cdot \cancel{17}}{2 \cdot \cancel{17}} = \frac{3}{2} \quad \text{reduced}$$

$$\begin{array}{r} 1 \\ 2\overline{)3} \\ \underline{2} \\ 1 \end{array} \qquad \frac{3}{2} = 1\frac{1}{2} \quad \text{as a mixed number}$$

☐ **NOW WORK EXERCISES 5–8 IN THE MARGIN.**

Answers to Marginal Exercises

1. $\dfrac{57}{8}$ **2.** $\dfrac{47}{4}$ **3.** $\dfrac{73}{9}$ **4.** $\dfrac{11}{2}$ **5.** $2\dfrac{4}{7}$ **6.** $1\dfrac{1}{4}$ **7.** $12\dfrac{1}{2}$ **8.** $2\dfrac{1}{3}$

NAME _____ SECTION _____ DATE _____

EXERCISES **4.1** _____

Reduce to lowest terms.

1. $\dfrac{24}{18}$ 2. $\dfrac{25}{10}$ 3. $\dfrac{16}{12}$ 4. $\dfrac{10}{8}$

5. $\dfrac{39}{26}$ 6. $\dfrac{48}{32}$ 7. $\dfrac{35}{25}$ 8. $\dfrac{18}{16}$

9. $\dfrac{80}{64}$ 10. $\dfrac{75}{60}$

Change to mixed numbers with the fraction part reduced.

11. $\dfrac{100}{24}$ 12. $\dfrac{25}{10}$ 13. $\dfrac{16}{12}$ 14. $\dfrac{10}{8}$

15. $\dfrac{39}{26}$ 16. $\dfrac{42}{8}$ 17. $\dfrac{43}{7}$ 18. $\dfrac{34}{16}$

19. $\dfrac{45}{6}$ 20. $\dfrac{75}{12}$ 21. $\dfrac{56}{18}$ 22. $\dfrac{31}{15}$

Answers

1. _____
2. _____
3. _____
4. _____
5. _____
6. _____
7. _____
8. _____
9. _____
10. _____
11. _____
12. _____
13. _____
14. _____
15. _____
16. _____
17. _____
18. _____
19. _____
20. _____
21. _____
22. _____

Answers

23. _____

24. _____

25. _____

26. _____

27. _____

28. _____

29. _____

30. _____

31. _____

32. _____

33. _____

34. _____

35. _____

36. _____

37. _____

38. _____

39. _____

40. _____

41. _____

42. _____

43. _____

44. _____

45. _____

46. _____

47. _____

48. _____

49. _____

50. _____

23. $\dfrac{36}{12}$ **24.** $\dfrac{48}{16}$ **25.** $\dfrac{72}{16}$ **26.** $\dfrac{70}{34}$

27. $\dfrac{45}{15}$ **28.** $\dfrac{60}{36}$ **29.** $\dfrac{35}{20}$ **30.** $\dfrac{185}{100}$

Change to improper fraction form and reduce.

31. $4\dfrac{5}{8}$ **32.** $3\dfrac{3}{4}$ **33.** $5\dfrac{1}{15}$ **34.** $1\dfrac{3}{5}$

35. $4\dfrac{2}{11}$ **36.** $2\dfrac{11}{44}$ **37.** $2\dfrac{9}{27}$ **38.** $4\dfrac{6}{7}$

39. $10\dfrac{8}{12}$ **40.** $11\dfrac{3}{8}$ **41.** $6\dfrac{8}{10}$ **42.** $14\dfrac{1}{5}$

43. $16\dfrac{2}{3}$ **44.** $12\dfrac{4}{8}$ **45.** $20\dfrac{3}{15}$ **46.** $9\dfrac{4}{10}$

47. $13\dfrac{1}{7}$ **48.** $49\dfrac{0}{12}$ **49.** $17\dfrac{0}{3}$ **50.** $3\dfrac{1}{50}$

4.2
ADDING MIXED NUMBERS

From Section 4.1, we know that a mixed number is the sum of a whole number and a fraction. This means that the sum of mixed numbers is the sum of whole numbers and fractions. With this relationship in mind, we add mixed numbers by treating the whole numbers and fraction parts separately.

> To add mixed numbers,
>
> 1. add the fraction parts;
> 2. add the whole numbers;
> 3. write the mixed number so that the fraction part is less than 1.

Example 1 $4\frac{2}{7} + 6\frac{3}{7}$

$$4\frac{2}{7} + 6\frac{3}{7} = 4 + \frac{2}{7} + 6 + \frac{3}{7}$$

$$= (4 + 6) + \left(\frac{2}{7} + \frac{3}{7}\right)$$

$$= 10 + \frac{5}{7}$$

$$= 10\frac{5}{7}$$

Or, vertically

$$4\frac{2}{7}$$
$$+\ 6\frac{3}{7}$$
$$\overline{10\frac{5}{7}}$$

Find each sum.

1. $2\dfrac{1}{2} + 3\dfrac{1}{3} =$

2. $14\dfrac{7}{10} + 22\dfrac{1}{5} =$

3. $4\dfrac{5}{8} + 10\dfrac{7}{8} =$

4. 8
  $+4\dfrac{3}{10}$
  $\overline{\phantom{+4\dfrac{3}{10}}}$

Example 2 $25\dfrac{1}{6} + 3\dfrac{7}{18}$

$$25\dfrac{1}{6} + 3\dfrac{7}{18} = 25 + \dfrac{1}{6} + 3 + \dfrac{7}{18}$$

$$= (25 + 3) + \left(\dfrac{1}{6} + \dfrac{7}{18}\right)$$

$$= 28 + \left(\dfrac{1}{6} \cdot \dfrac{3}{3} + \dfrac{7}{18}\right)$$

$$= 28 + \left(\dfrac{3}{18} + \dfrac{7}{18}\right)$$

$$= 28\dfrac{10}{18} = 28\dfrac{5}{9}$$

Or, vertically,

$$25\,\dfrac{1}{6} = 25\,\dfrac{1}{6} \cdot \dfrac{3}{3} = 25\dfrac{3}{18} \quad \text{LCD} = 18$$

$$+\ 3\dfrac{7}{18} = \ 3\ \ \dfrac{7}{18}\ \ = \ 3\dfrac{7}{18}$$

$$\overline{}$$

$$28\dfrac{10}{18} = 28\dfrac{5}{9}$$

Example 3 $7\dfrac{2}{3} + 9\dfrac{4}{5}$

$$7\dfrac{2}{3} = 7\,\dfrac{2}{3} \cdot \dfrac{5}{5} = \ 7\dfrac{10}{15} \quad \text{LCD} = 15$$

$$+9\dfrac{4}{5} = 9\,\dfrac{4}{5} \cdot \dfrac{3}{3} = \ 9\dfrac{12}{15}$$

$$\overline{}$$

$$16\dfrac{22}{15} = 16 + 1\dfrac{7}{15} = 17\dfrac{7}{15}$$

$$\uparrow \qquad\qquad\quad \uparrow$$

Fraction part is   Change it to a
greater than 1.   mixed number.

☐ **NOW WORK EXERCISES 1–4 IN THE MARGIN.**

Answers to Marginal Exercises

1. $5\dfrac{5}{6}$ **2.** $36\dfrac{9}{10}$ **3.** $15\dfrac{1}{2}$ **4.** $12\dfrac{3}{10}$

NAME _____ SECTION _____ DATE _____

EXERCISES **4.2** _____

Find each sum.

1. 10

  $+ \; 4\frac{5}{7}$
  _____

2. $6\frac{1}{2}$

  $+3\frac{1}{2}$
  _____

3. $5\frac{1}{3}$

  $+2\frac{2}{3}$
  _____

4. 8

  $+7\frac{3}{11}$
  _____

5. 12

  $+ \; \frac{3}{4}$
  _____

6. 15

  $+ \; \frac{9}{10}$
  _____

7. $3\;\frac{1}{4}$

  $+5\frac{5}{12}$
  _____

8. $5\frac{11}{12}$

  $+3\;\frac{5}{6}$
  _____

9. $21\;\frac{1}{3}$

  $+13\frac{1}{18}$
  _____

10. $9\frac{2}{3}$

  $+ \; \frac{5}{6}$
  _____

11. $4\frac{1}{2} + 3\frac{1}{6}$

12. $3\frac{1}{4} + 7\frac{1}{8}$

13. $25\frac{1}{10} + 17\frac{1}{4}$

14. $5\frac{1}{7} + 3\frac{1}{3}$

15. $6\frac{5}{12} + 4\frac{1}{3}$

1. _____

2. _____

3. _____

4. _____

5. _____

6. _____

7. _____

8. _____

9. _____

10. _____

11. _____

12. _____

13. _____

14. _____

15. _____

Answers

16. _____

17. _____

18. _____

19. _____

20. _____

21. _____

22. _____

23. _____

24. _____

25. _____

26. _____

27. _____

16. $5\frac{3}{10} + 2\frac{1}{14}$

17. $8\frac{2}{9} + 4\frac{1}{27}$

18. $11\frac{3}{4} + 2\frac{5}{16}$

19. $6\frac{4}{9} + 12\frac{1}{15}$

20. $4\frac{1}{6} + 13\frac{9}{10}$

21. $21\frac{3}{4} + 6\frac{3}{4}$

22. $3\frac{5}{8} + 3\frac{5}{8}$

23. $7\frac{3}{5} + 2\frac{1}{8}$

24. $9\frac{1}{8} + 3\frac{7}{12}$

25. $3\frac{1}{3} + 4\frac{1}{4} + 5\frac{1}{5}$

26. $\frac{3}{7} + 2\frac{1}{14} + 2\frac{1}{6}$

27. $20\frac{5}{8} + 42\frac{5}{6}$

28. $25\dfrac{2}{3} + 1\dfrac{1}{16}$

29. $32\dfrac{1}{64} + 4\dfrac{1}{24} + 17\dfrac{3}{8}$

30. $3\dfrac{1}{20} + 7\dfrac{1}{15} + 2\dfrac{3}{10}$

31.
$$\begin{array}{r} 4\dfrac{1}{3} \\[4pt] 8\dfrac{3}{8} \\[4pt] +6\dfrac{1}{6} \\ \hline \end{array}$$

32.
$$\begin{array}{r} 3\dfrac{2}{3} \\[4pt] 14\dfrac{1}{10} \\[4pt] +5\dfrac{1}{5} \\ \hline \end{array}$$

33.
$$\begin{array}{r} 13\dfrac{5}{8} \\[4pt] 13\dfrac{1}{12} \\[4pt] +10\dfrac{1}{4} \\ \hline \end{array}$$

34.
$$\begin{array}{r} 5\dfrac{1}{8} \\[4pt] 1\dfrac{1}{5} \\[4pt] +3\dfrac{1}{40} \\ \hline \end{array}$$

35.
$$\begin{array}{r} 27\dfrac{2}{3} \\[4pt] 30\dfrac{5}{8} \\[4pt] +31\dfrac{5}{6} \\ \hline \end{array}$$

Answers

28. _____

29. _____

30. _____

31. _____

32. _____

33. _____

34. _____

35. _____

Answers

36. _____

37. _____

38. _____

39. _____

40. _____

36. A bus trip is made in three parts. The first part takes $2\frac{1}{3}$ hours, the second part takes $2\frac{1}{2}$ hours, and the third part takes $3\frac{3}{4}$ hours. How many hours does the trip take?

37. A construction company built three sections of highway. One section was $20\frac{7}{10}$ kilometers, the second section was $3\frac{4}{10}$ kilometers, and the third section was $11\frac{6}{10}$ kilometers. What was the total length of highway built?

38. A triangle (three-sided figure) has sides measuring $42\frac{3}{4}$ feet, $23\frac{1}{2}$ feet, and $22\frac{7}{8}$ feet. What is the perimeter (total distance around) of the triangle?

39. A quadrilateral (four-sided figure) has sides measuring $3\frac{1}{2}$ inches, $2\frac{1}{4}$ inches, $3\frac{5}{8}$ inches, and $2\frac{3}{4}$ inches. What is the perimeter (total distance around) of the quadrilateral?

40. During one week, Sam played three rounds of golf. The first round took $4\frac{1}{2}$ hours, the second round took $3\frac{3}{4}$ hours, and the third round took $4\frac{3}{4}$ hours. How much time did Sam spend playing golf that week?

4.3

SUBTRACTING MIXED NUMBERS

Subtraction with mixed numbers also involves working with the whole number and fraction parts separately. With subtraction, though, we have another concern related to the fraction parts. We will see that, in some cases, a fraction part must be changed to an improper fraction by **borrowing** the whole number 1.

> To subtract mixed numbers,
> 1. subtract the fraction parts;
> 2. subtract the whole numbers.

Example 1

$$4\frac{3}{5} - 1\frac{2}{5} = (4 - 1) + \left(\frac{3}{5} - \frac{2}{5}\right)$$

$$= 3 + \frac{1}{5}$$

$$= 3\frac{1}{5}$$

Or, vertically,

$$
\begin{array}{r}
4\frac{3}{5} \\
-1\frac{2}{5} \\
\hline
3\frac{1}{5}
\end{array}
$$

Example 2 $5\frac{6}{7} - 3\frac{2}{7}$

$$
\begin{array}{r}
5\frac{6}{7} \\
-3\frac{2}{7} \\
\hline
2\frac{4}{7}
\end{array}
$$

Example 3 $10\frac{3}{5} - 6\frac{3}{20}$

$$
\begin{array}{rl}
10\frac{3}{5} = & 10\frac{12}{20} \quad \text{LCD = 20} \\
-6\frac{3}{20} = & 6\frac{3}{20} \\
\hline
& 4\frac{9}{20}
\end{array}
$$

☐ **NOW WORK EXERCISES 1–3 IN THE MARGIN.**

Find each difference.

1. $9\frac{7}{10} - 3\frac{1}{5} =$

2. $15\frac{2}{3}$
$\quad -\ 7\frac{1}{10}$

3. $6\frac{3}{4} - 2\frac{3}{14} =$

> If the fraction part being subtracted is larger than the first fraction,
>
> 1. **borrow** the whole number 1 from the first whole number;
>
> 2. add this 1 to the first fraction in the form $\frac{k}{k}$ where k is the denominator of the fraction;
>
> 3. subtract.

Example 4 $7\frac{2}{5} - 4\frac{3}{5}$

Solution

Since $\frac{3}{5}$ is larger than $\frac{2}{5}$, **borrow** the whole number 1 from 7. We write 7 as $6 + 1$; then write $1 + \frac{2}{5}$ as $\frac{7}{5}$.

$$7\frac{2}{5} = 6 + 1 + \frac{2}{5} = 6 + 1\frac{2}{5} = 6\frac{7}{5}$$
$$-4\frac{3}{5} = 4 + \frac{3}{5} = 4 + \frac{3}{5} = 4\frac{3}{5}$$
$$2\frac{4}{5}$$

Example 5 $19\frac{2}{3} - 5\frac{3}{4}$

Solution

First change the fraction parts so they have the same denominator, 12. Then **borrow** the whole number 1 from 19.

$$19\frac{2}{3} = 19\frac{8}{12} = 18\frac{20}{12} \qquad 1 + \frac{8}{12} = \frac{12}{12} + \frac{8}{12} = \frac{20}{12}$$
$$-5\frac{3}{4} = 5\frac{9}{12} = 5\frac{9}{12}$$
$$13\frac{11}{12}$$

Example 6 $7 - 4\frac{2}{3}$

Solution

In this case, the whole number 7 has no fraction part. We still **borrow** the whole number 1. We write 1 as $\frac{3}{3}$ so that its denominator will be the same as that of the other fraction part, $\frac{2}{3}$.

$$7 = 6\frac{3}{3} \qquad \text{Here, } 7 = 6 + 1 = 6 + \frac{3}{3} = 6\frac{3}{3}.$$
$$-4\frac{2}{3} = 4\frac{2}{3}$$
$$2\frac{1}{3}$$

Completion Example 7 $10 - 6\dfrac{5}{8}$

The whole number 10 has no fraction part. We still need to **borrow** the whole number 1 from 10.

$$10 \quad = 9 \underline{\qquad}$$
$$- 6\dfrac{5}{8} = 6 \quad \dfrac{5}{8} \underline{\qquad}$$

Completion Example 8

$$14\dfrac{3}{4} = 14 \underline{\qquad} = 13 \underline{\qquad}$$
$$- 5\dfrac{9}{10} = 5 \underline{\qquad} = 5 \underline{\qquad}$$
$$\underline{\qquad}$$

☐ **NOW WORK EXERCISES 4 AND 5 IN THE MARGIN.**

Subtract.

4. $17 - 9\dfrac{6}{11} =$

5. $\quad 13\dfrac{1}{2}$
$$\quad - 8\dfrac{5}{6}$$

Completion Example Answers

7. $10 \quad = 9\dfrac{8}{8}$
$$- 6\dfrac{5}{8} = 6\dfrac{5}{8}$$
$$\overline{\quad 3\dfrac{3}{8}}$$

8. $14\dfrac{3}{4} = 14\dfrac{15}{20} = 13\dfrac{35}{20}$
$$- 5\dfrac{9}{10} = \quad 5\dfrac{18}{20} = \quad 5\dfrac{18}{20}$$
$$\overline{\quad 8\dfrac{17}{20}}$$

Answers to Marginal Exercises

1. $6\dfrac{1}{2}$ **2.** $8\dfrac{17}{30}$ **3.** $4\dfrac{15}{28}$ **4.** $7\dfrac{5}{11}$ **5.** $4\dfrac{2}{3}$

NAME _____ SECTION _____ DATE _____

EXERCISES **4.3** _____

Find the differences.

1. $5\frac{1}{2}$
    -1

2. $7\frac{3}{4}$
    -2

3. $4\frac{5}{12}$
    -3

4. $3\frac{5}{8}$
    -2

5. $6\frac{1}{2}$
    $-2\frac{1}{2}$

6. $9\frac{1}{4}$
    $-5\frac{1}{4}$

7. $5\frac{3}{4}$
    $-2\frac{1}{4}$

8. $7\frac{9}{10}$
    $-3\frac{3}{10}$

9. $14\frac{5}{8}$
    $-11\frac{3}{8}$

10. $20\frac{7}{16}$
    $-15\frac{5}{16}$

11. $4\frac{7}{8}$
    $-1\frac{1}{4}$

12. $9\frac{5}{16}$
    $-2\frac{1}{4}$

13. $5\frac{11}{12}$
    $-1\frac{1}{4}$

14. $10\frac{5}{6}$
    $-4\frac{2}{3}$

15. $8\frac{5}{6}$
    $-2\frac{1}{4}$

16. $15\frac{5}{8}$
    $-11\frac{3}{4}$

17. $14\frac{6}{10}$
    $-3\frac{4}{5}$

18. $8\frac{3}{32}$
    $-4\frac{3}{16}$

19. $12\frac{3}{4}$
    $-7\frac{1}{6}$

20. $8\frac{11}{12}$
    $-5\frac{9}{10}$

1. _____
2. _____
3. _____
4. _____
5. _____
6. _____
7. _____
8. _____
9. _____
10. _____
11. _____
12. _____
13. _____
14. _____
15. _____
16. _____
17. _____
18. _____
19. _____
20. _____

Answers

21. _____

22. _____

23. _____

24. _____

25. _____

26. _____

27. _____

28. _____

29. _____

30. _____

31. _____

32. _____

33. _____

34. _____

21. $4\dfrac{7}{16} - 3$

22. $5\dfrac{9}{10} - 2$

23. $7 - 6\dfrac{2}{3}$

24. $12 - 4\dfrac{1}{5}$

25. $2 - 1\dfrac{3}{8}$

26. $75 - 17\dfrac{5}{6}$

27. $4\dfrac{9}{16} - 2\dfrac{7}{8}$

28. $3\dfrac{7}{10} - 2\dfrac{5}{6}$

29. $15\dfrac{11}{16} - 13\dfrac{7}{8}$

30. $20\dfrac{3}{6} - 3\dfrac{4}{8}$

31. $17\dfrac{3}{12} - 12\dfrac{2}{8}$

32. $18\dfrac{2}{7} - 4\dfrac{1}{3}$

33. $13\dfrac{5}{8} - 6\dfrac{11}{20}$

34. $18\dfrac{7}{8} - 2\dfrac{2}{3}$

NAME _____ SECTION _____ DATE _____

35. $10\frac{3}{10} - 2\frac{1}{2}$

35. _____

36. Sara can paint a room in $3\frac{3}{5}$ hours, and Emily can paint a room of the same size in $4\frac{1}{5}$ hours. (a) How many hours are saved by having Sara paint the room? (b) How many minutes are saved?

36. (a) _____

(b) _____

37. _____

37. A teacher graded two sets of test papers. The first set took $3\frac{3}{4}$ hours to grade, and the second set took $2\frac{3}{5}$ hours. How much faster did the teacher grade the second set?

38. _____

39. _____

38. Mike takes $1\frac{1}{2}$ hours to clean a pool, and Tom takes $2\frac{1}{3}$ hours to clean the same pool. How much longer does Tom take?

39. A long-distance runner was in training. He ran ten miles in $50\frac{3}{10}$ minutes. Three months later, he ran the same ten miles in $43\frac{7}{10}$ minutes. By how much did his time improve?

Answers

40. _____

41. _____

42. _____

43. _____

44. _____

40. Certain shares of stock were selling for $43\frac{7}{8}$ dollars per share. One month later, the same stock was selling for $48\frac{1}{2}$ dollars per share. By how much did the stock increase in price?

41. You want to lose 10 pounds. If you weigh 180 pounds now and you lose $3\frac{1}{4}$ pounds during the first week and $3\frac{1}{2}$ pounds during the second week, how much more weight do you need to lose?

42. Mr. Johnson originally weighed 240 pounds. During each week of six weeks of dieting, he lost $5\frac{1}{2}$ pounds, $2\frac{3}{4}$ pounds, $4\frac{5}{16}$ pounds, $1\frac{3}{4}$ pounds, $2\frac{5}{8}$ pounds, and $3\frac{1}{4}$ pounds. If he was 35 years old, what did he weigh at the end of the six weeks?

43. A board is 10 feet long. If a piece $6\frac{3}{4}$ feet is cut off, what is the length of the remaining piece?

44. A salesman drove $5\frac{3}{4}$ hours one day and then $6\frac{1}{2}$ hours the next day. How much more time did he spend driving on the second day?

4.4
MULTIPLYING MIXED NUMBERS

The fact that a mixed number is the sum of a whole number and a fraction does not help in understanding multiplication (or division) with mixed numbers. The simplest way to multiply mixed numbers is to change each number to an improper fraction before multiplying. Just as with multiplying fractions in Chapter 3, we can use prime factorizations and reduce before actually multiplying.

> To multiply mixed numbers,
>
> 1. change each mixed number to improper fraction form;
>
> 2. multiply these fractions using prime factorizations; then reduce;
>
> 3. change the answer to a mixed number or leave it in fraction form.

Example 1 $\left(1\frac{2}{3}\right)\left(2\frac{3}{4}\right) = \left(\frac{5}{3}\right)\left(\frac{11}{4}\right) = \frac{5\cdot 11}{3\cdot 4} = \frac{55}{12}$ or $4\frac{7}{12}$

Example 2 $\frac{5}{6}\cdot 3\frac{3}{10} = \frac{5}{6}\cdot\frac{33}{10} = \frac{\cancel{5}\cdot\cancel{3}\cdot 11}{2\cdot\cancel{3}\cdot 2\cdot\cancel{5}} = \frac{11}{4}$ or $2\frac{3}{4}$

Example 3 $4\frac{1}{2}\cdot 1\frac{1}{6}\cdot 3\frac{1}{3} = \frac{9}{2}\cdot\frac{7}{6}\cdot\frac{10}{3} = \frac{\cancel{3}\cdot\cancel{3}\cdot 7\cdot\cancel{2}\cdot 5}{\cancel{2}\cdot 2\cdot\cancel{3}\cdot\cancel{3}} = \frac{35}{2}$ or $17\frac{1}{2}$

Large mixed numbers can be multiplied in the same way.

Example 4 $24\frac{3}{8}\cdot 45\frac{1}{4}$

(a) Change to improper fractions.

$$
\begin{array}{ccc}
\begin{array}{r} 24 \\ \times\ 8 \\ \hline 192 \end{array} &
\begin{array}{r} 192 \\ +\ \ 3 \\ \hline 195 \end{array} &
24\frac{3}{8} = \frac{195}{8}
\end{array}
$$

$$
\begin{array}{ccc}
\begin{array}{r} 45 \\ \times\ 4 \\ \hline 180 \end{array} &
\begin{array}{r} 180 \\ +\ \ 1 \\ \hline 181 \end{array} &
45\frac{1}{4} = \frac{181}{4}
\end{array}
$$

Find each product.

1. $\left(2\dfrac{1}{2}\right)\left(5\dfrac{3}{5}\right) =$

2. $4\dfrac{1}{5} \cdot 2\dfrac{3}{4} \cdot \dfrac{10}{33} =$

3. $10\dfrac{1}{12} \cdot 18\dfrac{3}{4} =$

(b) Now multiply.

$$24\dfrac{3}{8} \cdot 45\dfrac{1}{4} = \dfrac{195}{8} \cdot \dfrac{181}{4} = \dfrac{35{,}295}{32} = 1102\dfrac{31}{32}$$

$$
\begin{array}{r}
195 \\
\times\ 181 \\
\hline
195 \\
15\ 60 \\
19\ 5 \\
\hline
35{,}295
\end{array}
\qquad
\begin{array}{r}
1102\dfrac{31}{32} \\
32\overline{)35{,}295} \\
32 \\
\hline
3\ 2 \\
3\ 2 \\
\hline
09 \\
0 \\
\hline
95 \\
64 \\
\hline
31
\end{array}
$$

Completion Example 5

$$2\dfrac{1}{16} \cdot 1\dfrac{1}{7} \cdot 1\dfrac{3}{11} = \dfrac{33}{16} \cdot \underline{\hspace{1cm}} \cdot \underline{\hspace{1cm}}$$

$$= \dfrac{3 \cdot 11 \cdot \underline{\hspace{0.5cm}}}{2 \cdot 2 \cdot 2 \cdot 2 \cdot \underline{\hspace{0.5cm}}} = \underline{\hspace{1cm}}$$

Completion Example 6

$$\left(4\dfrac{3}{8}\right)\left(3\dfrac{1}{5}\right)\left(\dfrac{1}{34}\right) = \dfrac{35}{8} \cdot \underline{\hspace{1cm}} \cdot \underline{\hspace{1cm}}$$

$$= \dfrac{5 \cdot 7 \cdot \underline{\hspace{0.5cm}}}{2 \cdot 2 \cdot 2 \cdot \underline{\hspace{0.5cm}}} = \underline{\hspace{1cm}}$$

Completion Example 7

$$20\dfrac{3}{5} \cdot 18\dfrac{1}{4} = \dfrac{103}{5} \cdot \underline{\hspace{1cm}}$$

$$= \underline{\hspace{1.5cm}} = \underline{\hspace{1.5cm}}$$

☐ **NOW WORK EXERCISES 1–3 IN THE MARGIN.**

Completion Example Answers

5. $\dfrac{33}{16} \cdot \dfrac{8}{7} \cdot \dfrac{14}{11} = \dfrac{3 \cdot \cancel{11} \cdot \cancel{2} \cdot \cancel{2} \cdot \cancel{2} \cdot \cancel{2} \cdot \cancel{7}}{\cancel{2} \cdot \cancel{2} \cdot \cancel{2} \cdot \cancel{2} \cdot \cancel{7} \cdot \cancel{11}} = 3$

6. $\dfrac{35}{8} \cdot \dfrac{16}{5} \cdot \dfrac{1}{34} = \dfrac{\cancel{5} \cdot 7 \cdot \cancel{2} \cdot \cancel{2} \cdot \cancel{2} \cdot \cancel{2} \cdot 1}{\cancel{2} \cdot \cancel{2} \cdot \cancel{2} \cdot \cancel{5} \cdot \cancel{2} \cdot 17} = \dfrac{7}{17}$

7. $\dfrac{103}{5} \cdot \dfrac{73}{4} = \dfrac{7519}{20} = 375\dfrac{19}{20}$

A fraction **of** a number means to **multiply** the number by the fraction.

Example 8   Find $\dfrac{3}{4}$ of 40.

$$\frac{3}{4} \cdot 40 = \frac{3}{\cancel{4}\,_1} \cdot \frac{\cancel{40}^{\,10}}{1} = 30$$

Example 9   Find $\dfrac{3}{4}$ of 80.

$$\frac{3}{4} \cdot 80 = \frac{3}{\cancel{4}\,_1} \cdot \frac{\cancel{80}^{\,20}}{1} = 60$$

Example 10   Find $\dfrac{2}{3}$ of $\dfrac{9}{10}$.

$$\frac{2}{3} \cdot \frac{9}{10} = \frac{\cancel{2} \cdot \cancel{3} \cdot 3}{\cancel{3} \cdot \cancel{2} \cdot 5} = \frac{3}{5}$$

NAME _____ SECTION _____ DATE _____

EXERCISES **4.4** _____

Answers

Find the indicated products.

1. $\left(2\frac{1}{3}\right)\left(3\frac{1}{4}\right)$

2. $\left(1\frac{1}{5}\right)\left(1\frac{1}{7}\right)$

3. $4\frac{1}{2}\left(2\frac{1}{3}\right)$

4. $3\frac{1}{3}\left(2\frac{1}{5}\right)$

5. $6\frac{1}{4}\left(3\frac{3}{5}\right)$

6. $5\frac{1}{3}\left(2\frac{1}{4}\right)$

7. $\left(8\frac{1}{2}\right)\left(3\frac{2}{3}\right)$

8. $\left(9\frac{1}{3}\right)2\frac{1}{7}$

9. $\left(6\frac{2}{7}\right)1\frac{3}{11}$

10. $\left(11\frac{1}{4}\right)1\frac{1}{15}$

11. $6\frac{2}{3}\cdot4\frac{1}{2}$

12. $4\frac{3}{8}\cdot2\frac{2}{7}$

13. $9\frac{3}{4}\cdot2\frac{6}{26}$

14. $7\frac{1}{2}\cdot\frac{2}{15}$

15. $\frac{3}{4}\cdot1\frac{1}{3}$

1. _____
2. _____
3. _____
4. _____
5. _____
6. _____
7. _____
8. _____
9. _____
10. _____
11. _____
12. _____
13. _____
14. _____
15. _____

Answers

16. _____

17. _____

18. _____

19. _____

20. _____

21. _____

22. _____

23. _____

24. _____

25. _____

26. _____

27. _____

28. _____

16. $3\frac{4}{5} \cdot 2\frac{1}{7}$

17. $12\frac{1}{2} \cdot 2\frac{1}{5}$

18. $9\frac{3}{5} \cdot 1\frac{1}{16}$

19. $6\frac{1}{8} \cdot 3\frac{1}{7}$

20. $5\frac{1}{4} \cdot 11\frac{1}{3}$

21. $\frac{1}{4} \cdot \frac{2}{3} \cdot \frac{6}{7}$

22. $\frac{7}{8} \cdot \frac{24}{25} \cdot \frac{5}{21}$

23. $\frac{3}{16} \cdot \frac{8}{9} \cdot \frac{3}{5}$

24. $\frac{2}{5} \cdot \frac{1}{5} \cdot \frac{4}{7}$

25. $\frac{6}{7} \cdot \frac{2}{11} \cdot \frac{3}{5}$

26. $\left(3\frac{1}{2}\right)\left(2\frac{1}{7}\right)\left(5\frac{1}{4}\right)$

27. $\left(4\frac{3}{8}\right)\left(2\frac{1}{5}\right)\left(1\frac{1}{7}\right)$

28. $\left(6\frac{3}{16}\right)\left(2\frac{1}{11}\right)\left(5\frac{3}{5}\right)$

NAME _____ SECTION _____ DATE _____

29. $7\frac{1}{3} \cdot 5\frac{1}{4} \cdot 6\frac{2}{7}$

30. $2\frac{5}{8} \cdot 3\frac{2}{5} \cdot 1\frac{3}{4}$

31. $2\frac{1}{16} \cdot 4\frac{1}{3} \cdot 1\frac{3}{11}$

32. $5\frac{1}{10} \cdot 3\frac{1}{7} \cdot 2\frac{1}{17}$

33. $2\frac{1}{4} \cdot 6\frac{3}{8} \cdot 1\frac{5}{27}$

34. $1\frac{3}{32} \cdot 1\frac{1}{7} \cdot 1\frac{1}{25}$

35. $1\frac{5}{16} \cdot 1\frac{1}{3} \cdot 1\frac{1}{5}$

36. $24\frac{1}{5} \cdot 35\frac{1}{6}$

Answers

29. _____

30. _____

31. _____

32. _____

33. _____

34. _____

35. _____

36. _____

Answers

37. _____

38. _____

39. _____

40. _____

41. _____

42. _____

43. _____

44. _____

45. _____

46. _____

37. $72\dfrac{3}{5} \cdot 25\dfrac{1}{6}$

38. $42\dfrac{5}{6} \cdot 30\dfrac{1}{7}$

39. $75\dfrac{1}{3} \cdot 40\dfrac{1}{25}$

40. $36\dfrac{3}{4} \cdot 17\dfrac{5}{12}$

41. Find $\dfrac{2}{3}$ of 60.

42. Find $\dfrac{1}{4}$ of 80.

43. Find $\dfrac{1}{5}$ of 100.

44. Find $\dfrac{3}{5}$ of 100.

45. Find $\dfrac{3}{4}$ of 100.

46. Find $\dfrac{5}{6}$ of 120.

NAME _____ SECTION _____ DATE _____

47. Find $\dfrac{1}{2}$ of $\dfrac{5}{8}$.

48. Find $\dfrac{1}{6}$ of $\dfrac{3}{4}$.

49. Find $\dfrac{9}{10}$ of $\dfrac{15}{21}$.

50. Find $\dfrac{7}{8}$ of $\dfrac{8}{10}$.

Answers

47. _____

48. _____

49. _____

50. _____

51. _____

52. _____

51. A telephone pole is 32 feet long. If $\dfrac{5}{16}$ of the pole must be underground and $\dfrac{11}{16}$ of the pole aboveground, how much of the pole is underground? How much is aboveground?

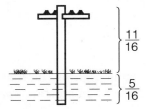

52. The total distance around a square (its perimeter) is found by multiplying the length of one side by 4. Find the perimeter of a square if the length of one side is $5\dfrac{1}{16}$ inches.

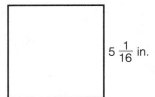

$5\dfrac{1}{16}$ in.

53. A man drives $17\frac{7}{10}$ miles one way to work, five days a week. How many miles does he drive each week going to and from work?

54. A length of pipe is $27\frac{3}{4}$ feet. What would be the total length if $36\frac{1}{2}$ of these pipe sections were laid end to end?

55. A woman reads $\frac{1}{6}$ of a book in 3 hours. If the book contains 540 pages, how many pages will she read in 3 hours? How long will she take to read the entire book?

56. Three towns (A, B, and C) are located on the same highway (assume a straight section of highway). Towns A and B are 53 kilometers apart. Town C is $45\frac{9}{10}$ kilometers from town B. How far apart are towns A and C? (The sketch shows that there are two possible situations to consider.)

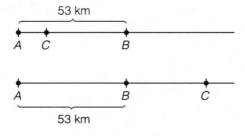

4.5
DIVIDING MIXED NUMBERS _____

Division with mixed numbers is the same as division with fractions. We change each mixed number to an improper fraction; then multiply by the **reciprocal** of the divisor. Remember from Section 3.4 that the **reciprocal** of $\frac{a}{b}$ is $\frac{b}{a}$ ($a \neq 0$ and $b \neq 0$), and $\frac{a}{b} \cdot \frac{b}{a} = 1$.

Note: There is no reciprocal for 0.

> To divide mixed numbers,
>
> 1. change each mixed number to an improper fraction;
> 2. write the reciprocal of the divisor;
> 3. multiply by using prime factorizations; then reduce.

Example 1 $3\frac{1}{4} \div 7\frac{4}{5} = \frac{13}{4} \div \frac{39}{5} = \frac{13}{4} \cdot \boxed{\frac{5}{39}}$ Note that the divisor is $\frac{39}{5}$, and we multiply by its reciprocal, $\frac{5}{39}$.

$$= \frac{\cancel{13} \cdot 5}{4 \cdot \cancel{13} \cdot 3} = \frac{5}{12}$$

Example 2 $\dfrac{2\frac{1}{4}}{4\frac{1}{2}} = \dfrac{\frac{9}{4}}{\frac{9}{2}} = \dfrac{\cancel{9}}{\cancel{4}_2} \cdot \dfrac{\cancel{2}^1}{\cancel{9}} = \dfrac{1}{2}$ Note that the divisor is $\frac{9}{2}$, and we multiply by its reciprocal, $\frac{2}{9}$.

Example 3 $6 \div 7\frac{7}{8} = 6 \div \frac{63}{8} = \frac{6}{1} \cdot \boxed{\frac{8}{63}}$

$$= \frac{\cancel{3} \cdot 2 \cdot 8}{\cancel{3} \cdot 3 \cdot 7} = \frac{16}{21}$$

Example 4 $7\frac{7}{8} \div 6 = \frac{63}{8} \div \frac{6}{1} = \frac{63}{8} \cdot \boxed{\frac{1}{6}}$

$$= \frac{\cancel{3} \cdot 3 \cdot 7}{8 \cdot 2 \cdot \cancel{3}} = \frac{21}{16} \quad \text{or} \quad 1\frac{5}{16}$$

Find each quotient.

1. $3\dfrac{3}{4} \div 2\dfrac{2}{5} =$

2. $4\dfrac{2}{3} \div 4\dfrac{2}{3} =$

3. $16 \div 2\dfrac{2}{3} =$

Completion Example 5 $10\dfrac{1}{2} \div 3\dfrac{3}{4} = \dfrac{21}{2} \div \underline{\hspace{1cm}} = \dfrac{21}{2} \cdot \underline{\hspace{1cm}}$

$$= \dfrac{3 \cdot 7 \cdot \underline{\hspace{0.4cm}}}{2 \cdot \underline{\hspace{0.3cm}}} = \underline{\hspace{1cm}} \quad \text{or} \quad \underline{\hspace{1cm}}$$

Completion Example 6 $6\dfrac{2}{3} \div 3\dfrac{1}{5} = \dfrac{20}{3} \div \underline{\hspace{1cm}} = \dfrac{20}{3} \cdot \underline{\hspace{1cm}}$

$$= \dfrac{2 \cdot 2 \cdot 5 \cdot \underline{\hspace{0.4cm}}}{3 \cdot \underline{\hspace{0.3cm}}} = \underline{\hspace{1cm}} \quad \text{or} \quad \underline{\hspace{1cm}}$$

☐ **NOW WORK EXERCISES 1–3 IN THE MARGIN.**

Example 7   The product of $2\dfrac{1}{3}$ with another number is $5\dfrac{1}{6}$. What is the other number?

Solution

$$2\dfrac{1}{3} \cdot \,? = 5\dfrac{1}{6}$$

To find the missing number, *divide* $5\dfrac{1}{6}$ by $2\dfrac{1}{3}$.

$$5\dfrac{1}{6} \div 2\dfrac{1}{3} = \dfrac{31}{6} \div \dfrac{7}{3} = \dfrac{31}{\cancel{6}_2} \cdot \dfrac{\cancel{3}^1}{7} = \dfrac{31}{14} \text{ or } 2\dfrac{3}{14}$$

The other number is $2\dfrac{3}{14}$.

CHECK: $2\dfrac{1}{3} \cdot 2\dfrac{3}{14} = \dfrac{\cancel{7}}{3} \cdot \dfrac{31}{\cancel{14}_2} = \dfrac{31}{6} = 5\dfrac{1}{6}$

Completion Example Answers

5. $\dfrac{21}{2} \div \dfrac{15}{4} = \dfrac{21}{2} \cdot \dfrac{4}{15} = \dfrac{\cancel{3} \cdot 7 \cdot \cancel{2} \cdot 2}{2 \cdot \cancel{3} \cdot 5} = \dfrac{14}{5}$ or $2\dfrac{4}{5}$

6. $\dfrac{20}{3} \div \dfrac{16}{5} = \dfrac{20}{3} \cdot \dfrac{5}{16} = \dfrac{\cancel{2} \cdot \cancel{2} \cdot 5 \cdot 5}{3 \cdot \cancel{2} \cdot \cancel{2} \cdot 2 \cdot 2} = \dfrac{25}{12}$ or $2\dfrac{1}{12}$

Example 8 Find the sum of $\frac{7}{8}, \frac{9}{16}$, and $\frac{7}{16}$, then divide the sum by 3.

Solution

Add the three fractions, then divide by 3.

$$\left(\frac{7}{8} + \frac{9}{16} + \frac{7}{16}\right) \div 3 = \left(\frac{14}{16} + \frac{9}{16} + \frac{7}{16}\right) \div 3$$

$$= \left(\frac{30}{16}\right) \div 3 = \frac{30}{16} \cdot \frac{1}{3}$$

$$= \frac{\cancel{3} \cdot \cancel{2} \cdot 5}{\cancel{2} \cdot 8 \cdot \cancel{3}} = \frac{5}{8}$$

In Example 8, both of the operations of addition and division were involved. The directions were given as to which operation to perform first. When there are no specific directions and an expression involves more than one operation, we use the rules for order of operations. These rules were stated in Sections 2.2 and 3.7 and are restated here for convenience and emphasis.

RULES FOR ORDER OF OPERATIONS

1. First, simplify within grouping symbols, such as parentheses ( ), brackets [ ], or braces { }. Start with the innermost grouping.
2. Second, find any powers indicated by exponents.
3. Third, moving from **left to right,** perform any multiplications or divisions in the order they appear.
4. Fourth, moving from **left to right,** perform any additions or subtractions in the order they appear.

Example 9 Use the rules for order of operations to simplify the following expression.

$$2\frac{1}{2} \cdot 1\frac{1}{6} + 7 \div \frac{3}{4} = \frac{5}{2} \cdot \frac{7}{6} + \frac{7}{1} \cdot \frac{4}{3} \quad \text{Multiply and divide from left to right.}$$

$$= \frac{35}{12} + \frac{28}{3} \quad \text{Now add.}$$

$$= \frac{35}{12} + \frac{28}{3} \cdot \boxed{\frac{4}{4}}$$

$$= \frac{35}{12} + \frac{112}{12}$$

$$= \frac{147}{12} = 12\frac{3}{12} = 12\frac{1}{4}$$

Or, working with separate parts, we can write

$$2\frac{1}{2} \cdot 1\frac{1}{6} = \frac{5}{2} \cdot \frac{7}{6} = \frac{35}{12} = 2\frac{11}{12} \quad \text{multiplying}$$

$$7 \div \frac{3}{4} = \frac{7}{1} \cdot \frac{4}{3} = \frac{28}{3} = 9\frac{1}{3} \quad \text{dividing}$$

$$
\begin{array}{rl}
2\dfrac{11}{12} = & 2\dfrac{11}{12} \quad \text{adding the results}\\[2mm]
+9\dfrac{1}{3} = & 9\dfrac{4}{12}\\[2mm]
\hline
& 11\dfrac{15}{12} = 12\dfrac{3}{12} = 12\dfrac{1}{4}
\end{array}
$$

Answers to Marginal Exercises

1. $\frac{25}{16}$ or $1\frac{9}{16}$ 2. 1   3. 6

NAME _____ SECTION _____ DATE _____

EXERCISES **4.5** _____

Find the indicated quotients.

1. $\dfrac{2}{21} \div \dfrac{2}{7}$

2. $\dfrac{9}{32} \div \dfrac{5}{8}$

3. $\dfrac{5}{12} \div \dfrac{3}{4}$

4. $\dfrac{6}{17} \div \dfrac{6}{17}$

5. $\dfrac{5}{6} \div 3\dfrac{1}{4}$

6. $\dfrac{7}{8} \div 7\dfrac{1}{2}$

7. $\dfrac{29}{50} \div 3\dfrac{1}{10}$

8. $4\dfrac{1}{5} \div 3\dfrac{1}{3}$

9. $2\dfrac{1}{17} \div 1\dfrac{1}{4}$

10. $5\dfrac{1}{6} \div 3\dfrac{1}{4}$

11. $2\dfrac{2}{49} \div 3\dfrac{1}{14}$

12. $6\dfrac{5}{6} \div 2$

13. $4\dfrac{1}{5} \div 3$

14. $4\dfrac{5}{8} \div 4$

15. $6\dfrac{5}{6} \div \dfrac{1}{2}$

Answers

1. _____

2. _____

3. _____

4. _____

5. _____

6. _____

7. _____

8. _____

9. _____

10. _____

11. _____

12. _____

13. _____

14. _____

15. _____

Answers

16. _____

17. _____

18. _____

19. _____

20. _____

21. _____

22. _____

23. _____

24. _____

25. _____

26. _____

27. _____

28. _____

29. _____

30. _____

16. $4\dfrac{5}{8} \div \dfrac{1}{4}$ 17. $4\dfrac{1}{5} \div \dfrac{1}{3}$ 18. $1\dfrac{1}{32} \div 3\dfrac{2}{3}$

19. $7\dfrac{5}{11} \div 4\dfrac{1}{10}$ 20. $13\dfrac{1}{7} \div 4\dfrac{2}{11}$

Evaluate the following expressions using the rules for order of operations.

21. $\dfrac{3}{5} \cdot \dfrac{1}{6} + \dfrac{1}{5} \div 2$ 22. $\dfrac{1}{2} \div \dfrac{1}{2} + 1 - \dfrac{2}{3} \cdot 3$

23. $3\dfrac{1}{2} \cdot 5\dfrac{1}{3} + \dfrac{5}{12} \div \dfrac{15}{16}$ 24. $2\dfrac{1}{4} + 1\dfrac{1}{5} + 2 \div \dfrac{20}{21}$

25. $\dfrac{5}{8} - \dfrac{1}{3} \cdot \dfrac{2}{5} + 6\dfrac{1}{10}$ 26. $1\dfrac{1}{6} \cdot 1\dfrac{2}{19} \div \dfrac{7}{8} + \dfrac{1}{38}$

27. $\dfrac{3}{10} + \dfrac{5}{6} \div \dfrac{1}{4} \cdot \dfrac{1}{8} - \dfrac{7}{60}$ 28. $5\dfrac{1}{7} \div (2 + 1)$

29. $\left(2 - \dfrac{1}{3}\right) \div \left(1 - \dfrac{1}{3}\right)$ 30. $\left(2\dfrac{4}{9} + 1\dfrac{1}{18}\right) \div \left(1\dfrac{2}{9} - \dfrac{1}{6}\right)$

31. Find the sum of the numbers $\frac{7}{8}$, $\frac{9}{10}$, and $1\frac{3}{4}$.

32. Find the sum of the numbers $\frac{5}{6}$ and $\frac{1}{15}$; then subtract $\frac{17}{30}$.

33. The product of $\frac{9}{10}$ with another number is $1\frac{2}{3}$. What is the other number?

34. The result of multiplying two numbers is $10\frac{1}{3}$. If one of the numbers is $7\frac{1}{6}$, what is the other number?

35. An airplane is carrying 150 passengers. This is $\frac{6}{7}$ of its capacity. What is the capacity of the airplane?

Answers

31. _____

32. _____

33. _____

34. _____

35. _____

Answers

36. _____

37. _____

38. _____

39. _____

40. _____

36. The sale price of a coat is \$135. This is $\frac{3}{4}$ of the original price. What was the original price?

37. If the product of $5\frac{1}{2}$ and $2\frac{1}{4}$ is added to the quotient of $\frac{9}{10}$ and $\frac{3}{4}$, what is the sum?

38. If the quotient of $\frac{5}{8}$ and $\frac{1}{2}$ is subtracted from the product of $2\frac{1}{4}$ and $3\frac{1}{5}$, what is the difference?

39. If $\frac{9}{10}$ of 70 is divided by $\frac{3}{4}$ of 10, what is the quotient?

40. If $\frac{2}{3}$ of $4\frac{1}{4}$ is added to $\frac{5}{8}$ of $6\frac{1}{3}$, what is the sum?

NAME _____ SECTION _____ DATE _____

CHAPTER **4** TEST _____

Answers

Reduce all fractions to lowest terms.

1. Change $\dfrac{17}{5}$ to a mixed number.

2. Change $7\dfrac{5}{8}$ to an improper fraction.

In Problems 3–20, all improper fractions should be expressed as whole or mixed numbers in your answers.

3. Find the sum of $2\dfrac{3}{4}$ and $3\dfrac{5}{6}$.

4. Find the difference between $7\dfrac{1}{10}$ and $4\dfrac{4}{15}$.

5. Find the product of $3\dfrac{3}{5}$ and $6\dfrac{2}{3}$.

6. Find the quotient of $2\dfrac{1}{3}$ and 14.

1. _____

2. _____

3. _____

4. _____

5. _____

6. _____

Answers

In Problems 7–18, perform the indicated operations.

7. _____

8. _____

9. _____

7. $9\dfrac{3}{8}$

$+\ \dfrac{5}{6}$

8. 7

$-5\dfrac{3}{5}$

9. $4\dfrac{3}{10} + 2\dfrac{3}{4}$

10. _____

11. _____

10. $6\dfrac{1}{6} - 4\dfrac{2}{7}$

11. $\dfrac{3}{5} + 1\dfrac{7}{10} + 2\dfrac{1}{8}$

12. $5\dfrac{5}{9} - 4\dfrac{2}{3}$

12. _____

13. _____

13. $2\dfrac{2}{3} \cdot \dfrac{5}{8}$

14. $\dfrac{5}{6} \div 2\dfrac{1}{2}$

15. $6\dfrac{2}{5} \cdot 3\dfrac{1}{8}$

14. _____

15. _____

16. _____

16. $4\dfrac{7}{8} \div 3\dfrac{1}{4}$

17. $\left(3\dfrac{1}{4}\right)\left(1\dfrac{2}{13}\right)\left(3\dfrac{1}{9}\right)$

18. $\dfrac{4}{5} \div 1\dfrac{3}{5} + \dfrac{2}{15} \cdot \dfrac{10}{7}$

17. _____

18. _____

19. _____

19. Find the difference between $3\dfrac{1}{2}$ and $2\dfrac{5}{8}$; then add $3\dfrac{1}{4}$ to the difference.

20. _____

20. How many shelves $3\dfrac{1}{3}$ feet long can be cut from a board 25 feet long?

Decimal Numbers

5

5.1
READING AND WRITING DECIMAL NUMBERS

As we discussed in Section 1.1, whole numbers can be related to a **place value system.** To be able to read and write whole numbers, we need to know the value of each place. The **decimal numbers** are an extension of the whole numbers. A **decimal point** is placed to the right of the ones place, and places are considered to the right of the decimal point. Each place to the right represents a fraction with a denominator that is a power of 10. Just as with whole numbers, to be able to read and write decimal numbers, we need to know the value of each place.

> **Powers of 10:** The numbers 1, 10, 100, 1000, 10,000, 100,000, 1,000,000, and so on

> **Decimal Number:** Mixed number or fraction in which the denominator of the fraction is a power of 10

Example 1 $\frac{3}{10}, \frac{41}{100}, \frac{53}{10,000}, \frac{89}{10}, \frac{7}{1}, 11\frac{4}{10}$, and $5\frac{36}{100}$ are all decimal numbers written in fraction form or mixed number form.

Decimal Notation: An extension of the place value system using a decimal point, with whole numbers written to the left of the decimal point and fractions to the right of the decimal point

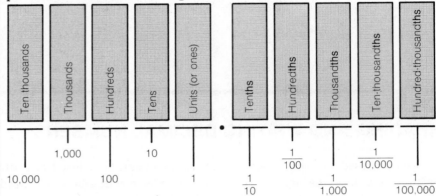

Note that **th** is used to indicate the fraction parts. You should memorize the value of each position.

To read or write a decimal number,

1. read (or write) the whole number as before;
2. read (or write) **and** in place of the decimal point;
3. read (or write) the fraction part as a whole number with the name of the place of the last digit.

Example 2

$$4\ 8 \ . \ 6 \qquad \text{the same as } 48\frac{6}{10}$$

forty-eight **and** six tenths

And indicates the decimal point and the digit 6 is in the tenths position.

Example 3

$$5 \ . \ 3\ 9\ 8 \qquad \text{the same as } 5\frac{398}{1000}$$

five **and** three hundred ninety-eight thousandths

And indicates the decimal point and the digit 8 is in the thousandths position.

Example 4

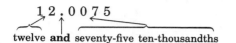

$$12.0075$$

twelve **and** seventy-five ten-thousandths

the same as $12\frac{75}{10,000}$

And indicates the decimal point and the digit 5 is in the ten-thousandths position.

> **SPECIAL NOTES**
>
> 1. The **th** at the end of a word indicates a fraction part (a part to the right of the decimal point).
>
>   eight hundred = 800
>   eight hundred**ths** = 0.08
>
> 2. The hyphen (-) indicates one word.
>
>   eight hundred thousand = 800,000
>   eight hundred-thousand**ths** = 0.00008

☐ **NOW WORK EXERCISES 1–4 IN THE MARGIN.**

Example 5 Write $13\frac{506}{1000}$ in decimal notation.

13.506

Example 6 Write fifteen thousandths in decimal notation.

0.015

Note that the digit 5 is in the thousandths position.

Example 7 Write sixty-two and forty-three hundredths in decimal notation.

62.43

Example 8 Write six hundred and five thousandths in decimal notation.

600.005

Example 9 Write six hundred five thousandths in decimal notation.

0.605

Write each decimal number in words.

1. 20.7

2. 9.04

3. 18.651

4. 2.0008

Write each number in decimal notation.

5. ten and eleven hundredths

6. four and five thousandths

7. eight hundred and three tenths

8. one thousand six hundred and two hundred sixty-four thousandths

9. four hundred and seven thousandths

10. four hundred seven thousandths

Note carefully how the use of **and** in Example 8 gives a completely different result from the result in Example 9.

☐ **NOW WORK EXERCISES 5–10 IN THE MARGIN.**

> As a safety measure when writing a check, the amount is written in number form and in word form to avoid problems of spelling and poor penmanship. If there is a discrepancy between the number and the written form, the bank will honor the written form.

Example 10

HOWARD TOWNSEND
JULIA TOWNSEND
123 ELM STREET 123-4569
SMALLTOWN, ILLINOIS 60001

No. **125**

70-1876
———
711

January 1 19 *90*

Pay to the
order of *Acme Management Corporation* $ *350⁰⁰*

Three hundred fifty and ⁿᵒ/100 ———— Dollars

SUBURBAN BANK
SMALLTOWN, ILLINOIS
60001

Julia Townsend

Answers to Marginal Exercises

1. twenty and seven tenths   **2.** nine and four hundredths
3. eighteen and six hundred fifty-one thousandths   **4.** two and eight ten-thousandths
5. 10.11   **6.** 4.005   **7.** 800.3   **8.** 1600.264   **9.** 400.007   **10.** 0.407

NAME _____ SECTION _____ DATE _____

EXERCISES **5.1** _____

Write the following mixed numbers in decimal notation.

1. $37\dfrac{498}{1000}$ **2.** $18\dfrac{76}{100}$ **3.** $4\dfrac{11}{100}$ **4.** $56\dfrac{3}{100}$

5. $87\dfrac{3}{1000}$ **6.** $95\dfrac{2}{10}$ **7.** $62\dfrac{7}{10}$ **8.** $100\dfrac{25}{100}$

9. $100\dfrac{38}{100}$ **10.** $250\dfrac{623}{1000}$

Write the following decimal numbers in mixed number form.

11. 82.56 **12.** 93.07 **13.** 10.576 **14.** 100.6

15. 65.003

Answers

1. _____

2. _____

3. _____

4. _____

5. _____

6. _____

7. _____

8. _____

9. _____

10. _____

11. _____

12. _____

13. _____

14. _____

15. _____

Answers

16. _____

17. _____

18. _____

19. _____

20. _____

21. _____

22. _____

23. _____

24. _____

25. _____

Write the following numbers in decimal notation.

16. three tenths

17. fourteen thousandths

18. seventeen hundredths

19. six and twenty-eight hundredths

20. sixty and twenty-eight thousandths

21. seventy-two and three hundred ninety-two thousandths

22. eight hundred fifty and thirty-six ten-thousandths

23. seven hundred and seventy-seven hundredths

24. eight thousand four hundred ninety-two and two hundred sixty-three thousandths

25. six hundred thousand, five hundred and four hundred two thousandths

Write the following decimal numbers in words.

| | | |

Answers

26. 0.5   **27.** 0.93

26. _____

27. _____

28. 5.06   **29.** 32.58

28. _____

29. _____

30. 71.06   **31.** 35.078

30. _____

31. _____

32. 7.003   **33.** 18.102

32. _____

33. _____

34. 50.008   **35.** 607.607

34. _____

35. _____

36. 800.006   **37.** 0.806

36. _____

37. _____

38. 4700.617   **39.** 5000.005

38. _____

39. _____

40. 603.0065   **41.** 900.4638

40. _____

41. _____

Answers

Write sample checks for the amounts indicated.

42. $356.45

42. _____

HOWARD TOWNSEND
JULIA TOWNSEND
123 ELM STREET 123-4569
SMALLTOWN, ILLINOIS 60001

No. 125
70-1876
711

January 1 19 *90*

Pay to the
order of *Acme Management Corporation* $ _____

_____ Dollars

SUBURBAN BANK
SMALLTOWN, ILLINOIS
60001

Julia Townsend

43. _____

43. $651.50

HOWARD TOWNSEND
JULIA TOWNSEND
123 ELM STREET 123-4569
SMALLTOWN, ILLINOIS 60001

No. 125
70-1876
711

January 1 19 *90*

Pay to the
order of *Acme Management Corporation* $ _____

_____ Dollars

SUBURBAN BANK
SMALLTOWN, ILLINOIS
60001

Julia Townsend

44. _____

44. $2506.64

HOWARD TOWNSEND
JULIA TOWNSEND
123 ELM STREET 123-4569
SMALLTOWN, ILLINOIS 60001

No. 125
70-1876
711

January 1 19 *90*

Pay to the
order of *Acme Management Corporation* $ _____

_____ Dollars

SUBURBAN BANK
SMALLTOWN, ILLINOIS
60001

Julia Townsend

5.2
ROUNDING OFF _____

As we discussed in Section 1.4 when rounding off whole numbers, there are several methods used for rounding off decimal numbers, particularly when different operations (adding, subtracting, multiplying, or dividing) are involved. For example, in one method, if the last digit is 5, then the number is rounded off to the nearest even digit. The method we will use for rounding off decimal numbers is the same as the one we used in Section 1.4 with whole numbers.

> **Rounding Off:** Rounding off a given number means to find another number close to the given number. The desired place of accuracy must be stated.

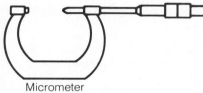

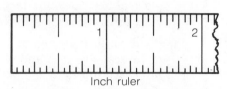

(a) The micrometer is marked to give approximate measures of small objects.

(b) The ruler is marked to give approximate measures of lengths in fourths, eighths, and sixteenths of an inch.

Example 1   Round off 5.76 to the nearest tenth.

Solution

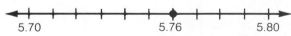

We can see from the number line that 5.76 is closer to 5.80. So, 5.76 rounds off to 5.80 or 5.8 (to the nearest tenth).

Example 2   Round off 0.153 to the nearest hundredth.

Solution

0.153 is between 0.150 and 0.160.

Looking at the numbers on the number line, we can see that 0.153 is closer to 0.150. So, 0.153 rounds off to 0.150 or 0.15 (to the nearest hundredth).

Round off each number to the place indicated. Draw a number line as an aid.

1. 8.63 (nearest tenth)

2. 5.042 (nearest hundredth)

3. 0.0179 (nearest thousandth)

Example 3   Round off 0.0637 to the nearest thousandth.

Solution

0.0637 is between 0.0630 and 0.0640.

0.0637 is closer to 0.0640. So, 0.0637 rounds off to 0.0640 or 0.064 (to the nearest thousandth).

☐ **NOW WORK EXERCISES 1–3 IN THE MARGIN.**

RULES FOR ROUNDING OFF DECIMAL NUMBERS

1. Look at the single digit just to the right of the place of desired accuracy.
2. If this digit is 5 or greater, make the digit in the desired place of accuracy one larger and replace all digits to the right with zeros. All digits to the left remain the same.
3. If this digit is less than 5, leave the digit in the desired place of accuracy as it is and replace all digits to the right with zeros. All digits to the left remain the same.

Note: Trailing 0's to the right of the decimal point may be dropped.

Example 4   Round off 6.749 to the nearest tenth.

Solution

(a)   6 . 7 4 9

        7 is in the tenths position. The next digit is 4.

(b) Since 4 is less than 5, leave the 7 and replace 4 and 9 with 0's.

(c) 6.749 rounds off to 6.700 or 6.7 to the nearest tenth.

Note: 6.700 and 6.7 are both correct. We have a choice of dropping trailing 0's to the right of the decimal point.

Example 5   Round off 3.45196 to the nearest thousandth.

Solution

(a)   3 . 4 5 1 9 6

       1 is in the thousandths position. The next digit is 9.

(b) Since 9 is greater than 5, change 1 to 2 and replace 9 and 6 with 0's.

(c) 3.45196 rounds off to 3.45200 or 3.452 to the nearest thousandth.

Example 6 Round off 4.9838 to the nearest tenth.

Solution

(a)   4 . 9 8 3 8

     9 is in the tenths position. 8 is the next digit.

(b) Since 8 is greater than 5, make 9 one larger and replace 8 and 3 and 8 with 0's. In this case, the digit 4 is also affected.

(c) 4.9838 rounds off to 5.0000 or 5.0 to the nearest tenth.

Note: The 0 in the tenths position is written to indicate the position of rounding off.

Example 7 Round off 17.835 to the nearest hundredth.

Solution

(a)   1 7 . 8 3 5

     hundredths position   next digit

(b) Since the next digit is 5, make 3 one larger, change 3 to 4, and replace 5 with 0.

(c) 17.835 rounds off to 17.840 or 17.84 to the nearest hundredth.

Completion Example 8 Round off 8472 to the nearest hundred.

Solution

(a) The decimal point is understood to be to the right of _____ .

(b) The digit in the hundreds position is _____ .

(c) The next digit is _____ .

(d) Since _____ is greater than 5, change the _____ to _____ and replace _____ and _____ with 0's.

(e) So, 8472 rounds off to _____ (to the nearest hundred).

Special Note: The 0's must **not** be dropped in a whole number. They can only be dropped if they are trailing 0's to the right of the decimal point.

Completion Example Answer

8. (a) 2

    (b) 4

    (c) 7

    (d) 7, 4 to 5, 7 and 2

    (e) 8500

Round off each number as indicated.

4. 3.349 (nearest tenth)

5. 0.07921 (nearest thousandth)

6. 239.53 (nearest unit)

7. 7558 (nearest thousand)

8. 0.0006873 (nearest hundred-thousandth)

Completion Example 9   Round off 1.00643 to the nearest ten-thousandth.

Solution

(a) The digit in the ten-thousandths position is _____ .

(b) The next digit is _____ .

(c) Since _____ is less than 5, leave _____ as it is and replace _____ with a 0.

(d) 1.00643 rounds off to _____ to the nearest _____ .

☐ **NOW WORK EXERCISES 4–8 IN THE MARGIN.**

Completion Example Answer

9. (a) 4

  (b) 3

  (c) 3, 4, 3

  (d) 1.00640 or 1.0064 to the nearest ten-thousandth

Answers to Marginal Exercises

1.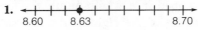
8.60   8.63   8.70

Answer: 8.60 or 8.6

2.
5.040 5.042   5.050

Answer: 5.040 or 5.04

3.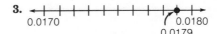
0.0170   0.0180
                     0.0179

Answer: 0.0180 or 0.018

4. 3.300 or 3.3

5. 0.07900 or 0.079

6. 240.00 or 240

7. 8000

8. 0.0006900 or 0.00069

NAME _____ SECTION _____ DATE _____

EXERCISES **5.2** _____

Round off each of the following decimal numbers as indicated.

To the nearest tenth:

1. 4.763   **2.** 5.031   **3.** 76.349   **4.** 76.352

5. 89.015   **6.** 7.555   **7.** 18.009   **8.** 37.666

9. 14.33382   **10.** 0.0364

To the nearest hundredth:

11. 0.385   **12.** 0.296   **13.** 5.7226   **14.** 8.9874

15. 6.99613   **16.** 13.13465   **17.** 0.0782   **18.** 6.0035

19. 5.7092   **20.** 2.8347

Answers

1. _____
2. _____
3. _____
4. _____
5. _____
6. _____
7. _____
8. _____
9. _____
10. _____
11. _____
12. _____
13. _____
14. _____
15. _____
16. _____
17. _____
18. _____
19. _____
20. _____

Answers

21. _____

22. _____

23. _____

24. _____

25. _____

26. _____

27. _____

28. _____

29. _____

30. _____

31. _____

32. _____

33. _____

34. _____

35. _____

36. _____

37. _____

38. _____

39. _____

40. _____

41. _____

42. _____

43. _____

44. _____

To the nearest thousandth:

21. 0.0672   **22.** 0.05550   **23.** 0.6338   **24.** 7.6666

25. 32.4785   **26.** 9.4302   **27.** 17.36371   **28.** 4.44449

29. 0.00191   **30.** 20.76962

To the nearest whole number (or nearest unit):

31. 479.23   **32.** 6.8   **33.** 17.5   **34.** 19.999

35. 382.48   **36.** 649.66   **37.** 439.78   **38.** 701.413

39. 6333.11   **40.** 8122.825

To the nearest ten:

41. 5163   **42.** 6475   **43.** 495   **44.** 572.5

NAME _____ SECTION _____ DATE _____

45. 998.5   **46.** 378.92   **47.** 5476.2   **48.** 76,523.1

49. 92,540.9   **50.** 7007.7

To the nearest thousand:

51. 7398   **52.** 62,275   **53.** 47,823.4   **54.** 103,499

55. 217,480.2   **56.** 9872.5   **57.** 379,500   **58.** 4,500,762

59. 7,305,438   **60.** 573,333.3

61. 0.0005783 (nearest hundred-thousandth)

Answers

45. _____

46. _____

47. _____

48. _____

49. _____

50. _____

51. _____

52. _____

53. _____

54. _____

55. _____

56. _____

57. _____

58. _____

59. _____

60. _____

61. _____

Answers

62. _____

63. _____

64. _____

65. _____

66. _____

67. _____

68. _____

69. _____

70. _____

62. 0.5449 (nearest hundredth)

63. 473.8 (nearest ten)

64. 5.00632 (nearest thousandth)

65. 473.8 (nearest hundred)

66. 5750 (nearest thousand)

67. 3.2296 (nearest thousandth)

68. 15.548 (nearest tenth)

69. 78,419 (nearest ten thousand)

70. 78,419 (nearest ten)

5.3

ADDING AND SUBTRACTING DECIMAL NUMBERS; ESTIMATING

> To add decimal numbers,
>
> 1. write the numbers one under the other;
> 2. keep the decimal points in line;
> 3. keep digits with the same place value in line (zeros may be filled in as aids);
> 4. add the numbers, placing the decimal point in the answer in line with the other decimal points.

Example 1 Find the sum $6.3 + 5.42 + 14.07$.

Solution

$$
\begin{array}{r}
6.30 \\
5.42 \\
+14.07 \\
\hline
25.79
\end{array}
$$

← 0 may be filled in to help keep the digits in line.

Example 2 Find the sum $9 + 4.86 + 37.479 + 0.6$.

Solution

$$
\begin{array}{r}
9.000 \\
4.860 \\
37.479 \\
+ 0.600 \\
\hline
51.939
\end{array}
$$

The decimal point is understood to be to the right of 9, as 9.0.

0's are filled in to help keep the digits in line.

Example 3 Add.

$$
\begin{array}{r}
56.2 \\
85.75 \\
+29.001
\end{array}
$$

You can write

$$
\begin{array}{r}
56.200 \\
85.750 \\
+ 29.001 \\
\hline
170.951
\end{array}
$$

☐ **NOW WORK EXERCISES 1–3 IN THE MARGIN.**

Find each sum.

1. $45.2 + 2.08 + 3.5 =$

2. $4 + 5.7 + 0.63 =$

3.
$$
\begin{array}{r}
17 \\
8.61 \\
5.004 \\
+29.19
\end{array}
$$

Find each difference.

4. $50.036 - 47.58 =$

5. $14 - 4.176 =$

> To subtract decimal numbers,
>
> 1. write the numbers one under the other.
> 2. line up the decimal points.
> 3. keep digits with the same place value in line (zeros may be filled in).
> 4. subtract, placing the decimal point in the answer in line with the other decimal points.

Example 4　Find the difference $16.715 - 4.823$.

Solution

$$
\begin{array}{r}
16.715 \\
-\ 4.823 \\
\hline
11.892
\end{array}
$$

Example 5　Find the difference $21.2 - 13.716$.

Solution

$$
\begin{array}{r}
21.200 \quad \leftarrow \text{fill in 0's.} \\
-13.716 \\
\hline
7.484
\end{array}
$$

☐ **NOW WORK EXERCISES 4 AND 5 IN THE MARGIN.**

Estimating

By rounding off each number to the place of the last nonzero digit on the left and adding (or subtracting) these rounded-off numbers, we can estimate (or approximate) the answer before the actual calculations are done. This technique of estimating answers is especially helpful when working with decimal numbers where the placement of the decimal point is so important.

Example 6　First estimate the sum $74 + 3.529 + 52.61$; then find the sum.

Solution

(a) Estimate by adding rounded-off numbers.

$$
\begin{array}{r}
70 \\
4 \\
+\ 50 \\
\hline
124 \leftarrow \text{estimate}
\end{array}
$$

(b) Find the actual sum.

$$
\begin{array}{r}
74.000 \\
3.529 \\
+\ 52.610 \\
\hline
130.139 \leftarrow \text{actual sum}
\end{array}
$$

The actual sum and the estimated sum are close (difference of about 6). Had a decimal number been misplaced in one of the original numbers, the difference could be as much as 50 to 100. For example, a common error is to write 74 as 0.74 instead of 74.0. This error would lead to a sum as follows:

$$
\begin{array}{r}
\text{wrong} \\
\text{place}
\end{array}
\quad
\begin{array}{r}
0.74 \\
3.529 \\
+52.610 \\
\hline
56.879 \leftarrow \text{wrong} \\
\text{sum}
\end{array}
$$

The difference $124 - 56 = 68$ is large in relation to the given numbers and indicates some kind of error.

Example 7 Estimate the difference $22.418 - 17.526$; then find the difference.

Solution

(a) Estimate the difference. In this case, both numbers round to 20.

$$
\begin{array}{r}
20 \\
-20 \\
\hline
0 \leftarrow \text{estimate}
\end{array}
$$

(b) Find the difference.

$$
\begin{array}{r}
{}^{1}\,{}^{1}\,{}^{1}{}^{3}\,1 \\
2\!\!\!/\,2\!\!\!/.\,4\!\!\!/\,18 \\
-1\,7.5\,26 \\
\hline
4.8\,92 \leftarrow \text{actual difference}
\end{array}
$$

We find that the actual difference is less than 5 and that is the reason the estimated difference was 0.

Example 8 Mrs. Finn went to the local store and bought a pair of shoes for $42.50, a blouse for $25.60, and a shirt for $37.55. How much did she spend? (Tax was included in the prices.) Estimate her expenses mentally before calculating her actual expenses.

Solution

$$
\begin{array}{r}
\$42.50 \\
25.60 \\
+\quad 37.55 \\
\hline
\$105.65
\end{array}
$$

She spent $105.65.

6. Brian got a haircut for $10.00 (including a shampoo) and a shave for $3.50. If he tipped the barber $2.00, how much change did he receive from a $20 bill?

Completion Example 9 Joe decided he needed some new fishing equipment. He bought a new rod for $55, a rod (on sale) for $22.50, and some fishing line for $2.70. If tax totaled $4.82, how much change did he receive from a $100 bill? Try to do a mental estimate first before calculating the actual amount of change he will get.

Solution

(a) Find the total of his expenses including tax.

$$\begin{array}{r} \$55.50 \\ 22.50 \\ 2.70 \\ +\quad 4.82 \\ \hline \$ \end{array}$$

(b) Subtract the answer in part (a) from $100.

$$\begin{array}{r} \$100.00 \\ -\qquad \\ \hline \$ \end{array}$$

☐ **NOW WORK EXERCISE 6 IN THE MARGIN.**

Completion Example Answer
9. (a) $85.02
   (b) $14.98

Answers to Marginal Exercises
1. 50.78   2. 10.33   3. 59.804   4. 2.456   5. 9.824   6. $4.50 change

NAME _____ SECTION _____ DATE _____

EXERCISES **5.3** _____

Find each of the indicated sums. Estimate your answers, either mentally or on paper, before doing the actual calculations. Check to see that your sums are close to the estimated values.

1. 0.6 + 0.4 + 1.3

2. 5 + 6.1 + 0.4

3. 0.59 + 6.91 + 0.05

4. 3.488 + 16.593 + 25.002

5. 37.02 + 25 + 6.4 + 3.89

6. 4.0086 + 0.034 + 0.6 + 0.05

7. 43.766 + 9.33 + 17 + 206

8. 52.3 + 6 + 21.01 + 4.005

9.
```
   75.2
   3.682
 +14.995
```

10.
```
   107.39
     5.061
    23.54
 + 64.9801
```

11.
```
   34.967
   50.6
   8.562
 + 9.3
```

12.
```
   4.156
   3.7
  25.682
 +13.405
```

1. _____

2. _____

3. _____

4. _____

5. _____

6. _____

7. _____

8. _____

9. _____

10. _____

11. _____

12. _____

Answers

13. _____

14. _____

15. _____

16. _____

17. _____

18. _____

19. _____

20. _____

21. _____

22. _____

23. _____

Find each of the indicated differences. First estimate the differences mentally.

13. 29.5 − 13.61

14. 1.0057 − 0.03

15. 78.015 − 13.068

16.   4.8
    −0.0026

17.   31.009
     − 0.534

18.   4
    −1.0566

19.   40.718
     − 6.532

20. Theresa got a haircut for $30.00 and a manicure for $10.50. If she tipped the stylist $5, how much change did she receive from a $50 bill?

21. The inside radius of a pipe is 2.38 inches, and the outside radius is 2.63 inches as shown in the figure. What is the thickness of the pipe. (Note: The radius of a circle is the distance from the center of the circle to a point on the circle.)

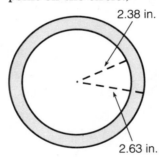
2.38 in.
2.63 in.

22. Mr. Johnson bought the following items at a department store: slacks, $32.50; shoes, $43.75; shirt, $18.60. How much did he spend? What was his change if he gave the clerk a $100 bill? (Tax was included in the prices.)

23. An architect's scale drawing shows a rectangular lot 2.38 inches on one side and 3.76 inches on the other side. What was the perimeter (distance around) of the rectangle on the drawing?

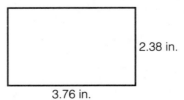
2.38 in.
3.76 in.

5.4
MULTIPLYING DECIMAL NUMBERS; ESTIMATING

Multiplying decimal numbers is almost the same as multiplying whole numbers. The major concern is correct placement of the decimal point in the product. If we remember that decimals are fractions, the following rule for multiplying seems reasonable.

> To multiply decimal numbers,
>
> 1. multiply the two numbers as if they were whole numbers;
> 2. count the total number of places to the right of the decimal points in both numbers being multiplied;
> 3. this sum is the number of places to the right of the decimal point in the product.

To illustrate how to place the decimal point in the product and the close relationship with fractions, several products are shown in both fraction and decimal notation.

FRACTIONS

DECIMALS

$$\frac{1}{10} \cdot \frac{1}{100} = \frac{1}{1000}$$

$$\begin{array}{r} .1 \\ \times\ .01 \\ \hline .001 \end{array}$$ 3 places (or thousandths)

$$\frac{3}{10} \cdot \frac{5}{100} = \frac{15}{1000}$$

$$\begin{array}{r} .3 \\ \times\ .05 \\ \hline .015 \end{array}$$ 3 places (or thousandths)
3 places (or thousandths)

$$\frac{6}{100} \cdot \frac{4}{1000} = \frac{24}{100,000}$$

$$\begin{array}{r} .004 \\ \times\ \ .06 \\ \hline .00024 \end{array}$$ 5 places (or hundred-thousandths)
5 places (or hundred-thousandths)

As these illustrations indicate, **there is no need to keep the decimal points lined up for multiplication.**

Find each product.

1. $(0.8)(0.2) =$

2. $(5.6)(0.04) =$

3. $\begin{array}{r} 3.781 \\ \times\ 3.01 \\ \hline \end{array}$

Example 1

$$\begin{array}{r} 2.432 \quad \leftarrow 3\ \text{places} \\ \times \quad 5.1 \quad \leftarrow 1\ \text{place} \\ \hline 2432 \\ 12\ 160 \\ \hline 12.4032 \quad \leftarrow 4\ \text{places in the product} \end{array}$$

total of 4 places

Example 2   Multiply 4.35×12.6.

$$\begin{array}{r} 4.35 \quad \leftarrow 2\ \text{places} \\ \times\ 12.6 \quad \leftarrow 1\ \text{place} \\ \hline 2\ 610 \\ 8\ 70 \\ 43\ 5 \\ \hline 54.810 \quad \leftarrow 3\ \text{places in the product} \end{array}$$

total of 3 places

Example 3   Multiply $(0.046)(0.007)$.

$$\begin{array}{r} 0.046 \quad \leftarrow 3\ \text{places} \\ \times\ 0.007 \quad \leftarrow 3\ \text{places} \\ \hline 0.000322 \quad \leftarrow 6\ \text{places in the product} \end{array}$$

total of 6 places

This means that three 0's had to be inserted between the 3 and the decimal point.

Completion Example 4

$$\begin{array}{r} 3.4 \quad \leftarrow \underline{\quad}\ \text{place} \\ \times \quad 5.8 \quad \leftarrow \underline{\quad}\ \text{place} \\ \hline 2\ 72 \\ 17\ 0 \\ \hline \end{array}$$

total of ____ places

_____ ____ places in the product

Completion Example 5   Multiply 0.003×0.03.

$$\begin{array}{r} 0.003 \quad \leftarrow \underline{\quad}\ \text{places} \\ \times\ 0.03 \quad \leftarrow \underline{\quad}\ \text{places} \\ \hline \end{array}$$

total of ____ places

_____ ____ places in the product

____ 0's had to be inserted to the decimal point.

☐ **NOW WORK EXERCISES 1–3 IN THE MARGIN.**

Completion Example Answers

4. $\begin{array}{r} 3.4 \\ \times\ 5.8 \\ \hline 2\ 72 \\ 17\ 0 \\ \hline 19.72 \end{array}$ 1 place   total of
           1 place   2 places

      2 places in
      the product

5. $\begin{array}{r} 0.003 \\ \times\ 0.03 \\ \hline 0.00009 \end{array}$ 3 places   total of
           2 places   5 places
      5 places in the product

4 0's had to be inserted
to the decimal point.

Estimating

Estimating products can be done by rounding off each number to the place of the last nonzero digit on the left and multiplying these rounded-off numbers. This technique is particularly helpful in correctly placing the decimal point in the actual product.

Example 6 First estimate the product (0.356)(6.1); then find the product and use the estimation to help place the decimal point.

Solution

(a) Estimate by multiplying rounded-off numbers.

$$
\begin{array}{r}
0.4 \quad \text{(0.356 rounded off)} \\
\times \ \ 6 \quad \text{(6.1 rounded off)} \\
\hline
2.4 \quad \leftarrow \text{estimate}
\end{array}
$$

(b) Find the actual product.

$$
\begin{array}{r}
0.356 \\
\times \ \ \ 6.1 \\
\hline
356 \\
2\,136 \ \ \ \\
\hline
2.1716 \quad \leftarrow \text{actual product}
\end{array}
$$

The estimated product 2.4 helps place the decimal point correctly in the product 2.1716.

Example 7 First find the product (19.9)(2.3); and then estimate the product and use this estimation as a check.

Solution

(a) Find the actual product.

$$
\begin{array}{r}
19.9 \\
\times \ \ 2.3 \\
\hline
5\,97 \\
39\,8 \ \ \\
\hline
45.77 \quad \leftarrow \text{actual product}
\end{array}
$$

(b) Estimate the product and use this estimation to check that the actual product is reasonable.

$$
\begin{array}{r}
20.0 \quad \text{(19.9 rounded off)} \\
\times \ \ \ 2 \quad \text{(2.3 rounded off)} \\
\hline
40.0 \quad \leftarrow \text{estimated product}
\end{array}
$$

The estimated product of 40.0 indicates that the actual product is reasonable and the placement of the decimal point is correct.

Note that, even though estimated products (or estimated sums or estimated differences) are close to the actual value, they do not guarantee the absolute accuracy of this actual value. The estimations help only in placing decimal points and checking the "reasonableness" of the answer.

Multiply by moving the decimal point the appropriate number of places.

4. $8.75 \times 10 =$

5. $100(6.3) =$

6. $1000 \times 0.1894 =$

> To multiply a decimal number by a power of 10,
>
> move the decimal point **to the right** the same number of places as the number of 0's in the power of 10.
>
> Multiplication by 10 moves the decimal point one place **to the right.**
>
> Multiplication by 100 moves the decimal point two places **to the right.**
>
> Multiplication by 1000 moves the decimal point three places **to the right,**
>
> and so on.

See Section 1.5 for multiplication of whole numbers by powers of 10.

Example 8 $10(2.68) = 26.8$ Move decimal point 1 place to the right.

Example 9 $100(2.68) = 268.$ Move decimal point 2 places to the right.

Example 10 $1000(0.9531) = 953.1$ Move decimal point 3 places to the right.

Example 11 $1000(7.24) = 7240.$ Move decimal point 3 places to the right. Note that a 0 had to be inserted.

☐ **NOW WORK EXERCISES 4–6 IN THE MARGIN.**

Word problems can involve several operations. The words do not usually say directly to add, subtract, multiply, or divide. We must reason from experience what to do with the numbers.

The following example illustrates a situation involving a choice between paying cash and making a down payment plus several monthly payments. In real life, such factors as the rate of interest, the cash available, and what other investments might be made with the cash, help in making a decision on how to proceed. These factors are not considered in Example 12.

Completion Example 12 You can buy a car for $7500 cash or you can make a down payment of $1500 and then pay $566.67 each month for twelve months. How much can you save by paying cash?

Solution

(a) Find the amount paid in monthly payments.

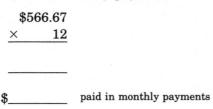

$ \underline{\hspace{2cm}} $ paid in monthly payments

(b) Find the total amount paid by adding the down payment to the answer in part (a)

$ \underline{\hspace{2cm}} $ total paid

(c) Find the savings by subtracting $7500 from the answer in part (b).

$$
\begin{array}{r}
\$ \quad\quad\quad \\
- \ 7500.00 \\
\hline
\end{array}
$$

$ \underline{\hspace{2cm}} $ savings by paying cash

Completion Example Answer

12. (a) $6800.04 paid in monthly payments

(b) $6800.04; $8300.04 total paid

(c) $8300.04; $800.04 savings by paying cash

Answers to Marginal Exercises

1. 0.16   **2.** 0.224   **3.** 11.38081   **4.** 87.5   **5.** 630.   **6.** 189.4

NAME _____ SECTION _____ DATE _____

EXERCISES **5.4** _____

Find each of the indicated products.

1. (0.6)(0.7)   **2.** (0.3)(0.8)   **3.** (0.2)(0.2)

4. (0.3)(0.3)   **5.** 8(2.7)   **6.** 4(9.6)

7. 1.4(0.3)   **8.** 1.5(0.6)   **9.** (0.2)(0.02)

10. (0.3)(0.03)   **11.** 5.4(0.02)   **12.** 7.3(0.01)

13. 0.23×0.12 **14.** 0.15×0.15 **15.** 8.1×0.006

Answers

1. _____

2. _____

3. _____

4. _____

5. _____

6. _____

7. _____

8. _____

9. _____

10. _____

11. _____

12. _____

13. _____

14. _____

15. _____

Answers

16. 7.1 × 0.008   **17.** 0.06 × 0.01   **18.** 0.25 × 0.01

16. _____

17. _____

18. _____

19. 3(0.125)   **20.** 4(0.375)   **21.** 1.6(0.875)

19. _____

20. _____

21. _____

22. 5.3(0.75)   **23.** 6.9(0.25)   **24.** 4.8(0.25)

22. _____

23. _____

24. _____

25. _____

25. 0.83(6.1)   **26.** 0.27(0.24)   **27.** 0.16(0.5)

26. _____

27. _____

28. _____

29. _____

28. 0.28(0.5)   **29.** 3.29(0.01)   **30.** 5.78(0.02)

30. _____

NAME _____ SECTION _____ DATE _____

31.   0.005
        ×0.009

32.   0.006
        ×0.004

33.   0.137
        × 0.06

31. _____

32. _____

33. _____

34.   0.106
        × 0.09

35.   1.07
        × 0.5

36.   5.08
        × 0.4

34. _____

35. _____

36. _____

In Exercises 37–45, first estimate the product; then find the actual product. Use the estimation to help in correctly placing the decimal point in the product.

37. _____

37.   0.0106
        × 0.087

38.   0.0213
        × 0.065

39.   83.105
        × 0.111

38. _____

39. _____

40. _____

41. _____

40.   17.002
        × 0.101

41.   86.1
        ×0.057

42.   7.83
        ×0.18

42. _____

Answers

43. _____

44. _____

45. _____

46. _____

47. _____

48. _____

49. _____

50. _____

51. _____

52. _____

53. _____

54. _____

55. _____

56. _____

57. _____

58. _____

59. _____

60. _____

61. _____

43. 95.62
    × 0.57

44. 6.002
    × 0.57

45. 8.034
    × 0.29

46. 100(3.46)

47. 100(20.57)

48. 100(7.82)

49. 100(6.93)

50. 100(16.1)

51. 100(38.2)

52. 10(0.435)

53. 10(0.719)

54. 10(1.86)

55. 1000(4.1782)

56. 1000(0.38)

57. 1000(0.47)

58. 10,000 × 0.005

59. 10,000 × 0.00615

60. 10,000 × 7.4

61. If an architect makes a drawing to scale so that 1 inch represents 6.75 feet, what distance is represented by 5.5 inches?

NAME _____ SECTION _____ DATE _____

62. To buy a car, you can pay $2036.50 cash, or put $400 down and make 18 monthly payments of $104.30. How much is saved by paying cash?

63. An automobile dealer makes $150.70 on each used car he sells and $425.30 on each new car he sells. (a) Estimate how much he made the month he sold 11 used and 6 new cars. (b) Find the amount he made. (c) Find the difference between the estimation and the actual income.

64. Suppose a tax assessor figures the tax at 0.07 of the assessed value of a home. (a) If the assessed value is determined at a rate of 0.32 of the market value, what taxes are paid on a home with a market value of $136,500? (b) Estimate these taxes and check to see that the actual taxes are close to the estimate.

65. (a) If the sale price of a new refrigerator is $583 and sales tax is figured at 0.06 times the price, approximately what amount is paid for the refrigerator? (b) What is the exact amount paid for the refrigerator?

Answers

62. _____

63. (a) _____

(b) _____

(c) _____

64. (a) _____

(b) _____

65. (a) _____

(b) _____

66. (a) _____

    (b) _____

67. _____

68. (a) _____

    (b) _____

69. _____

70. _____

66. You drive south at 57.6 miles per hour for 3 hours, then west at 52.4 miles per hour for 4 hours. (a) About how far have you driven in the 7 hours? (b) How far did you actually drive?

67. Multiply the numbers 2.456 and 3.16, then round off the product to the nearest tenth. Next, round off each of the factors to the nearest tenth and then multiply and round off this product to the nearest tenth. Did you get the same answer?

68. You paid $2500 as a down payment on a new car and made thirty-six monthly payments of $275.50. (a) About how much did you pay for the car? (b) Exactly each how much did you pay for the car?

69. If you were paid a salary of $350 per week and $13.75 for each hour you worked over 40 hours in a week, how much would you make if you worked 45 hours in one week?

70. If you bought a magazine for $2.50, a candy bar for $.65, a milk shake for $1.75, french fries for $.85, and you had to pay a tax of 0.06 times the total, how much change would you get from a $10 bill?

5.5
DIVIDING DECIMAL NUMBERS; ESTIMATING

Division with whole numbers gives a quotient and possibly a remainder. For example,

$$
\begin{array}{r}
27 \quad \text{quotient} \\
35\overline{)959} \\
\underline{70} \\
259 \\
\underline{245} \\
14 \quad \text{remainder}
\end{array}
$$

Now that we have decimal numbers, we can continue to divide and get a decimal quotient. Zeros are added on to the dividend if they are needed.

$$
\begin{array}{r}
27.4 \quad \text{quotient} \\
35\overline{)959.0} \quad \leftarrow \text{0 added on} \\
\underline{70} \\
259 \\
\underline{245} \\
14\,0 \\
\underline{14\,0} \\
0
\end{array}
$$

If the divisor has a decimal point, multiply both the divisor and dividend by a power of 10 to make the divisor a whole number. For example, we can write

$$
4.9\overline{)51.45} \quad \text{as} \quad \frac{51.45}{4.9} \times \frac{10}{10} = \frac{514.5}{49}
$$

This means that

$$
4.9\overline{)51.45} \quad \text{is the same as} \quad 49\overline{)514.5}
$$

Also, we can write

$$
1.36\overline{)5.1} \quad \text{as} \quad \frac{5.1}{1.36} \times \frac{100}{100} = \frac{510}{136} \quad \text{or} \quad 136\overline{)510}
$$

This discussion leads to the following procedure for dividing decimal numbers.

> To divide decimal numbers,
>
> 1. move the decimal point in the divisor so it is a whole number;
> 2. move the decimal point in the dividend the same number of places to the right;
> 3. place the decimal point in the quotient directly above the new decimal point in the dividend;
> 4. divide just as with whole numbers.

Note: In Step 3, to ensure that the decimal point is placed correctly, we place the decimal point in the quotient before dividing in Step 4.

Example 1   Divide $51.45 \div 4.9$.

Solution

(a) Write the problem as follows:

$$4.9\overline{)51.45}$$

(b) To move the decimal point so that the divisor is a whole number, move each decimal point 1 place. This makes the whole number 49 the divisor. Place the decimal point in the quotient before dividing.

$$\overset{\text{.}\curvearrowleft\text{decimal point in quotient}}{4.\underset{\curvearrowright}{9,}\overline{)51.\underset{\curvearrowright}{4,}5}}$$

(c) Divide as with whole numbers.

$$
\begin{array}{r}
10.5 \\
49.\overline{)514.5} \\
\underline{49} \\
24 \\
\underline{0} \\
24\,5 \\
\underline{24\,5} \\
0
\end{array}
$$

Example 2   Divide $5.1 \div 1.36$

Solution

(a) Write the problem as follows:

$$1.36\overline{)5.1}$$

(b) Move the decimal points so the divisor is a whole number. Add 0's in the dividend if needed. Place the decimal point in the quotient before dividing.

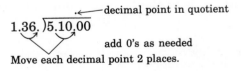

$$\overset{\text{.}\leftarrow\text{decimal point in quotient}}{1.\underset{\curvearrowright}{36,}\overline{)5.\underset{\curvearrowright}{10,}00}}$$

add 0's as needed

Move each decimal point 2 places.

(c) Divide.

```
              3.75
        136.)510.00
              408
              102 0
               95 2
                6 80
                6 80 ·
                    0
```

Example 3   Divide $6.3252 \div 6.3$.

Solution

```
              1.004
        6.3.)6.3.252
              6 3
               0 2
                 0
                 25
                  0
                 252
                 252
                   0
```

Note: There **must** be a digit to the right of the decimal point in the quotient above every digit to the right of the decimal point in the dividend.

Completion Example 4   Divide $24.225 \div 4.25$.

```
                 5.
      4.25.)24.22.5

         ____

         ____
```

☐ **NOW WORK EXERCISES 1–3 IN THE MARGIN.**

> When the remainder is not zero,
>
> 1. decide first how many decimal places are to be in the quotient;
> 2. divide until the quotient is **one digit past the place of desired accuracy;**
> 3. using this last digit, round off the quotient to the desired place of accuracy.

Completion Example Answer

```
              5.7
   4.  4.25.)24.22.5
              21 25
               2 97 5
               2 97 5
                   0
```

Find each quotient.

1. $9.87 \div 2.1$

2. $9.393 \div 9.3$

3. $10.01 \div 2.86$

Example 5 Find $8.24 \div 2.9$ to the nearest tenth.

Solution

Divide until the quotient is in hundredths (one more place than tenths), then round off to tenths.

$$
\begin{array}{r}
\overset{\text{hundredths}}{}\overset{\text{read \textbf{approximately}}}{} \\
2.84 \approx 2.8 \quad \text{rounded off to tenths} \\
2.9\,\overline{)8.2\,40} \\
\underline{58} \\
2\,44 \\
\underline{2\,32} \\
1\,20 \\
\underline{1\,16} \\
4
\end{array}
$$

$8.24 \div 2.9 \approx 2.8$ accurate to the nearest tenth

Example 6 Find $1.83 \div 4.1$ to the nearest hundredth.

Solution

Divide until the quotient is in thousandths (one more place than hundredths), then round off to hundredths.

$$
\begin{array}{r}
\overset{\text{read \textbf{approximately}}}{} \\
0.446 \approx 0.45 \\
4.1\,\overline{)1.8\,300} \quad \text{Add as many 0's as needed.} \\
\underline{1\,64} \\
1\,90 \\
\underline{1\,64} \\
260 \\
\underline{246} \\
14
\end{array}
$$

$1.83 \div 4.1 \approx 0.45$ accurate to the nearest hundredth

Example 7 Find $17 \div 3.3$ to the nearest thousandth.

Solution

Divide until the quotient is in ten-thousandths, then round off to thousandths.

$$\begin{array}{r} 5.1515 \approx 5.152 \\ 3.3\overline{)17.0\,0000} \\ \underline{16\,5} \\ 50 \\ \underline{3\,3} \\ 1\,70 \\ \underline{1\,65} \\ 50 \\ \underline{3\,3} \\ 170 \\ \underline{165} \\ 5 \end{array}$$

Add as many 0's as needed.

$17 \div 3.3 \approx 5.152$ accurate to thousandths

☐ **NOW WORK EXERCISES 4 AND 5 IN THE MARGIN.**

> To divide a decimal number by a power of 10,
>
> move the decimal point to the left the same number of places as the number of 0's in the power of 10.
>
> Division by 10 moves the decimal point **one** place **to the left.**
>
> Division by 100 moves the decimal point **two** places **to the left.**
>
> Division by 1000 moves the decimal point **three** places **to the left,**
>
> and so on

Example 8 $4.16 \div 100 = \dfrac{4.16}{100} = 0.0416$

Example 9 $782 \div 10 = \dfrac{782}{10} = 78.2$

Example 10 $\dfrac{593.3}{1000} = 0.5933$

Example 11 $\dfrac{186.4}{100} = 1.864$

☐ **NOW WORK EXERCISES 6–8 IN THE MARGIN.**

The term **average** will be discussed in detail in Chapter 9. In this section, we will use the word average in the sense of "an average speed of 55 miles per hour" and "the average number of miles per gallon of gas."

---

4. Find $1.83 \div 7$ to the nearest hundredth.

5. Find $43.721 \div 0.06$ to the nearest thousandth.

Find each quotient.

6. $42.31 \div 100 =$

7. $328 \div 10 =$

8. $\dfrac{57.6}{10,000} =$

If you ride your bicycle at an average speed of 15 miles per hour, you may sometimes be riding as slowly as 10 mph and at other times as fast as 22 mph. Whatever your actual speed at any time, the distance you travel can be found by multiplying your average speed by the time you spend riding.

Example 12   If you ride your bicycle at an average speed of 15.2 miles per hour, how far will you ride in 3.5 hours?

Solution

Multiply the average speed by the number of hours.

$$
\begin{array}{r}
15.2 \text{ mph} \\
\underline{3.5 \text{ hr}} \\
7\ 60 \\
\underline{45\ 6\ \ } \\
53.20 \text{ miles}
\end{array}
$$

You will ride 53.2 miles in 3.5 hours.

If we know the total number of units of a quantity (distance, dollars, gallons of gas, etc.), we can find the average amount per unit by dividing the amount by the number of units.

Example 13   Bud bought five shirts for a total price of $62.50. What was the average price per shirt?

Solution

Find the average price per shirt by dividing the total price by 5.

$$
\begin{array}{r}
12.50 \quad \text{average price per shirt} \\
5\overline{)62.50} \\
\underline{5} \\
12 \\
\underline{10} \\
2\ 5 \\
\underline{2\ 5} \\
00 \\
\underline{00} \\
0
\end{array}
$$

He paid an average price of $12.50 per shirt.

Estimating

As with addition, subtraction, and multiplication, we can use estimating with division to help in placing the decimal point in the quotient and to verify the "reasonableness" of the quotient. The technique is to round off both the divisor and dividend to the place of the last nonzero digit on the left and then divide with these rounded-off values.

Example 14 First estimate the quotient 6.2 ÷ 0.302, and then find the quotient to the nearest tenth.

Solution

(a) Estimate the quotient using 6.2 ≈ 6 and 0.302 ≈ 0.3.

$$
\begin{array}{r}
20. \\
0.3)\overline{6.0} \\
\underline{6} \\
0\ 0 \\
\underline{0} \\
0
\end{array}
$$

(b) Find the quotient to the nearest tenth.

$$
\begin{array}{r}
20.52 \approx 20.5 \\
0.302)\overline{6.200,00} \\
\underline{6\ 04} \\
160 \\
\underline{0} \\
160\ 0 \\
\underline{151\ 0} \\
9\ 00 \\
\underline{6\ 04}
\end{array}
$$

The estimated value 20 is very close to the rounded-off quotient 20.5.

Example 15 The gas tank of a car holds 17 gallons of gasoline. Approximately how many miles per gallon does the car average if it will go 450 miles on one tank of gas?

Solution

The question calls only for an approximate answer. We can use rounded-off values:

$$17 \approx 20 \text{ gal and } 450 \approx 500 \text{ miles}$$

$$
\begin{array}{r}
25 \text{ miles per gallon} \\
20)\overline{500} \\
\underline{40} \\
100 \\
\underline{100} \\
0
\end{array}
$$

The car averages about 25 miles per gallon.

NAME _____   SECTION _____   DATE _____

EXERCISES **5.5** _____

Divide.

1. $4.68 \div 2$

2. $1.71 \div 3$

3. $4.95 \div 5$

4. $1.62 \div 9$

5. $0.064 \div 0.8$

6. $0.63 \div 0.7$

7. $82.24 \div 0.04$

8. $16.02 \div 0.03$

9. $48 \div 2.4$

10. $28 \div 5.6$

Answers

1. _____

2. _____

3. _____

4. _____

5. _____

6. _____

7. _____

8. _____

9. _____

10. _____

Answers

11. _____

12. _____

13. _____

14. _____

15. _____

16. _____

17. _____

18. _____

19. _____

20. _____

21. _____

22. _____

23. _____

24. _____

25. _____

26. _____

27. _____

28. _____

Find each quotient to the nearest tenth.

11. $8\overline{)455}$ **12.** $4\overline{)263}$ **13.** $9.4\overline{)6.538}$

14. $4.6\overline{)5}$ **15.** $7.05\overline{)0.4977}$ **16.** $0.37\overline{)4.683}$

17. $0.23\overline{)65.226}$ **18.** $1.62\overline{)34}$ **19.** $1.33\overline{)75}$

Find each quotient to the nearest hundredth.

20. $24\overline{)0.1463}$ **21.** $1.23\overline{)14.91129}$ **22.** $0.075\overline{)0.42753}$

23. $2.7\overline{)2.583}$ **24.** $23\overline{)62.949}$ **25.** $9\overline{)2}$

26. $26\overline{)5.729}$ **27.** $13\overline{)65.476}$ **28.** $3.181\overline{)6}$

NAME _____   SECTION _____   DATE _____

In Exercises 29–37, first estimate each quotient; then find the quotient to the nearest thousandth.

29. $23\overline{)71}$

30. $69\overline{)293}$

31. $85.3\overline{)24.31}$

32. $2.57\overline{)0.4961}$

33. $13\overline{)1.029}$

34. $14\overline{)4.073}$

35. $16.2\overline{)0.11623}$

36. $25.7\overline{)6.27}$

37. $0.23\overline{)45.221}$

Answers

29. _____

30. _____

31. _____

32. _____

33. _____

34. _____

35. _____

36. _____

37. _____

Answers

38. _____

39. _____

40. _____

41. _____

42. _____

43. _____

44. _____

45. _____

46. _____

47. _____

48. _____

49. _____

50. _____

51. (a) _____

   (b) _____

Divide as indicated.

38. $78.4 \div 100$

39. $16.4963 \div 100$

40. $50.36 \div 100$

41. $45.621 \div 1000$

42. $73.85 \div 1000$

43. $18.6 \div 1000$

44. $\dfrac{167}{10}$

45. $\dfrac{138.1}{10}$

46. $\dfrac{7.85}{10}$

47. $\dfrac{1.54}{10,000}$

48. $\dfrac{169.9}{10,000}$

49. $\dfrac{10.413}{10,000}$

50. Sam bought three pictures for his home at an average price of $125.50 per picture. How much did he spend on the pictures altogether?

51. (a) If a car averages 24.6 miles per gallon, about how far will it go on 18 gallons of gas? (b) Exactly how many miles will it go on 18 gallons of gas?

52. If a motorcycle averages 32.4 miles per gallon, how many miles will it go on 7 gallons of gas?

Answers

52. _____

53. (a) _____

53. (a) If a car travels 300 miles on 16 gallons of gas, approximately how many miles does it travel per gallon? (b) Exactly how many miles does it travel per gallon?

(b) _____

54. (a) _____

(b) _____

54. (a) If a bicyclist rode 250.6 miles in 13.2 hours, about how fast did she ride per hour? (b) What was her average speed in miles per hour (to the nearest tenth)?

55. (a) _____

(b) _____

56. (a) _____

55. A quarter section of beef can be bought cheaper than the same amount of meat purchased a few pounds at a time. (a) Estimate the cost per pound if 150 pounds cost $187.50. (b) What is the cost per pound?

(b) _____

56. (a) If you drive 9.5 hours at an average speed of 52.2 miles per hour, about how far will you drive? (b) Exactly how far will you drive?

57. _____

58. _____

59. _____

60. _____

57. If new tires cost $56.50 per tire and tax is figured at 0.06 times the cost of each tire, what will you pay for 4 new tires?

58. If you bought 10 books for a total price of $225 plus tax, and tax is figured at 0.06 times the price, what average amount did you pay per book, including tax?

59. If the total price of a stereo was $312.70 including tax at 0.06 times the list price, you can find the list price by dividing the total price by 1.06. What was the list price? (**Note:** 1.06 represents the list price plus 0.06 times the list price.)

60. Suppose that the total interest paid on a 30-year mortgage for a home loan of $60,000 is going to be $189,570. What will be the payment each month if the payments are to pay off both the loan and the interest?

5.6
DECIMALS AND FRACTIONS

From Section 5.1 we know that decimal numbers are fractions with denominators that are powers of 10. From Section 5.5 we know how to divide with whole numbers and decimal numbers to get decimal numbers. In this section, we will discuss these two important relationships in more detail.

Changing from Decimals to Fractions

> A decimal number can be written in fraction form by writing a fraction with
>
> 1. Numerator: a whole number with all the digits of the decimal number; and
>
> 2. Denominator: the power of ten that names the rightmost digit.

For example,

$$0.25 = \frac{25}{100} \quad \text{and} \quad 0.025 = \frac{25}{1000}$$

The rightmost digit, 5, is in the hundredths position.   The rightmost digit, 5, is in the thousandths position.

We may or may not want to reduce some fractions. If we choose to reduce, factoring works just as it did in Chapter 3.

In the following examples, each decimal number is changed to fraction form, and then reduced.

Example 1 $0.25 = \frac{25}{100} = \frac{\cancel{5} \cdot \cancel{5} \cdot 1}{2 \cdot \cancel{5} \cdot 2 \cdot \cancel{5}} = \frac{1}{4}$

↑
hundredths

Example 2 $0.32 = \frac{32}{100} = \frac{\cancel{4} \cdot 8}{\cancel{4} \cdot 25} = \frac{8}{25}$

↑
hundredths

Example 3 $0.131 = \frac{131}{1000}$

↑
thousandths

Example 4 $0.075 = \frac{75}{1000} = \frac{\cancel{25} \cdot 3}{\cancel{25} \cdot 40} = \frac{3}{40}$

↑
thousandths

Example 5 $2.6 = \dfrac{26}{10} = \dfrac{\cancel{2} \cdot 13}{\cancel{2} \cdot 5} = \dfrac{13}{5}$

       ↑
   tenths

or, as a mixed number,

$$2.6 = 2\frac{6}{10} = 2\frac{3}{5}$$

Example 6 $1.42 = \dfrac{142}{100} = \dfrac{\cancel{2} \cdot 71}{\cancel{2} \cdot 50} = \dfrac{71}{50}$

    ↑
 hundredths

or, as a mixed number,

$$1.42 = 1\frac{42}{100} = 1\frac{21}{50}$$

Changing from Fractions to Decimals

> To change a fraction into a decimal number, divide the numerator by the denominator.
>
> 1. If the remainder is 0, the decimal is said to be **terminating.**
> 2. If the remainder is not 0, the decimal is said to be **nonterminating.**

The following examples illustrate fractions that convert to terminating decimals.

Example 7 $\dfrac{3}{8}$

$$
\begin{array}{r}
.375 \\
8\overline{)3.000} \\
\underline{2\,4} \\
60 \\
\underline{56} \\
40 \\
\underline{40} \\
0
\end{array}
$$

$\dfrac{3}{8} = 0.375$

Example 8 $\dfrac{3}{4}$

$$
\begin{array}{r}
.75 \\
4\overline{)3.00} \\
\underline{2\,8} \\
20 \\
\underline{20} \\
0
\end{array}
$$

$\dfrac{3}{4} = 0.75$

Example 9 $\dfrac{4}{5}$

$$
\begin{array}{r}
.8 \\
5\overline{)4.0} \\
\underline{4\,0} \\
0
\end{array}
$$

$\dfrac{4}{5} = 0.8$

Nonterminating decimals can be **repeating** or **nonrepeating.** A repeating nonterminating decimal has a repeating pattern to its digits. Every fraction that has a whole number numerator and nonzero denominator will be either a terminating decimal or a repeating decimal. Nonrepeating decimals will be discussed in detail in later courses.

The following examples illustrate nonterminating repeating decimals.

Example 10 $\dfrac{1}{3}$

$$
\begin{array}{r}
.333 \longleftarrow \text{The 3 will repeat without end.} \\
3\overline{)1.000} \\
\underline{9} \\
10 \\
\underline{9} \\
10 \\
\underline{9} \\
1 \longleftarrow \text{Continuing to divide will give a} \\
\text{remainder of 1 each time.}
\end{array}
$$

We write $\dfrac{1}{3} = 0.333\ldots$ The three dots mean "and so on" or to continue without stopping.

Example 11 $\dfrac{7}{12}$

$$
\begin{array}{r}
.5833 \longleftarrow \text{The 3 will repeat without end.} \\
12\overline{)7.0000} \\
\underline{6\,0} \\
1\,00 \\
\underline{96} \\
40 \\
\underline{36} \\
40 \\
\underline{36} \\
4 \quad \text{Continuing to divide will give a} \\
\text{remainder of 4 each time.}
\end{array}
$$

We write $\dfrac{7}{12} = 0.58333\ldots$

Example 12 $\dfrac{1}{7}$

$$
\begin{array}{r}
.142857 \longleftarrow \text{The six digits will repeat in the} \\
7\overline{)1.000000} \quad \text{the same pattern without end.} \\
\underline{7} \\
30 \\
\underline{28} \\
20 \\
\underline{14} \\
60 \\
\underline{56} \\
40 \\
\underline{35} \\
50 \\
\underline{49} \\
1
\end{array}
$$

The remainder will repeat in sequence 1, 3, 2, 6, 4, 5, 1, and so on. Therefore, the digits in the quotient will also repeat.

We write $\dfrac{1}{7} = 0.142857142857142857\ldots$

Another way to write repeating decimals is to write a **bar** over the repeating digits. Thus, in Examples 10, 11, and 12, we can write

$$\frac{1}{3} = 0.\overline{3} \quad \text{and} \quad \frac{7}{12} = 0.58\overline{3} \quad \text{and} \quad \frac{1}{7} = 0.\overline{142857}$$

We may choose to round off the quotient to some decimal place just as was done with division in Section 5.5. Perform the division one place past the desired round-off position.

Example 13 $\frac{5}{11}$

$$
\begin{array}{r}
.454 \approx 0.45 \quad \text{(nearest hundredth)} \\
11\overline{)5.000} \\
\underline{4\,4} \\
60 \\
\underline{55} \\
50 \\
\underline{44}
\end{array}
$$

Example 14 $\frac{5}{6}$

$$
\begin{array}{r}
.833 \approx 0.83 \quad \text{(nearest hundredth)} \\
6\overline{)5.000} \\
\underline{4\,8} \\
20 \\
\underline{18} \\
20 \\
\underline{18}
\end{array}
$$

As the following examples illustrate, we can perform operations with both fractions and decimal numbers by changing the fractions to decimal form. (**Note:** We could also change the decimal numbers to fraction form.)

Example 15　Find the sum $10\frac{1}{2} + 7.32 + 5\frac{3}{5}$ in decimal form.

Solution

$$
\begin{array}{rll}
10\frac{1}{2} & = 10.50 & \left(\frac{1}{2} = 0.50\right) \\
7.32 & = 7.32 & \\
+\ 5\frac{3}{5} & = \underline{5.60} & \left(\frac{3}{5} = 0.60\right) \\
& \ 23.42 &
\end{array}
$$

Example 16　Determine whether $\frac{3}{16}$ is larger than 0.18 by changing $\frac{3}{16}$ to decimal form and then comparing the two numbers. Find the difference.

Solution

$$
\begin{array}{r}
.1875 \\
16\overline{)3.0000} \\
\underline{1\,6} \\
1\,40 \\
\underline{1\,28} \\
120 \\
\underline{112} \\
80 \\
\underline{80} \\
0
\end{array}
$$

So, $\dfrac{3}{16} = 0.1875$.

$$
\begin{array}{r}
0.1875 \\
-0.1800 \\
\hline
0.0075 \quad \text{difference}
\end{array}
$$

Thus, $\dfrac{3}{16}$ is larger than 0.18 and their difference is 0.0075.

NAME _____ SECTION _____ DATE _____

EXERCISES **5.6** _____

Change each decimal to fraction form. Do not reduce.

1. 0.9   2. 0.3   3. 0.5   4. 0.8

5. 0.62   6. 0.38   7. 0.57   8. 0.41

9. 0.526   10. 0.625   11. 0.016   12. 0.012

13. 5.1   14. 7.2   15. 8.15   16. 6.35

Change each decimal to fraction form (or mixed number form) and reduce if possible.

17. 0.125   18. 0.36   19. 0.18   20. 0.375

21. 0.225   22. 0.455   23. 0.17   24. 0.029

25. 3.2   26. 1.25   27. 6.25   28. 2.75

1. _____   2. _____

3. _____   4. _____

5. _____   6. _____

7. _____   8. _____

9. _____   10. _____

11. _____   12. _____

13. _____   14. _____

15. _____   16. _____

17. _____   18. _____

19. _____   20. _____

21. _____   22. _____

23. _____   24. _____

25. _____   26. _____

27. _____   28. _____

Answers

29. _____

30. _____

31. _____

32. _____

33. _____

34. _____

35. _____

36. _____

37. _____

38. _____

39. _____

40. _____

41. _____

42. _____

43. _____

44. _____

45. _____

46. _____

47. _____

48. _____

Change each fraction to decimal form. If the decimal is nonterminating, write it with the bar notation over the repeating pattern of digits.

29. $\dfrac{2}{3}$ **30.** $\dfrac{5}{16}$ **31.** $\dfrac{5}{11}$ **32.** $\dfrac{3}{11}$

33. $\dfrac{11}{16}$ **34.** $\dfrac{9}{16}$ **35.** $\dfrac{3}{7}$ **36.** $\dfrac{5}{7}$

37. $\dfrac{1}{6}$ **38.** $\dfrac{5}{18}$ **39.** $\dfrac{5}{9}$ **40.** $\dfrac{2}{9}$

Change each fraction to decimal form rounded off to the nearest thousandth.

41. $\dfrac{7}{24}$ **42.** $\dfrac{16}{33}$ **43.** $\dfrac{5}{12}$ **44.** $\dfrac{13}{16}$

45. $\dfrac{1}{32}$ **46.** $\dfrac{1}{14}$ **47.** $\dfrac{16}{13}$ **48.** $\dfrac{20}{9}$

NAME _____ SECTION _____ DATE _____

49. $\dfrac{30}{32}$ **50.** $\dfrac{40}{3}$

49. _____

Perform the indicated operations by writing all the numbers in decimal form. Round off to the nearest thousandth if necessary.

50. _____

51. $\dfrac{1}{4} + 0.25 + \dfrac{1}{5}$ **52.** $\dfrac{3}{4} + \dfrac{1}{10} + 3.55$

51. _____

52. _____

53. _____

53. $\dfrac{5}{8} + \dfrac{3}{5} + 0.41$ **54.** $6 + 2\dfrac{37}{100} + 3\dfrac{11}{50}$

54. _____

55. _____

55. $2\dfrac{53}{100} + 5\dfrac{1}{10} + 7.35$ **56.** $37.02 + 25 + 6\dfrac{2}{5} + 3\dfrac{89}{100}$

56. _____

57. _____

58. _____

57. $1\dfrac{1}{4} - 0.125$ **58.** $2\dfrac{1}{2} - 1.75$

59. _____

59. $36.71 - 23\dfrac{1}{5}$ **60.** $3.1 - 2\dfrac{1}{100}$

60. _____

Answers

61. _____

62. _____

63. _____

64. _____

65. _____

66. _____

67. _____

68. _____

69. _____

70. _____

71. _____

72. _____

61. $\left(\dfrac{35}{100}\right)(0.73)$

62. $\left(5\dfrac{1}{10}\right)(2.25)$

63. $\left(1\dfrac{3}{8}\right)(3.1)(2.6)$

64. $\left(1\dfrac{3}{4}\right)\left(2\dfrac{1}{2}\right)(5.35)$

65. $5\dfrac{54}{100} \div 2.1$

66. $72.16 \div \dfrac{2}{5}$

67. $13.65 \div \dfrac{1}{2}$

68. $91.7 \div \dfrac{1}{4}$

In each of the following exercises, change any fraction to decimal form; then determine which number is larger. Find the difference.

69. $2\dfrac{1}{4},\ 2.3$

70. $\dfrac{7}{8},\ 0.878$

71. $0.28,\ \dfrac{3}{11}$

72. $\dfrac{1}{3},\ 0.3$

NAME _____ SECTION _____ DATE _____

CHAPTER **5** TEST _____

1. Write 0.056 as a fraction and reduce to lowest terms.

1. _____

2. _____

2. Write 21.021 in words.

3. _____

3. Write the following in decimal notation:

three thousand five and two hundred three ten-thousandths

4. _____

5. _____

In Problems 4–6, round off as indicated.

6. _____

4. 203.016 to the nearest hundredth

7. _____

8. _____

5. 1961.7035 to the nearest thousand

6. 0.0782 to the nearest tenth

In Problems 7–15, perform the indicated operations.

7. 75.817 + 16.943   **8.** 856 − 93.41

Answers

9. 82 + 14.91 + 25.2   **10.** 100.6 − 82.493

9. _____

10. _____
                                11. $13\frac{2}{5}$ + 6 + 17.913   **12.** $1\frac{1}{1000}$ − 0.09705

11. _____

12. _____
                                13. (0.35)(0.84)   **14.**   16.31   **15.** (1.92)(1000)
                                                                  ×0.785
13. _____

14. _____

15. _____   Find each quotient to the nearest hundredth.

16. _____   **16.** 82 ÷ 4.6   **17.** $0.13\overline{)8.617}$ **18.** 3.614 ÷ 100

17. _____

18. _____   **19.** Change 0.375 to fraction   **20.** Change $\frac{5}{16}$ to decimal form
                                   form and reduce.   rounded off to the nearest
19. _____   thousandth.

20. _____

                                21. (a) If an automobile gets 18.3 miles per gallon and its tank holds 21.4
21. (a) _____   gallons, approximately how far can the car travel on a full tank of gas?
                                        (b) How many miles can the car actually travel on a full tank of gas (to
                                        the nearest mile)?
   (b) _____

22. _____   **22.** If you make a down payment of $250.00 on a refrigerator and make
                                   twelve equal payments of $24.50, how much did you pay for the refrig-
                                   erator?

Ratio and Proportion

6

6.1

RATIO AND PROPORTION

We know that a fraction can be used to:

1. Indicate a part of a whole

$$\frac{3}{8} \quad \text{means} \quad \frac{3 \text{ pieces of cherry pie}}{8 \text{ pieces in the whole pie}}$$

2. Indicate division

$$\frac{3}{8} \quad \text{means} \quad 3 \div 8 \quad \text{or} \quad 8\overline{)3.000}$$

$$
\begin{array}{r}
.375 \\
8\overline{)3.000} \\
\underline{2\,4} \\
60 \\
\underline{56} \\
40 \\
\underline{40} \\
0
\end{array}
$$

A third use of fractions is to compare two numbers. For example,

$$\frac{6}{2} \quad \text{might mean} \quad \frac{6 \text{ minutes}}{2 \text{ hours}} \quad \text{or} \quad \frac{6 \text{ apples}}{2 \text{ oranges}}$$

In these comparisons, we must know the units. Such a comparison is called **a ratio.**

> **Ratio:** A comparison of two quantities by division
>
> **Notation:** The ratio of a to b can be written $\frac{a}{b}$ or $a : b$ or a to b

To avoid confusion, the units in a ratio should be written down or otherwise explained in the problem. If possible, the units should be the same in both the numerator and denominator. Generally, the ratio should also be reduced.

Example 1   Compare the quantities 30 students and 40 chairs as a ratio.

Solution

Since the units (students and chairs) are not the same, the units should be written in the ratio. We can write

(a) $\dfrac{30 \text{ students}}{40 \text{ chairs}}$

(b) 30 students : 40 chairs

(c) 30 students to 40 chairs

> **TWO CHARACTERISTICS OF RATIOS**
>
> 1. Ratios can be reduced just as fractions can be reduced.
> 2. Common units should be used in the numerator and denominator of a ratio whenever possible.

Example 2   In Example 1, we could write the ratio as

$$\dfrac{3 \text{ students}}{4 \text{ chairs}} \quad \text{since} \quad \dfrac{30}{40} = \dfrac{3}{4}$$

Example 3   Write the comparison of 2 feet to 3 yards as a ratio.

Solution

(a) We can write $\dfrac{2 \text{ feet}}{3 \text{ yards}}$.

(b) But a better procedure is to change to common units. Since there are 3 feet in one yard, 3 yards = 9 feet.

     The ratio $\dfrac{2 \text{ feet}}{3 \text{ yards}}$ can be written $\dfrac{2 \text{ feet}}{9 \text{ feet}}$ or 2 : 9.

Example 4   A baseball player has a hitting average of .250. Write his hitting average as a ratio of hits to times at bat.

Solution

$$.250 = \dfrac{250 \text{ hits}}{1000 \text{ times at bat}} = \dfrac{1 \text{ hit}}{4 \text{ times at bat}}$$

Example 5 What is the reduced ratio of 300 centimeters (cm) to 2 meters (m)? Centimeters and meters are units of length used in the metric system and will be discussed in detail in Chapter 10. (There are 100 centimeters in 1 meter.)

Solution

Since 1 m = 100 cm, we have 2 m = 200 cm.

The ratio is $$\frac{300 \text{ cm}}{2 \text{ m}} = \frac{300 \text{ cm}}{200 \text{ cm}} = \frac{3}{2}$$

The reduced ratio can also be written 3 : 2 or 3 to 2.

☐ **NOW WORK EXERCISES 1–3 IN THE MARGIN.**

> **Proportion:** A statement that two ratios are equal
>
> **Symbols:** $\frac{a}{b} = \frac{c}{d}$ is a proportion
>
> A proportion has four terms:
>
> first term — third term
> $$\frac{a}{b} = \frac{c}{d}$$
> second term — fourth term
>
> **Extremes:** Terms 1 and 4 (a and d)
> **Means:** Terms 2 and 3 (b and c)

Example 6 $\frac{3}{6} = \frac{4}{8}$ is a proportion.

3 and 8 are the extremes.

6 and 4 are the means.

Example 7 $\frac{8.4}{4.2} = \frac{10.2}{5.1}$ is a proportion.

8.4 and 5.1 are the extremes.

4.2 and 10.2 are the means.

Write each comparison as a ratio reduced to lowest terms.

1. 3 quarters to 1 dollar

2. 36 inches to 5 feet

3. Inventory shows 5000 washers and 4000 bolts. What is the ratio of washers to bolts?

Completion Example 8 $\dfrac{2\frac{1}{2}}{10} = \dfrac{3\frac{1}{4}}{13}$ is a _____.

_____ and _____ are the extremes.

_____ and _____ are the means.

> **A proportion is true if the product of the extremes equals the product of the means.**
>
> $$\frac{a}{b} = \frac{c}{d} \quad \text{if and only if} \quad a \cdot d = b \cdot c \quad \text{where } b \neq 0 \text{ and } d \neq 0.$$
>
> **Note:** The terms can be decimal numbers, fractions, mixed numbers, or whole numbers. The rules are the same.

Example 9   Is the proportion $\dfrac{9}{13} = \dfrac{4.5}{6.5}$ true or false?

Solution

$$
\begin{array}{cc}
6.5 & 4.5 \\
\times \ \ 9 & \times \ \ 13 \\
\hline
58.5 & 13\,5 \\
& 45 \\
\hline
& 58.5
\end{array}
$$

Since $9(6.5) = 13(4.5)$, the proportion is true.

Example 10   Is the proportion $\dfrac{5}{8} = \dfrac{7}{10}$ true or false?

Solution $5 \cdot 10 = 50$ and $8 \cdot 7 = 56$

Since $50 \neq 56$, the proportion is false.

Completion Example Answer

8. $\dfrac{2\frac{1}{2}}{10} = \dfrac{3\frac{1}{4}}{13}$ is a proportion.

$2\frac{1}{2}$ and 13 are the extremes.

10 and $3\frac{1}{4}$ are the means.

Completion Example 11 Is the proportion $\frac{1}{4} : \frac{2}{3} = 9 : 24$ true or false?

Solution

(a) The extremes are _____ and _____ .

The means are _____ and _____ .

(b) The product of the extremes is _____ .

The product of the means is _____ .

(c) The proportion is _____ , because

the products are _____ .

☐ **NOW WORK EXERCISES 4–6 IN THE MARGIN.**

Determine whether each proportion is true or false.

4. $\dfrac{15}{20} = \dfrac{21}{28}$

5. $\dfrac{5\frac{1}{2}}{2} = \dfrac{6\frac{1}{2}}{3}$

6. $\dfrac{1.3}{1.5} = \dfrac{1.82}{2.1}$

Completion Example Answer

11. (a) Extremes are $\frac{1}{4}$ and 24.

Means are $\frac{2}{3}$ and 9.

(b) Product of the extremes is $\frac{1}{4} \cdot 24 = 6$.

Product of the means is $\frac{2}{3} \cdot 9 = 6$.

(c) Proportion is true because the products are equal.

Answers to Marginal Exercises

1. $\frac{3}{4}$ or $3:4$ **2.** $\frac{3}{5}$ or $3:5$ **3.** $\dfrac{5 \text{ washers}}{4 \text{ bolts}}$ **4.** True, $15 \cdot 28 = 20 \cdot 21$

5. False, $5\frac{1}{2} \cdot 3 \neq 2 \cdot 6\frac{1}{2}$ **6.** True, $(1.3)(2.1) = (1.5)(1.82)$

NAME _____ SECTION _____ DATE _____

EXERCISES **6.1** _____

Write the following comparisons as ratios reduced to lowest terms. Use common units in the numerator and denominator whenever possible.

1. 1 dime to 3 nickels

2. 5 nickels to 3 quarters

3. 50 miles to 2 gallons of gas

4. 60 miles to 5 gallons of gas

5. 250 miles to 5 hours

6. 10 miles to 4 minutes

7. 30 chairs to 25 people

8. 40 people to 35 chairs

9. 18 inches to 2 feet

10. 10,560 feet to 3 miles
   (**Hint:** 5280 feet = 1 mile)

11. 10 centimeters to 1 meter
   (**Hint:** 100 centimeters = 1 meter)

12. 3 meters to 1000 millimeters
   (**Hint:** 1000 millimeters = 1 meter)

13. 8 days to 1 week

14. 3 weeks to 21 days

15. $200 in profit to $1000 invested

16. $80 in profit to $120 invested

1. _____

2. _____

3. _____

4. _____

5. _____

6. _____

7. _____

8. _____

9. _____

10. _____

11. _____

12. _____

13. _____

14. _____

15. _____

16. _____

Answers

Determine whether each proportion is true or false by comparing the products of the means and extremes.

17. _____

18. _____

19. _____

20. _____

21. _____

22. _____

23. _____

24. _____

25. _____

26. _____

27. _____

28. _____

29. _____

30. _____

31. _____

32. _____

33. _____

34. _____

35. _____

36. _____

37. _____

17. $\dfrac{5}{6} = \dfrac{10}{12}$ **18.** $\dfrac{2}{7} = \dfrac{5}{17}$ **19.** $\dfrac{7}{21} = \dfrac{4}{12}$

20. $\dfrac{3}{5} = \dfrac{60}{100}$ **21.** $\dfrac{125}{1000} = \dfrac{1}{8}$ **22.** $\dfrac{3}{8} = \dfrac{375}{1000}$

23. $\dfrac{1}{4} = \dfrac{25}{100}$ **24.** $\dfrac{7}{8} = \dfrac{875}{1000}$ **25.** $\dfrac{3}{16} = \dfrac{9}{48}$

26. $\dfrac{2}{3} = \dfrac{66}{100}$ **27.** $\dfrac{1}{3} = \dfrac{33}{100}$ **28.** $\dfrac{14}{6} = \dfrac{21}{8}$

29. $\dfrac{12}{18} = \dfrac{14}{21}$ **30.** $\dfrac{5}{6} = \dfrac{7}{8}$ **31.** $\dfrac{7.5}{10} = \dfrac{3}{4}$

32. $\dfrac{6.2}{3.1} = \dfrac{10.2}{5.1}$ **33.** $\dfrac{8\frac{1}{2}}{2\frac{1}{3}} = \dfrac{4\frac{1}{4}}{1\frac{1}{6}}$ **34.** $\dfrac{6\frac{1}{5}}{1\frac{1}{7}} = \dfrac{3\frac{1}{10}}{\frac{8}{14}}$

35. $\dfrac{6}{24} = \dfrac{10}{48}$ **36.** $\dfrac{7}{16} = \dfrac{3\frac{1}{2}}{8}$ **37.** $\dfrac{10}{17} = \dfrac{5}{8\frac{1}{2}}$

6.2
FINDING THE UNKNOWN TERM IN A PROPORTION

> To find the unknown term in a proportion,
>
> 1. represent the unknown term with some letter such as x, y, w, A, B, etc.
> 2. write an equation that sets the product of the extremes equal to the product of the means.
> 3. divide both sides of the equation by the number multiplying the unknown. (This number is called the **coefficient** of the unknown.)
>
> The resulting equation gives the missing value for the unknown.

Important Note: Write one equation under the other, just as in the example.

Example 1 Find x if $\dfrac{3}{6} = \dfrac{5}{x}$.

Solution

(a) Write the proportion:
$$\frac{3}{6} = \frac{5}{x}$$

(b) Write product of extremes equal to product of means:
$$3 \cdot x = 6 \cdot 5$$

(c) Divide both sides by 3, the coefficient of x:
$$\frac{3 \cdot x}{3} = \frac{30}{3}$$

(d) Reduce to find the solution:
$$\frac{\cancel{3} \cdot x}{\cancel{3}} = \frac{\cancel{30}^{10}}{\cancel{3}}$$
$$x = 10$$

Example 2 Find y if $\dfrac{6}{16} = \dfrac{y}{24}$.

Solution

(a) Write the proportion:

$$\frac{6}{16} = \frac{y}{24}$$

(b) Write product of extremes equal to product of means:

$$6 \cdot 24 = 16 \cdot y$$

(c) Divide both sides by 16, the coefficient of y:

$$\frac{6 \cdot 24}{16} = \frac{16 \cdot y}{16}$$

(d) Reduce to find the missing value:

$$\frac{\overset{3}{\cancel{6}} \cdot \overset{3}{\cancel{24}}}{\underset{2}{\cancel{16}}} = \frac{\cancel{16} \cdot y}{\cancel{16}}$$

$$9 = y$$

Second Solution

(a) Write the proportion:

$$\frac{6}{16} = \frac{y}{24}$$

(b) Reduce the fraction $\dfrac{6}{16}$ to $\dfrac{3}{8}$:

$$\frac{3}{8} = \frac{y}{24}$$

(c) Solve as before:

$$3 \cdot 24 = 8 \cdot y$$

$$\frac{72}{8} = \frac{\cancel{8} \cdot y}{\cancel{8}}$$

$$9 = y$$

Example 3 Find w if $\dfrac{w}{7} = \dfrac{20}{\frac{2}{3}}$.

Solution

(a) Write the proportion:

$$\frac{w}{7} = \frac{20}{\frac{2}{3}}$$

(b) Set product of extremes equal to product of means:

$$w \cdot \frac{2}{3} = 7 \cdot 20$$

(c) Divide both sides by $\dfrac{2}{3}$ $\left(\text{or multiply by } \dfrac{3}{2}, \text{ the reciprocal of } \dfrac{2}{3}\right)$:

$$\frac{w \cdot \frac{2}{3}}{\frac{2}{3}} = \frac{7 \cdot 20}{\frac{2}{3}} \quad \text{or} \quad w \cdot \frac{2}{3} \cdot \frac{3}{2} = 7 \cdot 20 \cdot \frac{3}{2}$$

(d) Reduce. Remember that to divide by a fraction, you multiply by its reciprocal:

$$\frac{w \cdot \dfrac{2}{3}}{\dfrac{2}{3}} = \frac{7 \cdot 20}{\dfrac{2}{3}} \quad \text{or} \quad w\frac{\cancel{2}}{\cancel{3}} \cdot \frac{\cancel{3}}{\cancel{2}} = \frac{7}{1} \cdot \frac{\overset{10}{\cancel{20}}}{1} \cdot \frac{3}{\cancel{2}}$$

$$w = 7 \cdot \overset{10}{\cancel{20}} \cdot \frac{3}{\cancel{2}} = 210, \qquad\qquad w = 210$$

Completion Example 4 Find A if $\dfrac{A}{3} = \dfrac{7.5}{6}$.

Solution

(a) $\dfrac{A}{3} = \dfrac{7.5}{6}$

(b) $6 \cdot A = $ _____

(c) $\dfrac{\cancel{6} \cdot A}{\cancel{6}} = $ _____

(d) $A = $ _____

Completion Example 5 Find R if $\dfrac{3}{5} = \dfrac{R}{100}$.

Solution

(a) $\qquad \dfrac{3}{5} = \dfrac{R}{100}$

(b) _____ $= 5 \cdot R$

(c) _____ $= \dfrac{\cancel{5} \cdot R}{\cancel{5}}$

(d) _____ $= R$

Completion Example Answers

4. (b) $6 \cdot A = 3(7.5)$

(c) $\dfrac{\cancel{6} \cdot A}{\cancel{6}} = \dfrac{22.5}{6}$

(d) $A = 3.75$

5. (b) $3 \cdot 100 = 5 \cdot R$

(c) $\dfrac{300}{5} = \dfrac{\cancel{5} \cdot R}{\cancel{5}}$

(d) $\quad 60 = R$

Solve each proportion for the unknown term.

1. $\dfrac{3}{10} = \dfrac{R}{100}$

2. $\dfrac{\frac{1}{2}}{6} = \dfrac{5}{x}$

3. $\dfrac{x}{1.50} = \dfrac{11}{2.75}$

Example 6 Find x if $\dfrac{x}{1\frac{1}{2}} = \dfrac{1\frac{2}{3}}{3\frac{1}{3}}$.

Solution

(a) $\dfrac{x}{1\frac{1}{2}} = \dfrac{1\frac{2}{3}}{3\frac{1}{3}}$

(b) $\dfrac{10}{3} \cdot x = \dfrac{\cancel{3}}{2} \cdot \dfrac{5}{\cancel{3}}$ Write each mixed number as an improper fraction.

(c) $\dfrac{\cancel{3}}{\cancel{10}} \cdot \dfrac{\cancel{10}}{\cancel{3}} \cdot x = \dfrac{3}{10} \cdot \dfrac{5}{2}$ Multiply each side by $\dfrac{3}{10}$.

(d) $x = \dfrac{3}{\cancel{10}} \cdot \dfrac{\cancel{5}}{2} = \dfrac{3}{4}$
$\qquad\qquad\quad 2$

Completion Example 7 Find y if $\dfrac{2\frac{1}{2}}{6} = \dfrac{3}{y}$.

Solution

(a) $\qquad \dfrac{2\frac{1}{2}}{6} = \dfrac{3}{y}$

(b) $\qquad \dfrac{5}{2} \cdot y = \underline{\qquad}$ $\left(2\frac{1}{2} = \dfrac{5}{2}\right)$

(c) $\dfrac{\cancel{2}}{\cancel{5}} \cdot \dfrac{\cancel{5}}{\cancel{2}} \cdot y = \dfrac{2}{5} \cdot \underline{\qquad}$

(d) $\qquad y = \underline{\qquad}$

▢ **NOW WORK EXERCISES 1–3 IN THE MARGIN.**

Completion Example Answer

7. (b) $\quad \dfrac{5}{2} \cdot y = 6 \cdot 3$

(c) $\dfrac{\cancel{2}}{\cancel{5}} \cdot \dfrac{\cancel{5}}{\cancel{2}} \cdot y = \dfrac{2}{5} \cdot 18$

(d) $\qquad y = \dfrac{36}{5}$ or $7\dfrac{1}{5}$

Answers to Marginal Exercises

1. $R = 30$ **2.** $x = 60$ **3.** $x = 6$

NAME _____ SECTION _____ DATE _____

EXERCISES **6.2** _____

Answers

Solve the following proportions.

1. $\dfrac{3}{6} = \dfrac{6}{x}$

2. $\dfrac{7}{21} = \dfrac{y}{6}$

3. $\dfrac{5}{7} = \dfrac{x}{28}$

4. $\dfrac{4}{10} = \dfrac{5}{x}$

5. $\dfrac{8}{B} = \dfrac{6}{30}$

6. $\dfrac{7}{B} = \dfrac{5}{15}$

7. $\dfrac{1}{2} = \dfrac{x}{100}$

8. $\dfrac{3}{4} = \dfrac{x}{100}$

9. $\dfrac{A}{3} = \dfrac{7}{2}$

10. $\dfrac{x}{100} = \dfrac{1}{20}$

11. $\dfrac{3}{5} = \dfrac{60}{D}$

12. $\dfrac{3}{16} = \dfrac{9}{x}$

1. _____

2. _____

3. _____

4. _____

5. _____

6. _____

7. _____

8. _____

9. _____

10. _____

11. _____

12. _____

13. _____

14. _____

15. _____

16. _____

17. _____

18. _____

19. _____

20. _____

21. _____

22. _____

23. _____

24. _____

25. _____

26. _____

27. _____

13. $\dfrac{\frac{1}{2}}{x} = \dfrac{5}{10}$

14. $\dfrac{\frac{2}{3}}{3} = \dfrac{y}{127}$

15. $\dfrac{\frac{1}{3}}{x} = \dfrac{5}{9}$

16. $\dfrac{\frac{3}{4}}{7} = \dfrac{3}{z}$

17. $\dfrac{\frac{1}{8}}{6} = \dfrac{\frac{1}{2}}{w}$

18. $\dfrac{\frac{1}{6}}{5} = \dfrac{5}{w}$

19. $\dfrac{1}{4} = \dfrac{1\frac{1}{2}}{y}$

20. $\dfrac{1}{5} = \dfrac{x}{2\frac{1}{2}}$

21. $\dfrac{1}{5} = \dfrac{x}{7\frac{1}{2}}$

22. $\dfrac{2}{5} = \dfrac{R}{100}$

23. $\dfrac{3}{5} = \dfrac{R}{100}$

24. $\dfrac{A}{4} = \dfrac{75}{100}$

25. $\dfrac{A}{4} = \dfrac{50}{100}$

26. $\dfrac{20}{B} = \dfrac{1}{4}$

27. $\dfrac{30}{B} = \dfrac{25}{100}$

NAME _____ SECTION _____ DATE _____

28. $\dfrac{A}{20} = \dfrac{15}{100}$ 　　　　**29.** $\dfrac{1}{3} = \dfrac{R}{100}$ 　　　　**30.** $\dfrac{2}{3} = \dfrac{R}{100}$

31. $\dfrac{9}{x} = \dfrac{4\frac{1}{2}}{11}$ 　　　　**32.** $\dfrac{y}{6} = \dfrac{2\frac{1}{2}}{12}$ 　　　　**33.** $\dfrac{x}{4} = \dfrac{1\frac{1}{4}}{5}$

34. $\dfrac{5}{x} = \dfrac{2\frac{1}{4}}{27}$ 　　　　**35.** $\dfrac{x}{3} = \dfrac{16}{3\frac{1}{5}}$ 　　　　**36.** $\dfrac{6.2}{5} = \dfrac{x}{15}$

Answers

28. _____

29. _____

30. _____

31. _____

32. _____

33. _____

34. _____

35. _____

36. _____

Answers

37. _____

38. _____

39. _____

40. _____

41. _____

42. _____

43. _____

44. _____

45. _____

46. _____

47. _____

48. _____

49. _____

50. _____

37. $\dfrac{3.5}{2.6} = \dfrac{10.5}{B}$

38. $\dfrac{4.1}{3.2} = \dfrac{x}{6.4}$

39. $\dfrac{7.8}{1.3} = \dfrac{x}{0.26}$

40. $\dfrac{7.2}{y} = \dfrac{4.8}{14.4}$

41. $\dfrac{150}{300} = \dfrac{R}{100}$

42. $\dfrac{19.2}{96} = \dfrac{R}{100}$

43. $\dfrac{12}{B} = \dfrac{25}{100}$

44. $\dfrac{13.5}{B} = \dfrac{15}{100}$

45. $\dfrac{A}{42} = \dfrac{65}{100}$

46. $\dfrac{A}{244} = \dfrac{18}{100}$

47. $\dfrac{A}{850} = \dfrac{30}{100}$

48. $\dfrac{A}{595} = \dfrac{6}{100}$

49. $\dfrac{5684}{B} = \dfrac{98}{100}$

50. $\dfrac{24}{27} = \dfrac{R}{100}$

6.3
SOLVING WORD PROBLEMS USING PROPORTIONS

To solve a word problem using proportions,

1. read the problem carefully (several times if necessary);
2. decide what is unknown (what you are trying to find) and assign some letter to represent it;
3. set up a proportion using Pattern A or Pattern B (both patterns will give the same answer);
4. solve the proportion.

Pattern A: Each ratio has different units but they are in the same order. For example,

$$\frac{500 \text{ miles}}{20 \text{ gallons}} = \frac{x \text{ miles}}{30 \text{ gallons}}$$

Pattern B: Each ratio has the same units, the numerators correspond, and the denominators correspond. For example,

$$\frac{500 \text{ miles}}{x \text{ miles}} = \frac{20 \text{ gallons}}{30 \text{ gallons}}$$

(500 miles corresponds to 20 gallons and x miles corresponds to 30 gallons.)

Example 1 You drove your car 500 miles and used 20 gallons of gasoline. How many miles would you expect to drive using 30 gallons of gasoline?

Solution

(a) Let x represent the unknown number of miles.

(b) Set up a proportion using either Pattern A or Pattern B. **Label the numerators and denominators to be sure the units are in the same order.**

$$\frac{500 \text{ miles}}{20 \text{ gallons}} = \frac{x \text{ miles}}{30 \text{ gallons}}$$

Pattern A Each ratio has different units but the numerators are the same units and the denominators are the same units.

(c) Solve the proportion.

$$\frac{500}{20} = \frac{x}{30}$$

$$500 \cdot 30 = 20 \cdot x$$

$$\frac{15,000}{20} = \frac{\cancel{20} \cdot x}{\cancel{20}}$$

$$750 = x$$

You would expect to drive 750 miles on 30 gallons of gas.

Note: *Any* of the following proportions would give the same answer.

$$\frac{20 \text{ gallons}}{500 \text{ miles}} = \frac{30 \text{ gallons}}{x \text{ miles}}$$ **Pattern A**   Each ratio has different units but the numerators are the same units and the denominators are the same units.

$$\frac{500 \text{ miles}}{x \text{ miles}} = \frac{20 \text{ gallons}}{30 \text{ gallons}}$$ **Pattern B**   Each ratio has the same units, numerators correspond, and denominators correspond.

$$\frac{x \text{ miles}}{500 \text{ miles}} = \frac{30 \text{ gallons}}{20 \text{ gallons}}$$ **Pattern B**   Each ratio has the same units, numerators correspond, and denominators correspond.

Example 2   An architect draws the plans for a building using a scale of $\frac{1}{2}$ inch to represent 10 feet. How many feet would 6 inches represent?

Solution

(a) Let y represent the unknown number of feet.

(b) Set up a proportion labeling the numerators and denominators.

$$\frac{\frac{1}{2} \text{ inch}}{6 \text{ inches}} = \frac{y \text{ feet}}{10 \text{ feet}} \qquad \textbf{WRONG}$$

This proportion is **WRONG** because $\frac{1}{2}$ inch does *not* correspond to y feet. $\frac{1}{2}$ inch corresponds to 10 feet.

$$\frac{\frac{1}{2} \text{ inch}}{6 \text{ inches}} = \frac{10 \text{ feet}}{y \text{ feet}} \qquad \textbf{RIGHT}$$

(c) Solve the **RIGHT** proportion.

$$\frac{\frac{1}{2}}{6} = \frac{10}{y}$$

$$\frac{1}{2} \cdot y = 6 \cdot 10$$

$$\frac{\frac{1}{2} \cdot y}{\frac{1}{2}} = \frac{60}{\frac{1}{2}}$$

$$y = \frac{60}{1} \cdot \frac{2}{1}$$

$$y = 120$$

6 inches would represent 120 feet.

Example 3   If 6 pencils cost $1.50, how many pencils could you buy with $2.75?

Solution

(a) Let P represent the unknown number of pencils.

(b) Set up a proportion.

$$\frac{6 \text{ pencils}}{\$1.50} = \frac{p \text{ pencils}}{\$2.75}$$

(c) Solve the proportion.

$$\frac{6}{1.50} = \frac{P}{2.75}$$

$$16.50 = 1.50 \cdot P$$

$$\frac{16.50}{1.50} = \frac{\cancel{1.50} \cdot P}{\cancel{1.50}}$$

$$11 = P$$

$2.75 will buy 11 pencils (not including tax).

1. One pound of candy costs $4.75. How many pounds of candy can be bought for $28.50?

Completion Example 4 A recommended mixture of weed killer is 3 capfuls for 2 gallons of water. How many capfuls should be mixed with 5 gallons of water?

Solution

(a) Let x = unknown number of capfuls of weed killer.

(b) $$\frac{x \text{ capfuls}}{5 \text{ gallons}} = \frac{3 \text{ capfuls}}{2 \text{ gallons}}$$

$$\underline{\hspace{1cm}} \cdot x = 5 \cdot \underline{\hspace{1cm}}$$

$$\frac{\underline{\hspace{1cm}} \cdot x}{?} = \frac{15}{?}$$

$$x = \underline{\hspace{1cm}}$$

$\underline{\hspace{1.5cm}}$ capfuls of weed killer should be mixed with 5 gallons of water.

2. Precinct 1 has 520 registered voters and Precinct 2 has 630 registered voters. If the ratio of actual voters to registered voters was the same in both precincts, how many voted in Precinct 2 in the last election if 104 voted in Precinct 1?

Completion Example 5 A jelly manufacturer puts 2.5 ounces of sugar into every 6-ounce jar of jelly. How many ounces of jelly can be made with 300 ounces of sugar?

Solution

(a) Let A = unknown amount of jelly.

(b) $$\frac{2.5 \text{ oz. sugar}}{6 \text{ oz. jelly}} = \frac{300 \text{ oz. sugar}}{A \text{ oz. jelly}}$$

$$\underline{\hspace{1cm}} \cdot \underline{\hspace{1cm}} = 1800$$

$$\frac{\underline{\hspace{1cm}} \cdot \underline{\hspace{1cm}}}{?} = \frac{1800}{?}$$

$$A = \underline{\hspace{1cm}}$$

300 ounces of sugar will make $\underline{\hspace{1.5cm}}$ ounces of jelly.

☐ **NOW WORK EXERCISES 1 AND 2 IN THE MARGIN.**

Completion Example Answers

4. $\frac{x}{5} = \frac{3}{2}$

$2 \cdot x = 5 \cdot 3$

$\frac{\cancel{2} \cdot x}{\cancel{2}} = \frac{15}{2}$

$x = 7\frac{1}{2}$

$7\frac{1}{2}$ capfuls of weed killer should be mixed with 5 gallons of water.

5. $\frac{2 \cdot 5}{6} = \frac{300}{A}$

$2.5 \cdot A = 1800$

$\frac{\cancel{2.5} \cdot A}{\cancel{2.5}} = \frac{1800}{2.5}$

$A = 720$

300 ounces of sugar will make 720 ounces of jelly.

Nurses and doctors work with proportions when prescribing medicine and giving injections. Medical texts write proportions in the form

$$2 : 40 :: x : 100$$

instead of

$$\frac{2}{40} = \frac{x}{100}$$

Using either notation, the solution is found by setting the product of the extremes equal to the product of the means and solving the equation for the unknown quantity.

Example 6 Solve the proportion.

$$2 \text{ ounces} : 40 \text{ grams} :: x \text{ ounces} : 100 \text{ grams}.$$

Solution

$$\overset{\text{means}}{2 : 40 :: x : 100}$$
$$\underset{\text{extremes}}{}$$

$$2 \cdot 100 = 40 \cdot x$$

$$\frac{200}{40} = \frac{\cancel{40} \cdot x}{\cancel{40}}$$

$$5 = x$$

The solution is $x = 5$ ounces.

Answers to Marginal Exercises

1. $x = 6$ pounds   **2.** $x = 126$ voted

NAME _____ SECTION _____ DATE _____

EXERCISES **6.3** _____

Solve the following word problems using proportions.

1. A map maker uses a scale of 2 inches to represent 30 miles. How many miles do 3 inches represent?

2. An investor thinks she should make $12 for every $100 she invests. How much would she expect to make on a $1500 investment?

3. If one dozen (12) eggs costs $1.09, what does three dozen eggs cost?

4. The price of a certain fabric is $1.75 per yard. How many yards can be bought with $35 (not including tax)?

5. A baseball team bought 8 bats for $96. What would they pay for 10 bats?

Answers

1. _____

2. _____

3. _____

4. _____

5. _____

6. _____

7. _____

8. _____

9. _____

10. _____

6. A store owner expects to make a profit of $2 on an item that sells for $10. How much profit will he expect to make on an item that sells for $60?

7. A saleswoman makes $8 for every $100 worth of the product she sells. What will she make if she sells $5000 worth of the product?

8. If property taxes are figured at $1.50 for every $100 in evaluation, what taxes will be paid on a home valued at $85,000?

9. A condominium owner pays property taxes of $2000 per year. If taxes are figured at a rate of $1.25 for every $100 in value, what is the value of his condominium?

10. Sales tax is figured at 6¢ for every $1.00 of merchandise purchased. What was the purchase price on an item that had a sales tax of $2.04?

NAME _____ SECTION _____ DATE _____

11. An architect drew plans for a city park using a scale of $\frac{1}{4}$ inch to represent 25 feet. How many feet would 2 inches represent?

12. A building 14 stories high casts a shadow of 30 feet at a certain time of day. What is the length of the shadow of a 20-story building at the same time of day in the same city?

13. Two numbers are in the ratio of 4 to 3. The number 10 is in that same ratio to a fourth number. What is the fourth number?

14. A salesman figured he drove 560 miles every two weeks. How far would he drive in three months (12 weeks)?

15. Driving steadily, a woman made a trip of 200 miles in $4\frac{1}{2}$ hours. How long would she take to drive 500 miles at the same rate of speed?

Answers

11. _____

12. _____

13. _____

14. _____

15. _____

16. _____

17. _____

18. _____

19. _____

20. _____

21. _____

16. If you can drive 286 miles in $5\frac{1}{2}$ hours, how long will it take you to drive 468 miles at the same rate of speed?

17. What will 21 gallons of gasoline cost if gasoline costs $1.25 per gallon?

18. If diesel fuel costs $1.10 per gallon, how much diesel fuel will $24.53 buy?

19. An electric fan makes 180 revolutions per minute. How many revolutions will the fan make if it runs for 24 hours?

20. An investor made $144 in one year on a $1000 investment. What would she have earned if her investment had been $4500?

21. A typist can type 8 pages of manuscript in 56 minutes. How long will this typist take to type 300 pages?

NAME _____ SECTION _____ DATE _____

22. On a map, $1\frac{1}{2}$ inches represents 40 miles. How many inches represent 50 miles?

23. In the metric system, there are 2.54 centimeters in one inch. How many centimeters are there in one foot?

24. If 40 pounds of fertilizer are used on 2400 square feet of lawn, how many pounds of fertilizer are needed for a lawn of 5400 square feet?

25. An English teacher must read and grade 27 essays. If the teacher takes 20 minutes to read and grade 3 essays, how much time will he need to grade all 27 essays?

26. If 2 cups of flour are needed to make 12 biscuits, how much flour will be needed to make 9 of the same kind of biscuits?

Answers

22. _____

23. _____

24. _____

25. _____

26. _____

27. If 2 cups of flour are needed to make 12 biscuits, how many of the same kind of biscuits can be made with 3 cups of flour?

27. _____

28. _____

The following exercises are examples of proportions in medicine. Solve for x and be sure to label your answers. (The abbreviations are from the metric system.)

28. 1 liter : 1000 mL :: x liter : 5000 mL   **29.** 1 kg : 1000 g :: x kg : 2700 g

29. _____

30. _____

30. 1 mg : 1000 mcg :: x mg : 4 mcg   **31.** 1 g : 1000 mg :: 0.5 g : x mg

31. _____

32. _____

32. 1 dram : 60 grains :: x dram : 90 grains

33. _____

34. _____

33. $\dfrac{1 \text{ ounce}}{8 \text{ drams}} = \dfrac{0.5 \text{ ounce}}{x}$ **34.** $\dfrac{1 \text{ dram}}{60 \text{ minims}} = \dfrac{x}{180 \text{ minims}}$

35. _____

36. _____

35. $\dfrac{1 \text{ ounce}}{480 \text{ minims}} = \dfrac{4 \text{ ounces}}{x}$ **36.** $\dfrac{1 \text{ g}}{15 \text{ grains}} = \dfrac{x}{60 \text{ grains}}$

37. _____

38. _____

37. $\dfrac{1 \text{ ounce}}{30 \text{ g}} = \dfrac{2 \text{ ounces}}{x}$ **38.** $\dfrac{1 \text{ tsp}}{4 \text{ mL}} = \dfrac{x}{12 \text{ mL}}$

39. _____

40. _____

39. $\dfrac{1 \text{ ounce}}{30 \text{ mL}} = \dfrac{x}{15 \text{ mL}}$ **40.** $\dfrac{1 \text{ pint}}{500 \text{ mL}} = \dfrac{1.5 \text{ pints}}{x}$

NAME _____ SECTION _____ DATE _____

CHAPTER **6** TEST _____

In Problems 1–4, complete each statement.

1. The ratio 7 to 6 can be expressed as the fraction _____ .

2. In the proportion $\dfrac{3}{4} = \dfrac{75}{100}$, 3 and 100 are called the _____ .

3. In the proportion 6 : 10 :: 3 : 5, 10 and 3 are called the

_____

4. Is the proportion $\dfrac{4}{6} = \dfrac{9}{14}$ true or false? Give a reason for your answer.

In Problems 5–7, write each comparison as a ratio reduced to lowest terms. Use common units whenever possible.

5. 15 inches to 1 yard

1. _____

2. _____

3. _____

4. _____

5. _____

Answers

6. _____

7. _____

8. _____

9. _____

10. _____

11. _____

12. _____

13. _____

14. _____

15. _____

6. 220 miles to 4 hours

7. 3 inches to 7.62 centimeters

In Problems 8–15, solve each proportion. Reduce all fractions to lowest terms and express all improper fractions as mixed numbers.

8. $\dfrac{9}{4} = \dfrac{x}{16}$

9. $\dfrac{5}{3} = \dfrac{75}{x}$

10. $\dfrac{x}{2.5} = \dfrac{1}{4}$

11. $\dfrac{\frac{1}{3}}{x} = \dfrac{5}{\frac{1}{6}}$

12. $\dfrac{x}{0.18} = \dfrac{17}{0.9}$

13. $\dfrac{\frac{3}{8}}{x} = \dfrac{9}{20}$

14. 17 is to 5 as 51 is to x

15. $9 : 4 :: x : 2$

NAME _____ SECTION _____ DATE _____

In Problems 16–24, use a proportion to solve each word problem.

Answers

16. How far can you drive on 5 gallons of gas if you can drive 120 miles on 3.5 gallons of gas at the same speed (to the nearest tenth of a mile)?

16. _____

17. _____

18. _____

17. On a certain map, 2 inches represents 15 miles. How far apart are two towns that are $3\frac{1}{5}$ inches apart on the map?

19. _____

20. _____

18. If you can buy 4 tires for \$236, what will it cost for 5 tires?

19. If you drive 165 miles in $3\frac{2}{3}$ hours, what is your average speed in miles per hour?

20. The ratio of boys to girls in a certain classroom is 3 to 4. How many boys are there in a class with 36 girls?

21. _____

22. _____

23. _____

24. _____

21. There are one thousand grams in one kilogram. How many grams are there in four and seven tenths kilograms?

22. An artist figures she can paint 3 portraits every two weeks. At this rate, how long will it take her to paint 18 portraits?

23. Two units of a certain gas weigh 175 grams. What is the weight of 5 units of this gas?

24. A manufacturing company expects to make a profit of $3 on a product that it sells for $8. How much profit does the company expect to make on a product that it sells for $20?

Percent

7.1
UNDERSTANDING PERCENT

The word **percent** comes from the Latin *per centum* meaning "per hundred." So, **percent** means **hundredths** or **the ratio of a number to 100.** The symbol % is called the **percent sign.** You can think of it as a rearrangement of the digits for 100 in the form 0/0. For example,

$$\frac{50}{100} = 50\% \quad \text{and} \quad \frac{27}{100} = 27\%$$

In Figure 7.1 there are 100 small squares and 27 are shaded. Thus, $\frac{27}{100}$ or 27% of the large square is shaded.

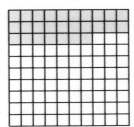 $\frac{27}{100} = 27\%$

FIGURE 7.1

> **Percent:** Percent means hundredths, or the ratio of a number to 100
> **Symbol:** % sign (used in place of hundredths)

In Examples 1–6, each fraction is changed to a percent.

Change each fraction to a percent.

1. $\dfrac{9}{100}$

Example 1 $\dfrac{30}{100} = 30\%$ Write the numerator and the % sign. Remember, % means hundredths.

Example 2 $\dfrac{45}{100} = 45\%$ Remember, % means hundredths.

Example 3 $\dfrac{60}{100} = 60\%$

2. $\dfrac{1.25}{100}$

Example 4 $\dfrac{8.6}{100} = 8.6\%$

Example 5 $\dfrac{17\frac{1}{2}}{100} = 17\frac{1}{2}\%$

3. $\dfrac{125}{100}$

Example 6 $\dfrac{200}{100} = 200\%$

☐ **NOW WORK EXERCISES 1–4 IN THE MARGIN.**

Example 7   The ratio of money made as profit to money invested can be called percent of profit. Based on percent of profit, is (a) or (b) the better investment?

(a) You make $150 profit by investing $300.

(b) You make $200 profit by investing $500.

4. $\dfrac{6\frac{1}{4}}{100}$

Solution

(a) $\dfrac{\$150 \text{ profit}}{\$300 \text{ invested}} = \dfrac{\cancel{3} \cdot 50}{\cancel{3} \cdot 100} = \dfrac{50}{100} = 50\%$

(b) $\dfrac{\$200 \text{ profit}}{\$500 \text{ invested}} = \dfrac{\cancel{5} \cdot 40}{\cancel{5} \cdot 100} = \dfrac{40}{100} = 40\%$

By writing both ratios of profit to investment as **hundredths,** we get **percent** and see that investment (a) is better than investment (b) since 50% is larger than 40%. Obviously, $200 profit is more than $150 profit, but the money risked ($500) is also greater. The use of percent gives an effective method for comparison because both ratios have the same denominator (100).

Completion Example 8   Which is the better investment?

(a) Making $40 profit by investing $200.

(b) Making $75 profit by investing $300.

Solution

Write each ratio as hundredths and compare the percents.

(a) $\dfrac{40}{200} = \dfrac{2\cdot}{2\cdot 100} = \dfrac{}{100} = \underline{\hspace{1cm}}\%$

(b) $\dfrac{75}{300} = \dfrac{3\cdot}{3\cdot 100} = \dfrac{}{100} = \underline{\hspace{1cm}}\%$

Investment $\underline{\hspace{1.5cm}}$ is better since $\underline{\hspace{1.5cm}}\%$ is larger than $\underline{\hspace{1.5cm}}\%$.

Completion Example Answer

8. (a) $\dfrac{40}{200} = \dfrac{\cancel{2}\cdot 20}{\cancel{2}\cdot 100} = \dfrac{20}{100} = 20\%$

(b) $\dfrac{75}{300} = \dfrac{\cancel{3}\cdot 25}{\cancel{3}\cdot 100} = \dfrac{25}{100} = 25\%$

Investment (b) is better since 25% is larger than 20%.

Answers to Marginal Exercises

1. 9%   2. 1.25%   3. 125%   4. $6\frac{1}{4}\%$

NAME _____ SECTION _____ DATE _____

EXERCISES **7.1** _____

What percent of each square is shaded?

Answers

1.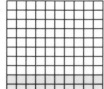
2.

3.

1. _____

2. _____

4.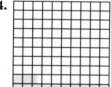
5.
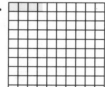
6.
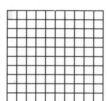

3. _____

4. _____

5. _____

6. _____

7.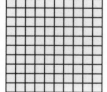
8.
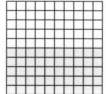

7. _____

8. _____

9. _____

9.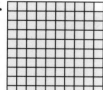
10.
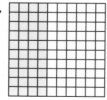

10. _____

Answers

Change the following fractions to percents.

11. _____

12. _____

13. _____

14. _____

15. _____

16. _____

17. _____

18. _____

19. _____

20. _____

21. _____

22. _____

23. _____

24. _____

25. _____

26. _____

27. _____

28. _____

29. _____

30. _____

31. _____

32. _____

33. _____

34. _____

11. $\dfrac{30}{100}$　　12. $\dfrac{20}{100}$　　13. $\dfrac{40}{100}$　　14. $\dfrac{50}{100}$

15. $\dfrac{7}{100}$　　16. $\dfrac{8}{100}$　　17. $\dfrac{90}{100}$　　18. $\dfrac{15}{100}$

19. $\dfrac{25}{100}$　　20. $\dfrac{35}{100}$　　21. $\dfrac{45}{100}$　　22. $\dfrac{65}{100}$

23. $\dfrac{75}{100}$　　24. $\dfrac{42}{100}$　　25. $\dfrac{53}{100}$　　26. $\dfrac{68}{100}$

27. $\dfrac{77}{100}$　　28. $\dfrac{48}{100}$　　29. $\dfrac{125}{100}$　　30. $\dfrac{110}{100}$

31. $\dfrac{150}{100}$　　32. $\dfrac{175}{100}$　　33. $\dfrac{200}{100}$　　34. $\dfrac{250}{100}$

NAME _____ SECTION _____ DATE _____

35. $\dfrac{236}{100}$ **36.** $\dfrac{120}{100}$ **37.** $\dfrac{16.3}{100}$ **38.** $\dfrac{27.2}{100}$

39. $\dfrac{13.4}{100}$ **40.** $\dfrac{38.6}{100}$ **41.** $\dfrac{20.25}{100}$ **42.** $\dfrac{93.5}{100}$

43. $\dfrac{0.5}{100}$ **44.** $\dfrac{1.5}{100}$ **45.** $\dfrac{0.25}{100}$ **46.** $\dfrac{3\frac{1}{2}}{100}$

47. $\dfrac{10\frac{1}{4}}{100}$ **48.** $\dfrac{1\frac{1}{4}}{100}$ **49.** $\dfrac{24\frac{1}{2}}{100}$ **50.** $\dfrac{17\frac{3}{4}}{100}$

Answers

35. _____

36. _____

37. _____

38. _____

39. _____

40. _____

41. _____

42. _____

43. _____

44. _____

45. _____

46. _____

47. _____

48. _____

49. _____

50. _____

Answers

51. _____

52. _____

53. _____

54. _____

55. _____

In each problem, write the ratio of profit to investment as hundredths and compare the percents. Tell which investment is better, (a) or (b).

51. (a) a profit of $36 on a $200 investment
(b) a profit of $51 on a $300 investment

52. (a) a profit of $40 on a $400 investment
(b) a profit of $60 on a $500 investment

53. (a) a profit of $70 on a $700 investment
(b) a profit of $30 on a $300 investment

54. (a) a profit of $150 on a $1500 investment
(b) a profit of $200 on a $2000 investment

55. (a) a profit of $300 on a $2000 investment
(b) a profit of $360 on a $3000 investment

7.2

DECIMALS AND PERCENTS _____

We know that percent means hundredths. So, by changing a decimal to a fraction with denominator 100, we effectively change the decimal to a percent. For example,

$$0.42 = \frac{42}{100} = 42\%$$

$$0.76 = \frac{76}{100} = 76\%$$

Not all decimals are hundredths. In these cases, we would proceed as follows:

$$0.253 = \frac{0.253 \times 100}{100} = \frac{25.3}{100} = 25.3\%$$

That is, by multiplying and dividing by 100, we can change the form of the decimal without changing its value. Again.

$$0.802 = \frac{0.802 \times 100}{100} = \frac{80.2}{100} = 80.2\%$$

By noting the way the decimal point is moved, the change can be done directly using the following rule.

> **To change a decimal to a percent,**
>
> 1. move the decimal point two places to the right.
> 2. write the % sign.

Example 1 $0.253 = 25.3\%$

decimal point moved   % sign added
two places to the
right

Example 2 $0.76 = 76\%$

decimal point moved   % sign added
two places to the
right

Example 3 $0.52 = \dfrac{52}{100} = 52\%$

writing the fraction in hundredths shows
why the decimal point is moved two
places to the right

Change each decimal to a percent.

1. 0.04

2. 0.35

3. $0.02\frac{1}{2}$

4. 0.936

5. 2.17

Example 4 $0.905 = \dfrac{90.5}{100} = 90.5\%$

Example 5 $0.005 = 0.5\%$

Example 6 $0.5 = 0.50 = 50\%$

Example 7 $1.5 = 1.50 = 150\%$

                 a number   is more
                 larger than 1   than 100%

Example 8 $2.35 = 235\%$

Example 9 $0.03 = 3\%$

Example 10 $0.3 = 30\%$

☐ **NOW WORK EXERCISES 1–5 IN THE MARGIN.**

To change percents to decimals, we simply reverse the procedure for changing decimals to percents.

> **To change a percent to a decimal,**
>
> 1. move the decimal point two places to the left;
> 2. drop the % sign.

Example 11   We can change 56% to a decimal using a fraction as

$$56\% = \dfrac{56}{100} = 0.56$$

or, using the rule for moving the decimal point,

$$56\% = 0.56$$

      understood   decimal point   % sign dropped
      decimal   moved two
      point   places left

Example 12 $38\% = 0.38$

Example 13 $18.5\% = 0.185$

Example 14 $12\frac{1}{4}\% = 0.12\frac{1}{4}$ or 0.1225

or, using the fraction form, we can write

$$12\frac{1}{4}\% = \frac{12\frac{1}{4}}{100} = \frac{12.25}{100} = 0.1225$$

Example 15 $0.6\% = 0.006$

Example 16 $0.23\% = 0.0023$

Example 17 $230\% = 2.30$

Example 18 $100\% = 1.0$

Example 19 $1.5\% = 0.015$

Example 20 $101\% = 1.01$

☐ **NOW WORK EXERCISES 6–10 IN THE MARGIN.**

Change each percent to a decimal.

6. 12%

7. $10\frac{1}{2}\%$

8. 18.2%

9. 0.7%

10. 165%

Answers to Marginal Exercises

1. 4% **2.** 35% **3.** $2\frac{1}{2}\%$ or 2.5% **4.** 93.6% **5.** 217% **6.** 0.12

7. $0.10\frac{1}{2}$ or 0.105 **8.** 0.182 **9.** 0.007 **10.** 1.65

NAME _____   SECTION _____   DATE _____

EXERCISES **7.2** _____

Answers

Change the following decimals to percents.

1. 0.02   **2.** 0.09   **3.** 0.1   **4.** 0.5

5. 0.7   **6.** 0.9   **7.** 0.36   **8.** 0.52

9. 0.83   **10.** 0.75   **11.** 0.25   **12.** 0.30

13. 0.40   **14.** 0.65   **15.** 0.025   **16.** 0.035

17. 0.046   **18.** 0.055   **19.** 0.003   **20.** 0.004

1. _____

2. _____

3. _____

4. _____

5. _____

6. _____

7. _____

8. _____

9. _____

10. _____

11. _____

12. _____

13. _____

14. _____

15. _____

16. _____

17. _____

18. _____

19. _____

20. _____

Answers

21. _____
22. _____
23. _____
24. _____
25. _____
26. _____
27. _____
28. _____
29. _____
30. _____
31. _____
32. _____
33. _____
34. _____
35. _____
36. _____
37. _____
38. _____
39. _____
40. _____
41. _____
42. _____
43. _____
44. _____
45. _____
46. _____
47. _____
48. _____
49. _____
50. _____

21. 1.10	**22.** 1.30	**23.** 1.25	**24.** 1.75
25. 2	**26.** 1.08	**27.** 1.05	**28.** 1.5
29. 2.3	**30.** 2.15		

Change the following percents to decimals.

31. 2%	**32.** 7%	**33.** 10%	**34.** 18%
35. 15%	**36.** 20%	**37.** 25%	**38.** 30%
39. 35%	**40.** 80%	**41.** 10.1%	**42.** 11.5%
43. 13.2%	**44.** 17.3%	**45.** $5\frac{1}{4}\%$	**46.** $6\frac{1}{2}\%$
47. $13\frac{3}{4}\%$	**48.** $15\frac{1}{4}\%$	**49.** $20\frac{1}{4}\%$	**50.** $18\frac{1}{2}\%$

51. 0.25%   **52.** 1.25%   **53.** 0.17%   **54.** 0.50%

55. 125%   **56.** 150%   **57.** 130%   **58.** 120%

59. 222%   **60.** 215%

61. Suppose sales tax is figured at 6%. Change 6% to a decimal.

62. The interest rate on a loan is 15%. Change 15% to a decimal.

63. The sales commission in a retail store is figured at $8\frac{1}{2}$% for the clerk. Change $8\frac{1}{2}$% to a decimal.

Answers

51. _____

52. _____

53. _____

54. _____

55. _____

56. _____

57. _____

58. _____

59. _____

60. _____

61. _____

62. _____

63. _____

Answers

64. _____

65. _____

66. _____

67. _____

68. _____

69. _____

70. _____

64. The discount during a special sale on dresses is 30%. Change 30% to a decimal.

65. The rate of profit based on sales price is 45%. Change 45% to a decimal.

66. A book store figures its profit to be $0.20 for every $1.00 in sales. Change 0.20 to a percent.

67. In calculating his sales commission, Mr. John multiplies by the decimal 0.12. Change 0.12 to a percent.

68. To calculate what your maximum house payment should be, a banker multiplied your income by 0.28. Change 0.28 to a percent.

69. The discount you earn by paying cash is found by multiplying the amount of your purchase by 0.02. Change 0.02 to a percent.

70. Suppose the state license fee is figured by multiplying the cost of your car by 0.005. Change 0.005 to a percent.

7.3

FRACTIONS AND PERCENTS
(Calculators Optional)

If a fraction has denominator 100, we can change it to a percent by writing the numerator and adding the % sign (Section 7.1). If the denominator is a factor of 100 (2, 4, 5, 10, 20, 25, or 50), we can write it in an equivalent form with denominator 100, and then change it to a percent. For example,

$$\frac{3}{4} = \frac{3}{4} \cdot \frac{25}{25} = \frac{75}{100} = 75\%$$

$$\frac{1}{2} = \frac{1}{2} \cdot \frac{50}{50} = \frac{50}{100} = 50\%$$

$$\frac{4}{5} = \frac{4}{5} \cdot \frac{20}{20} = \frac{80}{100} = 80\%$$

However, most fractions do not have denominators that are factors of 100. A more general approach (easily applied with calculators) is to change the fraction to decimal form (Section 5.6), and then change the decimal to a percent (Section 7.2).

> **To change a fraction to a percent,**
>
> 1. change the fraction to a decimal (divide the denominator into the numerator);
> 2. change the decimal to a percent.

Example 1   Change $\frac{3}{4}$ to a percent.

(a) Divide:

$$
\begin{array}{r}
0.75 \\
4\overline{)3.00} \\
2\,8 \\
\hline
20 \\
20 \\
\hline
0
\end{array}
$$

(b) Change 0.75 to a percent: $\frac{3}{4} = 0.75 = 75\%$

Change each fraction to a percent.

1. $\dfrac{13}{20}$

2. $1\dfrac{1}{2}$

3. $\dfrac{1}{4}$

Example 2 Change $\dfrac{5}{8}$ to a percent.

(a) Divide:

$$
\begin{array}{r}
0.625 \\
8\overline{)5.000} \\
\underline{4\ 8} \\
20 \\
\underline{16} \\
40 \\
\underline{40} \\
0
\end{array}
$$

(b) Change 0.625 to a percent: $\dfrac{5}{8} = 0.625 = 62.5\%$

Example 3 Change $\dfrac{18}{20}$ to a percent.

(a) Divide:

$$
\begin{array}{r}
0.9 \\
20\overline{)18.0} \\
\underline{18\ 0} \\
0
\end{array}
$$

(b) Change 0.9 to a percent: $\dfrac{18}{20} = 0.9 = 90\%$

Example 4 Change $2\dfrac{1}{4}$ to a percent.

(a) $2\dfrac{1}{4} = \dfrac{9}{4}$

$$
\begin{array}{r}
2.25 \\
4\overline{)9.00} \\
\underline{8} \\
1\ 0 \\
\underline{8} \\
20 \\
\underline{20} \\
0
\end{array}
$$

(b) $2\dfrac{1}{4} = \dfrac{9}{4} = 2.25 = 225\%$

☐ **NOW WORK EXERCISES 1–3 IN THE MARGIN.**

AGREEMENT FOR ROUNDING OFF DECIMAL QUOTIENTS

1. Decimal quotients that are exact with four decimal places will be written with four decimal places (or less).
2. Decimal quotients that are not exact will be divided to the fourth place and the quotient will be rounded off to the third place (thousandths).

> **USING A CALCULATOR**
>
> If your instructor agrees that this is the appropriate time, you may choose to use a calculator to do the long division when changing a fraction to a decimal. Most calculators give answers accurate to 8 digits, so you will need to round off your answers according to the preceding box, Agreement for Rounding Off Decimal Quotients.

Example 5 $\frac{1}{3} = 0.3333333$ using a calculator

(a) Rounding off the decimal quotient to the third decimal place:

$$\frac{1}{3} = 0.333 = 33.3\%$$ The answer is rounded off and not exact.

(b) Without a calculator, we can divide and use fractions:

$$
\begin{array}{r}
0.33\frac{1}{3} \\
3\overline{)1.00} \\
\underline{9} \\
10 \\
\underline{9} \\
1
\end{array}
\qquad
\frac{1}{3} = 0.33\frac{1}{3} = 33\frac{1}{3}\%
$$

$33\frac{1}{3}\%$ is exact and 33.3% is rounded off.

Both answers are acceptable, but be aware that 33.3% is a rounded-off answer.

Example 6 $\frac{2}{3} = 0.6666666$ using a calculator

(a) Rounding off the decimal quotient to the third decimal place:

$$\frac{2}{3} = 0.667 = 66.7\%$$ Remember, the answer is rounded off.

(b) Without a calculator, we can divide and use fractions:

$$
\begin{array}{r}
0.66\frac{2}{3} \\
3\overline{)2.00} \\
\underline{1\,8} \\
20 \\
\underline{18} \\
2
\end{array}
\qquad
\frac{2}{3} = 0.66\frac{2}{3} = 66\frac{2}{3}\%
$$

Both $66\frac{2}{3}\%$ and 66.7% are acceptable. Remember that 66.7% is a rounded-off answer.

Change each fraction to a percent. A calculator may be used.

4. $\dfrac{3}{8}$

5. $\dfrac{3}{40}$

6. $\dfrac{7}{9}$

Example 7 $\dfrac{1}{7} = 0.1428571$ using a calculator

(a) Rounding off the decimal quotient to the third decimal place:

$$\frac{1}{7} = 0.143 = 14.3\%$$

(b) Using long division, we can write the answer with a fraction or continue to divide and round off the decimal answer. The choice is yours.

$$\frac{0.14\frac{2}{7}}{7\overline{)1.00}} = 14\frac{2}{7}\%$$

$$\begin{array}{r} 7 \\ \hline 30 \\ 28 \\ \hline 2 \end{array}$$

or:

$$\frac{0.1428}{7\overline{)1.0000}} = 14.3\% \quad \text{by rounding off the decimal quotient to the third decimal place}$$

$$\begin{array}{r} 7 \\ \hline 30 \\ 28 \\ \hline 20 \\ 14 \\ \hline 60 \\ 56 \\ \hline 4 \end{array}$$

☐ **NOW WORK EXERCISES 4–6 IN THE MARGIN.**

> **To change a percent to a fraction or mixed number,**
> 1. write the percent as a fraction with denominator 100 and drop the % sign;
> 2. reduce the fraction.

Example 8   Change 60% to a fraction.

$$60\% = \frac{60}{100} = \frac{3 \cdot \cancel{20}}{5 \cdot \cancel{20}} = \frac{3}{5}$$

Example 9 Change 18% to a fraction.

$$18\% = \frac{18}{100} = \frac{9 \cdot \cancel{2}}{50 \cdot \cancel{2}} = \frac{9}{50}$$

Example 10 Change $7\frac{1}{4}\%$ to a fraction.

$$7\frac{1}{4}\% = \frac{7\frac{1}{4}}{100} = \frac{\frac{29}{4}}{100} = \frac{29}{4} \cdot \frac{1}{100} = \frac{29}{400}$$

Example 11 Change 130% to a mixed number.

$$130\% = \frac{130}{100} = \frac{13 \cdot \cancel{10}}{10 \cdot \cancel{10}} = \frac{13}{10} = 1\frac{3}{10}$$

☐ **NOW WORK EXERCISES 7–10 IN THE MARGIN.**

A COMMON MISUNDERSTANDING

The fractions $\frac{1}{4}$ and $\frac{1}{2}$ are often confused with the percents $\frac{1}{4}\%$ and $\frac{1}{2}\%$. The differences can be clarified by using decimals.

PERCENT	DECIMAL	FRACTION
$\frac{1}{4}\%$ (or 0.25%)	0.0025	$\frac{1}{400}$
$\frac{1}{2}\%$ (or 0.5%)	0.005	$\frac{1}{200}$
25%	0.25	$\frac{1}{4}$
50%	0.50	$\frac{1}{2}$

Thus,

$$\frac{1}{4} = 0.25 \quad \text{and} \quad \frac{1}{4}\% = 0.0025$$

$$0.25 \neq 0.0025$$

Similarly,

$$\frac{1}{2} = 0.50 \quad \text{and} \quad \frac{1}{2}\% = 0.005$$

$$0.50 \neq 0.005$$

You can think of $\frac{1}{4}$ as being one-fourth of a dollar (a quarter) and $\frac{1}{4}\%$ as being one-fourth of a penny. $\frac{1}{2}$ can be thought of as one-half of a dollar and $\frac{1}{2}\%$ as one-half of a penny.

Change each percent to a fraction or mixed number.

7. 80%

8. 16%

9. $5\frac{1}{2}\%$

10. 235%

Some percents are so common that their decimal and fraction equivalents should be memorized. Their fractional values are particularly easy to work with, and many times calculations involving these fractions can be done mentally.

COMMON PERCENT–DECIMAL–FRACTION EQUIVALENTS

$1\% = 0.01 = \dfrac{1}{100}$ $33\dfrac{1}{3}\% = 0.33\dfrac{1}{3} = \dfrac{1}{3}$ $12\dfrac{1}{2}\% = 0.125 = \dfrac{1}{8}$

$25\% = 0.25 = \dfrac{1}{4}$ $66\dfrac{2}{3}\% = 0.66\dfrac{2}{3} = \dfrac{2}{3}$ $37\dfrac{1}{2}\% = 0.375 = \dfrac{3}{8}$

$50\% = 0.50 = \dfrac{1}{2}$ $62\dfrac{1}{2}\% = 0.625 = \dfrac{5}{8}$

$75\% = 0.75 = \dfrac{3}{4}$ $87\dfrac{1}{2}\% = 0.875 = \dfrac{7}{8}$

$100\% = 1.00 = 1$

Answers to Marginal Exercises

1. 65%   **2.** 150%   **3.** 25%   **4.** 37.5%   **5.** 7.5%   **6.** 77.8% or $77\dfrac{7}{9}\%$

7. $\dfrac{4}{5}$ **8.** $\dfrac{4}{25}$ **9.** $\dfrac{11}{200}$ **10.** $2\dfrac{7}{20}$

NAME _____ SECTION _____ DATE _____

EXERCISES **7.3** _____

Change the following numbers to percents. A calculator may be used if your instructor thinks its use is appropriate.

1. $\dfrac{3}{100}$ 2. $\dfrac{16}{100}$ 3. $\dfrac{7}{100}$ 4. $\dfrac{29}{100}$

5. $\dfrac{1}{2}$ 6. $\dfrac{3}{4}$ 7. $\dfrac{1}{4}$ 8. $\dfrac{1}{20}$

9. $\dfrac{11}{20}$ 10. $\dfrac{7}{10}$ 11. $\dfrac{3}{10}$ 12. $\dfrac{3}{5}$

13. $\dfrac{1}{5}$ 14. $\dfrac{2}{5}$ 15. $\dfrac{4}{5}$ 16. $\dfrac{1}{50}$

Answers

1. _____

2. _____

3. _____

4. _____

5. _____

6. _____

7. _____

8. _____

9. _____

10. _____

11. _____

12. _____

13. _____

14. _____

15. _____

16. _____

Answers

17. _____

18. _____

19. _____

20. _____

21. _____

22. _____

23. _____

24. _____

25. _____

26. _____

27. _____

28. _____

29. _____

30. _____

31. _____

32. _____

33. _____

34. _____

35. _____

36. _____

17. $\dfrac{13}{50}$ **18.** $\dfrac{1}{25}$ **19.** $\dfrac{12}{25}$ **20.** $\dfrac{24}{25}$

21. $\dfrac{1}{8}$ **22.** $\dfrac{5}{8}$ **23.** $\dfrac{7}{8}$ **24.** $\dfrac{1}{9}$

25. $\dfrac{5}{9}$ **26.** $\dfrac{2}{7}$ **27.** $\dfrac{3}{7}$ **28.** $\dfrac{5}{6}$

29. $\dfrac{7}{11}$ **30.** $\dfrac{5}{11}$ **31.** $1\dfrac{1}{14}$ **32.** $1\dfrac{1}{6}$

33. $1\dfrac{1}{20}$ **34.** $1\dfrac{1}{4}$ **35.** $1\dfrac{3}{4}$ **36.** $1\dfrac{1}{5}$

NAME _____   SECTION _____   DATE _____

37. $1\frac{3}{8}$ **38.** $2\frac{1}{2}$ **39.** $2\frac{1}{10}$ **40.** $2\frac{1}{15}$

Change the following percents to fractions or mixed numbers.

41. 10%   **42.** 5%   **43.** 15%   **44.** 17%

45. 25%   **46.** 30%   **47.** 50%   **48.** $12\frac{1}{2}$%

49. $37\frac{1}{2}$%   **50.** $16\frac{2}{3}$%   **51.** $33\frac{1}{3}$%

Answers

37. _____

38. _____

39. _____

40. _____

41. _____

42. _____

43. _____

44. _____

45. _____

46. _____

47. _____

48. _____

49. _____

50. _____

51. _____

Answers

52. 66$\frac{2}{3}$%

53. 33%

54. $\frac{1}{2}$%

52. _____

53. _____

54. _____

55. $\frac{1}{4}$%

56. 1%

57. 100%

55. _____

56. _____

57. _____

58. _____

58. 125%

59. 120%

60. 150%

59. _____

60. _____

61. _____

61. 0.3%

62. 2.5%

63. 62.5%

62. _____

63. _____

64. _____

64. 0.2%

65. 0.75%

65. _____

7.4

TYPES OF PERCENT PROBLEMS
(Calculators Optional)

Consider the sentence

35% of 80 is 28.

We want to translate the sentence into an equation as follows:

$$
\begin{array}{ccccc}
35\% & \text{of} & 80 & \text{is} & 28 \\
\downarrow & \downarrow & \downarrow & \downarrow & \downarrow \\
\text{Rate} & \text{times} & \text{Base} & = & \text{Amount} \\
\text{or} & & & & \text{or} \\
\text{Percent} & & & & \text{Percentage}
\end{array}
$$

This basic relationship holds for the three types of problems related to percent.

> R = Rate or percent (as a decimal or fraction)
> B = Base (number we are finding a percent of)
> A = Amount or percentage (a part of the Base)
>   "of" means times (multiply)
>   "is" means =
> The relationship between R, B, and A is
>
> $$R \times B = A$$

The equation

$$R \times B = A$$

has three quantities in it. Finding the value of one quantity when the other two are known corresponds to one of the **three types of percent problems.**

> There are three basic types of problems using percent.
>
> **Type 1**   Finding an amount (or percentage) knowing the percent and the number:
>
> $$
> \begin{array}{ccccc}
> \text{What} & \text{is} & 45\% & \text{of} & 70? \\
> \downarrow & \downarrow & \downarrow & \downarrow & \downarrow \\
> A & = & 0.45 & \times & 70 \\
> A & = & R & \times & B
> \end{array}
> $$

Type 2   Finding a number knowing that a percent of that number is a certain amount:

30% of what number is 18?

$$0.30 \times \quad B \quad = 18$$
$$R \times \quad B \quad = A$$

Type 3   Finding the percent of a number represented by a certain amount:

What percent of 84 is 16.8?

$$R \quad \times 84 = 16.8$$
$$R \quad \times B = A$$

Remember,

(a) "of" means to multiply;

(b) "is" means =; and

(c) the percent is changed to decimal or fraction form.

PROBLEM TYPE 1

Example 1   What is 45% of 70?

$$A = 0.45 \times 70$$
$$A = 31.5$$

$R = 0.45$
$B = 70$

$$\begin{array}{r} 70 \\ \times 0.45 \\ \hline 3\,50 \\ 28\,0 \\ \hline 31.50 \end{array}$$

So, 31.5 is 45% of 70.

The multiplication can be done with a calculator or to the side as shown here.

Example 2   18% of 200 is what?

$$0.18 \times 200 = A$$
$$36 = A$$

$R = 0.18$
$B = 200$

$$\begin{array}{r} 0.18 \\ \times \;\; 200 \\ \hline 36.00 \end{array}$$

So, 18% of 200 is 36.

Problems of Type 1 are the most common and the easiest to solve. The rate (R) and base (B) are known. The unknown amount (A) is found by multiplying the rate times the base.

PROBLEM TYPE 2

Example 3   30% of what number is 18?

$$0.30 \times \quad B \quad = 18$$

$R = 0.30$
$A = 18$

Here the **coefficient** of B is 0.30 and both sides of the equation are to be divided by 0.30. This is the same procedure we used in Chapter 6 when solving proportions.

$$.30 \times B = 18$$

$$\frac{.\cancel{30} \times B}{.\cancel{30}} = \frac{18}{.30}$$

$$B = 60$$

$$\begin{array}{r} 60. \\ .30\overline{)18.00} \\ \underline{18\ 0} \\ 00 \\ \underline{0} \\ 0 \end{array}$$

So, 30% of 60 is 18.

Example 4 82% of ___ is 246?

$\quad\quad\quad\downarrow\ \downarrow\ \downarrow\ \downarrow\ \downarrow$

$\quad\quad\quad.82\ \times\ B\ =\ 246$

$R = 0.82$
$A = 246$

$$.82 \times B = 246$$

$$\frac{.\cancel{82} \times B}{.\cancel{82}} = \frac{246}{.82}$$

$$B = 300$$

$$\begin{array}{r} 300. \\ .82\overline{)246.00} \\ \underline{246} \\ 0\ 0 \\ \underline{0} \\ 0 \end{array}$$

So, 82% of 300 is 246.

PROBLEM TYPE 3

Example 5 What percent of 84 is 16.8?

$\quad\quad\quad\quad\downarrow\quad\quad\ \downarrow\ \downarrow\ \downarrow$

$\quad\quad\quad R\quad\quad \times\ 84\ =\ 16.8$

$B = 84$
$A = 16.8$

$$R \times 84 = 16.8$$

$$\frac{R \times \cancel{84}}{\cancel{84}} = \frac{16.8}{84}$$

$$R = .2 = 20\%$$

$$\begin{array}{r} 0.2 \\ 84\overline{)16.8} \\ \underline{16\ 8} \\ 0 \end{array}$$

So, 20% of 84 is 16.8.

Example 6 ___% of 85 is 27.2.

$\quad\quad\ \downarrow\ \ \downarrow\ \downarrow\ \downarrow\ \downarrow$

$\quad\quad\ R\ \ \times\ 85\ =\ 27.2$

$$R \times 85 = 27.2$$

$$\frac{R \times \cancel{85}}{\cancel{85}} = \frac{27.2}{85}$$

$$R = .32 = 32\%$$

$$\begin{array}{r} 0.32 \\ 85\overline{)27.20} \\ \underline{25\ 5} \\ 1\ 70 \\ \underline{1\ 70} \\ 0 \end{array}$$

So, 32% of 85 is 27.2.

The multiplication or division shown in these examples can easily be done with a calculator. But you should always write the equation down first so you know that you are performing the correct operation and you know what quantity is unknown.

Find the unknown quantity.

1. 10% of 90 is _____ .

☐ **NOW WORK EXERCISES 1–3 IN THE MARGIN.**

When working with applications, the wording of the problem is unlikely to be just as shown in Examples 1–6. The following examples show some alternative wording, but each problem is still one of the three types.

Example 7 Find 65% of 42.

Solution

2. 50% of _____ is 73.

Here the Rate = 65% and the Base = 42. This is a Type 1 Problem:

$$65\% \text{ of } 42 \text{ is } \underline{\hspace{1cm}}.$$

$$.65 \times 42 = A$$
$$27.3 = A$$

$$\begin{array}{r} .65 \\ \times\ \ 42 \\ \hline 1.30 \\ 26.0 \\ \hline 27.30 \end{array}$$

3. _____% of 56 is 67.2.

Example 8 What percent of 125 gives a result of 31.25?

Solution

Here the Rate is unknown. The Base = 125 and Amount = 31.25. This is a Type 3 Problem:

$$\underline{\hspace{1cm}}\% \text{ of } 125 \text{ is } 31.25.$$

$$R \times 125 = 31.25$$

4. Find 33% of 180.

$$\frac{R \times \cancel{125}}{\cancel{125}} = \frac{31.25}{125}$$

$$R = .25 = 25\%$$

$$\begin{array}{r} 0.25 \\ 125\overline{)31.25} \\ \underline{25\ 0} \\ 6\ 25 \\ \underline{6\ 25} \\ 0 \end{array}$$

☐ **NOW WORK EXERCISES 4 AND 5 IN THE MARGIN.**

5. Find 150% of 60.

The following examples show how to use fraction equivalents for percents to simplify your work.

Example 9 What is 25% of 24?

$$A = \frac{1}{\cancel{4}} \times \overset{6}{\cancel{24}} = 6 \qquad \textbf{Type 1 Problem}$$
$$25\% = \frac{1}{4}$$

Example 10 Find 75% of 36.

$$A = \frac{3}{\cancel{4}} \times \overset{9}{\cancel{36}} = 27$$

Type 1 Problem

$$75\% = \frac{3}{4}$$

Example 11 What is the value of $33\frac{1}{3}\%$ of 72?

$$A = \frac{1}{\cancel{3}} \times \overset{24}{\cancel{72}} = 24$$

Type 1 Problem

$$33\frac{1}{3}\% = \frac{1}{3}$$

Example 12 $37\frac{1}{2}\%$ of what number is 300?

$$\frac{3}{8} \times B = 300$$

Type 2 Problem

$$37\tfrac{1}{2}\% = \frac{3}{8}$$

$$\frac{\cancel{8}}{\cancel{3}} \times \frac{\cancel{3}}{\cancel{8}} \times B = \frac{8}{\cancel{3}} \times \overset{100}{\cancel{300}}$$

$\dfrac{8}{3}$ is the reciprocal of $\dfrac{3}{8}$.

$$B = 800$$

Remember that a percent is just another form of a fraction and that, when you find a percent of a number, the amount will be:

(a) smaller than the number if the percent is less than 100%;

(b) greater than the number if the percent is more than 100%.

NAME _____ SECTION _____ DATE _____

EXERCISES **7.4** _____

Solve each problem for the unknown quantity. You may use a calculator.

1. 10% of 70 is _____ .

2. 5% of 62 is _____ .

3. 15% of 60 is _____ .

4. 25% of 72 is _____ .

5. 75% of 12 is _____ .

6. 60% of 30 is _____ .

7. 100% of 36 is _____ .

8. 80% of 50 is _____ .

9. 2% of _____ is 3.

10. 20% of _____ is 17.

Answers

1. _____

2. _____

3. _____

4. _____

5. _____

6. _____

7. _____

8. _____

9. _____

10. _____

Answers

11. _____

12. _____

13. _____

14. _____

15. _____

16. _____

17. _____

18. _____

19. _____

20. _____

21. _____

22. _____

11. 3% of _____ is 21.

12. 30% of _____ is 21.

13. 100% of _____ is 75.

14. 50% of _____ is 42.

15. 150% of _____ is 63.

16. 110% of _____ is 330.

17. _____% of 60 is 90.

18. _____% of 150 is 60.

19. _____% of 75 is 15.

20. _____% of 12 is 4.

21. _____% of 34 is 17.

22. _____% of 30 is 6.

NAME _____ SECTION _____ DATE _____

23. _____% of 48 is 16.   **24.** _____% of 100 is 35.

Answers

23. _____

24. _____

25. _____

Each of the following problems is one of the three types discussed written in a different way. Remember that *B* is the number you are finding the percent of.

26. _____

25. _____ is 50% of 25.   **26.** _____ is 31% of 76.

27. _____

28. _____

29. _____

27. 22 is 20% of _____.   **28.** 86 is 100% of _____.

30. _____

29. 18 is _____% of 10.   **30.** 15 is _____% of 10.

Answers

31. _____

32. _____

33. _____

34. _____

35. _____

36. _____

37. _____

38. _____

39. _____

40. _____

31. 24 is $33\frac{1}{3}$% of _____ .

32. 92.1 is 15% of _____ .

33. 119.6 is 23% of _____ .

34. 9.5 is 25% of _____ .

35. 36 is _____% of 18.

36. 60 is _____% of 40.

37. _____ is 96% of 17.

38. _____ is 84% of 32.

39. _____ is 18% of 325.

40. _____ is 28% of 460.

NAME _____ SECTION _____ DATE _____

41. Find 18% of 244.

42. Find 15.2% of 75.

41. _____

42. _____

43. _____

43. Find 120% of 60.

44. What percent of 32 is 8?

44. _____

45. _____

45. 100 is 125% of what number?

46. _____

47. _____

Use fractions to solve the following problems. Do the work mentally if you can after you have set up the related equation.

46. Find 50% of 32.

47. Find $66\frac{2}{3}$% of 60.

Answers

48. _____

49. _____

50. _____

51. _____

52. _____

53. _____

54. _____

55. _____

48. What is $12\frac{1}{2}\%$ of 80?

49. What is $62\frac{1}{2}\%$ of 16?

50. $33\frac{1}{3}\%$ of 75 is what number?

51. 25% of 150 is what number?

52. 75% of what number is 21?

53. 50% of what number is 35?

54. $37\frac{1}{2}\%$ of what number is 61.2?

55. 100% of what number is 76.3?

7.5

APPLICATIONS WITH PERCENT (DISCOUNT, COMMISSION, SALES TAX, AND OTHERS) (Calculators Recommended)

You should be able to operate with decimals, round off decimals, change percents to decimals, and set up and solve the three basic types of percent problems. Since this section is concerned with the application of skills already learned, the author recommends the use of a calculator. A calculator is not meant to replace necessary skills and understanding but to enhance these abilities by providing answers rapidly and accurately. You must know from your own experience, knowledge, and understanding what numbers to work with and what operations to use.

> GENERAL PLAN FOR SOLVING PERCENT PROBLEMS
>
> 1. Read the problem carefully.
> 2. Decide what is unknown (A, B, or R).
> 3. Write down the values you do know for A, B, or R. (You must know two of them.)
> 4. Set up and solve the equation $R \times B = A$.
> 5. Check to see that the answer is reasonable. Rework the problem if the answer does not make sense to you.

In Examples 1–4 you may do any of the multiplying, dividing, adding, or subtracting with a calculator. When solving money problems, round off answers to the next higher cent.

Example 1   A refrigerator that regularly sells for $975 is on sale at a 30% discount. (a) What is the amount of the discount? (b) What is the sale price?

Solution

There are two problems here: (a) find the discount, and (b) find the sale price.

(a) Find the discount.
   30% of $975 is _____ .

$$0.30 \times 975 = A$$
$$292.50 = A$$

$$\begin{array}{r} \$975 \\ \times \quad .30 \\ \hline \$292.50 \end{array}$$

Type 1 Problem

discount

The discount is $292.50.

(b) Find the sale price by subtracting the discount from the original price.

$$
\begin{array}{r}
\$975.00 \\
-\ \ 292.50 \\
\hline
\$682.50 \quad \text{sale price}
\end{array}
$$

The sale price is $682.50.

Example 2   If the sales tax rate is 6%, what would be the final cost of the refrigerator in Example 1?

Solution

(a) Find the amount of the sales tax.

6% of $682.50 is _____ .

$$
\begin{array}{ll}
0.06 \times 682.50 = A & \$682.50 \quad \text{sale price} \\
 40.95 = A & \underline{\times .06} \quad \text{tax rate (as a decimal)} \\
& \$40.9500 \quad \text{sales tax}
\end{array}
$$

(b) Add this tax to the sale price to get the final cost.

(You should be able to relate these examples to your own experience with buying a sale item at a discount, or paying sales tax on a purchase.)

$$
\begin{array}{r}
\$682.50 \quad \text{sale price} \\
+\ \ \ 40.95 \quad \text{sales tax} \\
\hline
\$723.45 \quad \text{final cost}
\end{array}
$$

The final cost of the refrigerator is $723.45.

Example 3   An auto dealer paid $7566 for a large order of a special part. This was not the original price. He received a 3% discount off the original price because he paid cash. What was the original price?

Solution

Do **not** take 3% of $7566. We know that if 3% of something is gone, then 97% remains (100% − 3% = 97%). That is, $7566 represents 97% of the original price.

97% of _____ is $7566.   **Type 2 Problem**

$$.97 \times B = 7566$$

$$\frac{.97 \times B}{.97} = \frac{7566}{.97}$$

$$B = 7800$$

$$
\begin{array}{r}
7800. \\
.97\,)\overline{7566.00} \\
\underline{679} \\
776 \\
\underline{776} \\
0\,0 \\
\underline{0} \\
00 \\
\underline{0} \\
0
\end{array}
$$

The original price was $7800.

If you want to check this result, take 3% of $7800 and subtract this amount from $7800. You should get $7566.

Example 4 A saleswoman earns a fixed salary of $900 a month plus a commission of 8% on whatever she sells after she has sold $6500 in merchandise. What did she earn the month she sold $11,800 in merchandise?

Solution

(a) First find the amount of her commission. She earns 8% of what she sells over $6500.

$$
\begin{array}{rl}
\$11,800 & \text{total amount she sold} \\
-6,500 & \text{amount on which she does \textbf{not} earn a commission} \\
\hline
\$5300 & \text{base of commission}
\end{array}
$$

(b) Now find 8% of this base.

8% of $5300 is _____ .

Type 1 Problem

$$
\begin{array}{rl}
0.08 \times 5300 = A & \quad \$5300 \quad \text{base of commission} \\
424 = A & \times \quad .08 \quad \text{rate of commission} \\
& \overline{\$424.00} \quad \text{commission}
\end{array}
$$

(c) Her monthly pay is her salary plus her commission.

$$
\begin{array}{rl}
\$900.00 & \text{fixed salary} \\
+424.00 & \text{commission} \\
\hline
\$1324.00 & \text{total pay for the month}
\end{array}
$$

She earned $1324.

Profit: The difference between selling price and cost

$$(\text{profit} = \text{selling price} - \text{cost})$$

Percent of profit: There are two approaches to this:

1. Percent of profit based on cost is the ratio of profit to cost:

$$\frac{\text{profit}}{\text{cost}} \quad \% \text{ profit based on cost}$$

2. Percent of profit based on selling price is the ratio of profit to selling price:

$$\frac{\text{profit}}{\text{selling price}} \quad \% \text{ profit based on selling price}$$

Example 5 A company manufactures light fixtures that cost $21 each to produce and are sold for $28 each. (a) What is the profit on each fixture? (b) What is the percent of profit based on cost? (c) What is the percent of profit based on selling price?

Solution

(a)
$$\begin{array}{rl} \$28 & \text{selling price} \\ -\ 21 & \text{cost} \\ \hline \$\ 7 & \text{profit} \end{array}$$

The profit on each light fixture is $7. Using ratios, then changing the fractions to percents,

(b) $\dfrac{\$7 \text{ profit}}{\$21 \text{ cost}} = \dfrac{1}{3} = 0.33\dfrac{1}{3} = 33\dfrac{1}{3}\%$ profit based on cost

(c) $\dfrac{\$7 \text{ profit}}{\$28 \text{ selling price}} = \dfrac{1}{4} = 0.25 = 25\%$ profit based on selling price

Percent of profit **based on cost** is higher than percent of profit **based on selling price.** The business community reports whichever percent serves its purpose better. Your responsibility as an investor or consumer is to know which percent is reported and what it means to you.

Completion Example 6 Women's coats were on sale for $250. This was a discount of $100 from the original selling price. (a) What was the original selling price? (b) What was the rate of discount? (c) If the coats cost the store owner $200, what was his percent of profit based on cost? (d) What was his percent of profit based on the actual selling price?

Solution

(a) Find the original selling price.

$$
\begin{array}{ll}
\$250 & \text{sale price} \\
+\ 100 & \text{discount} \\
\hline
\$350 & \text{original selling price}
\end{array}
$$

(b) Find the rate of discount.

_____ % of $350 is $100.

$$R \times 350 = 100$$

$$\frac{R \times \cancel{350}}{\cancel{350}} = \frac{100}{350}$$

$$R = \underline{\hspace{1cm}} = \underline{\hspace{1cm}} \%$$

First find the profit and then find the percent of profit.

$$
\begin{array}{ll}
\$\underline{\hspace{1.5cm}} & \text{actual selling price} \\
-\qquad 200 & \text{cost} \\
\hline
\$\underline{\hspace{1.5cm}} & \text{profit}
\end{array}
$$

(c) $\dfrac{\underline{\ } \text{ profit}}{\underline{\ } \text{ cost}} = \underline{\hspace{1cm}} \%$ profit based on cost

(d) $\dfrac{\underline{\ } \text{ profit}}{\underline{\ } \text{ selling price}} = \underline{\hspace{1cm}} \%$ profit based on actual selling price

---

Completion Example Answer

6. (b) 28.6% of 350 is 100.

$R = 0.286 = 28.6\%$

$$
\begin{array}{ll}
\$250 & \text{actual selling price} \\
-\ 200 & \text{cost} \\
\hline
\$\ 50 & \text{profit}
\end{array}
$$

(c) $\dfrac{\$50 \text{ profit}}{\$200 \text{ cost}} = \dfrac{1}{4} = 25\%$ profit based on cost

(d) $\dfrac{\$50 \text{ profit}}{\$250 \text{ selling price}} = \dfrac{1}{5} = 20\%$ profit based on actual selling price

NAME _____ SECTION _____ DATE _____

EXERCISES **7.5** _____

Some of the following problems involve several calculations. Write down the known information and the basic set-up to work each problem. Calculators are recommended for these exercises.

1. A realtor works on a 6% commission. What is his commission on a house he sold for $95,000?

2. A store owner received a 3% discount from the manufacturer when she bought $6500 worth of dresses. (a) What was the amount of the discount? (b) What did she pay for the dresses?

3. A sales clerk receives a monthly salary of $500 plus a commission of 6% on all sales over $2500. What did the clerk earn the month she sold $6000 in merchandise?

Answers

1. _____

2. (a) _____

(b) _____

3. _____

Answers

4. _____

5. _____

6. _____

4. If a salesman works on a 10% commission only, how much will he have to sell to earn $1800 in one month?

5. A computer programmer was told she would be given a bonus of 5% of any money her programs could save the company. How much would she have to save the company to earn a bonus of $600?

6. The property taxes on a house were $750. What was the tax rate if the house was valued at $25,000?

7. Towels were on sale at a discount of 30%. If the sale price was $3.01, what was the original price?

Answers

7. _____

8. _____

9. _____

8. A shoe salesman worked on a fixed salary of $300 per month plus a 5% commission. How much did the salesman make during the month in which he sold $4500 worth of shoes?

9. A student missed 3 problems on a mathematics test and received a grade of 85%. If all the problems were of equal value, how many problems were on the test?

Answers

10. _____

11. (a) _____

  (b) _____

12. (a) _____

  (b) _____

  (c) _____

10. In one season, a basketball player missed 15% of his free throws. How many free throws did he make if he attempted 180 free throws?

11. A basketball player made 120 of 300 shots she attempted. (a) What percent of her shots did she make? (b) What percent did she miss?

12. The discount on a fur coat was $150. This was a 20% discount. (a) What was the original selling price of the coat? (b) What was the sale price? (c) What was the total amount paid for the coat if a 6% sales tax was added to the sale price?

NAME _____ SECTION _____ DATE _____

13. If sales tax in a state is figured at 6%, (a) what is the tax on a purchase of $30.20? (b) What is the total amount paid for the purchase?

14. Golf clubs were marked on sale for $320. This was a discount of 20% off the original selling price. (a) What was the original selling price? The clubs cost the golf pro $240. (b) What was his profit? (c) What was his percent of profit based on cost? (d) What was his percent of profit based on the sale price?

15. The discount on men's suits was $50, and they were on sale for $200. (a) What was the original selling price? (b) What was the rate of discount? (c) If the suits cost the store owner $150 each, what was his percent of profit based on cost? (d) What was his percent of profit based on the sale price?

Answers

13. (a) _____

(b) _____

14. (a) _____

(b) _____

(c) _____

(d) _____

15. (a) _____

(b) _____

(c) _____

(d) _____

16. (a) _____

    (b) _____

    (c) _____

17. (a) _____

    (b) _____

18. (a) _____

    (b) _____

    (c) _____

16. The author of a book was told that she would have to cut the number of pages by 12% in order for the book to sell at a popular price and still show a profit. (a) What percent of the pages were in the final book? (b) If the book contained 220 pages in final form, how many pages were in it originally? (c) How many pages were cut?

17. In order to get more subscribers, a book club offered three books for the single price of $7.02. The total selling price was originally $17.55 for all three books. (a) What was the amount of the discount? (b) Based on the original selling price, what was the rate of discount on these three books?

18. The cost of a television set to a store owner was $350, and he sold the set for $490. (a) What was his profit? (b) What was his percent of profit based on cost? (c) What was his percent of profit based on selling price?

NAME _____ SECTION _____ DATE _____

19. A car dealer bought a used car for $1500. He marked up the price so he would make a profit of 25% based on his cost. (a) What was the selling price? (b) If the customer paid 8% of the selling price in taxes and fees, what did the customer pay for the car?

Answers

19. (a) _____

(b) _____

20. (a) _____

(b) _____

21. _____

20. A man weighed 200 pounds. He lost 20 pounds in three months. Then he gained back 20 pounds two months later. (a) What percent of his weight did he lose in the first three months? (b) What percent did he gain? The loss and the gain are the same, but the two percents are different. Why?

21. An auto supply store received a shipment of auto parts with the bill for $845.30. Some of the parts were not as ordered and they were returned immediately. The value of the parts returned was $175.50. The terms of the billing provided the store with a 2% discount if it paid cash (for the parts it kept) within two weeks. What did the store pay for the parts it kept if it paid cash within two weeks?

Answers

22. _____

22. You want to purchase a new home for $98,000. The bank will loan you 80% of the purchase price. How much cash do you need if loan fees are 2% of the loan and other fees total $850?

NAME _____ SECTION _____ DATE _____

CHAPTER **7** TEST _____

Answers

In Problems 1–5, change each number to a percent.

1. $\dfrac{101}{100}$ **2.** 0.003   **3.** 0.173

1. _____

2. _____

3. _____

4. $\dfrac{4}{25}$ **5.** $1\dfrac{1}{8}$

4. _____

5. _____

In Problems 6–8, change each percent to a decimal.

6. _____

6. 30%   **7.** 325%   **8.** $4\dfrac{1}{2}\%$

7. _____

8. _____

In Problems 9–11, change each percent to a fraction or mixed number. Reduce to lowest terms.

9. _____

9. 140%   **10.** $24\dfrac{1}{2}\%$ **11.** 5.6%

10. _____

11. _____

In Problems 12–17, solve each problem for the unknown quantity.

12. _____

12. 10% of 36 is _____ .   **13.** 13 is _____% of 26.

13. _____

Answers

14. _____

15. _____

16. _____

17. _____

18. _____

19. _____

20. _____

21. _____

14. 33 is 110% of _____ .   15. _____ is 42% of 349.

16. 15.225 is 75% of _____ .   17. 16.55 is _____ % of 50.

18. A customer received a 2% discount on the purchase of a refrigerator because she paid cash. If she paid $735, what was the amount of the discount (disregarding sales tax)?

19. A salesman earns $750 each month plus a commission of 9% of his sales over $3500 each month. How much did he earn for the month in which his sales were $10,000?

20. If 24-carat gold is pure gold, what percent gold is 14-carat gold (to the nearest tenth of a percent)?

21. In one year, Mr. James earned $15,000. He spent $4800 on rent, $5250 on food, and $1800 on taxes. What percent of his income did he spend on each of those items?

Consumer Applications

8

(Calculators Recommended)

A NOTE ON CALCULATORS

You are encouraged to use a calculator for all the problems in this chapter. A calculator will save you considerable time because many of the problems have several steps. Accuracy to three decimal places will be acceptable.

Be aware that some answers will differ slightly if you round off at different places in a calculation involving multiplication and/or division.

For example, to calculate $\$500 \times 0.15 \times \frac{1}{3}$, you might write

$$\$500 \times 0.15 \times \frac{1}{3} = \frac{500 \times 0.15}{3} = \frac{75}{3} = \$25$$

or, using a decimal, $\frac{1}{3} = 0.333$ (to three decimal places),

$$\$500 \times 0.15 \times \frac{1}{3} = 500 \times 0.15 \times 0.333 = \$24.98$$

$25 is the correct answer, but if calculators are allowed, **round-off errors** must also be allowed, and $24.98 will be accepted. Be sure you understand that slight errors can and do occur with the use of calculators and rounded-off decimals.

8.1

SIMPLE INTEREST _____

Interest is money paid for the use of money. The money that is invested or borrowed is called the **principal.** The **rate** is the **percent of interest** and is almost always stated as **an annual (yearly) rate.**

Interest is either paid or earned, depending on whether you are the borrower or the lender. In either case, the calculations involved are the same. The purpose of this section is to explain how interest is calculated. The interest rates vary from year to year and from one area of the world to another, but the concept of interest is the same everywhere.

There are two kinds of interest, **simple interest** and **compound interest.** Compound interest involves interest paid on interest and will be discussed in Section 8.2. Many loans are based on simple interest. Only one payment (including the principal and the interest) is made at the end of the term of the loan. No monthly payments are made.

> **Interest:** Money paid for the use of money
>
> **Simple interest:** Interest calculated at a yearly rate based only on the original amount of money used

> FORMULA FOR CALCULATING SIMPLE INTEREST
>
> $$I = P \times R \times T$$
>
> where
>
> I = Interest
>
> P = Principal, the amount invested or borrowed
>
> R = Rate, the percent stated as an annual (yearly) rate
>
> T = Time, in years or fraction of a year
>
> **Note:** We will use 360 days in one year (30 days in a month), a common practice in business and banking.

The formula can be written $I = P \cdot R \cdot T$ using raised dots to indicate multiplication in place of the $\times$ sign. The $\times$ sign is used in this text to avoid confusion with decimal points.

Example 1 You want to borrow $2000 at 14% interest for 1 year. How much interest would you pay?

Solution

$$I = P \times R \times T$$
$$P = \$2000$$
$$R = 14\% = 0.14$$
$$T = 1 \text{ year}$$
$$I = \$2000 \times 0.14 \times 1 = \$280$$

You would pay $280 interest.

Example 2 You decided you might need the $2000 for only 90 days. How much interest would you pay? (The annual interest rate would still be stated as 14%.)

Solution

$$I = P \times R \times T$$
$$P = \$2000$$
$$R = 14\% = 0.14$$
$$T = 90 \text{ days} = \frac{90}{360} \text{ year} = \frac{1}{4} \text{ year}$$
$$I = \$2000 \times 0.14 \times \frac{1}{4} = \frac{280}{4} = \$70$$

or
$$I = \$2000 \times 0.14 \times 0.25 = \$70$$

You would pay $70 interest if you borrowed the money for 90 days.

Example 3 John loaned $500 to a friend for 6 months at an interest rate of 12%. How much will his friend pay him at the end of the 6 months?

Solution

John will be paid interest plus the original principal. Find the interest and then add the principal.

$$I = P \times R \times T$$
$$P = \$500$$
$$R = 12\% = 0.12$$
$$T = 6 \text{ months} = \frac{6}{12} \text{ year} = \frac{1}{2} \text{ year}$$
$$I = \$500 \times 0.12 \times \frac{1}{2} = \frac{60}{2} = \$30$$

Principal + Interest = $500 + $30 = $530

John will be paid $530.

1. Pat borrowed $3000 at 16% for one year. How much interest did she pay?

Completion Example 4   Sylvia borrowed $2500 at 10% interest for 30 days. How much interest did she have to pay?

Solution

$$P = \$2500$$

$$R = 10\% = \underline{\hspace{2cm}}$$

$$T = 30 \text{ days} = \frac{30}{360} \text{ year} = \underline{\hspace{2cm}} \text{ year}$$

$$I = P \times R \times T$$

$$I = \underline{\hspace{2cm}} \times \underline{\hspace{2cm}} \times \underline{\hspace{2cm}} = \underline{\hspace{2cm}}$$

☐ **NOW WORK EXERCISES 1 AND 2 IN THE MARGIN.**

2. Stacey loaned her uncle $1500 at 10% interest for 9 months. How much interest did she earn?

If you know any 3 of the values in the formula

$$I = P \times R \times T$$

you can find the fourth value.

To find *P*: $P = \dfrac{I}{R \times T}$

To find *R*: $R = \dfrac{I}{P \times T}$

To find *T*: $T = \dfrac{I}{P \times R}$

Remember, *T* is a part of a year.

Example 5   How much (what principal) would you need to invest if you wanted to make $100 in interest in 30 days and your investment would return 16% interest?

Completion Example Answer

4. $R = 10\% = 0.10$

$T = \dfrac{1}{12}$ year

$I = \$2500 \times 0.10 \times \dfrac{1}{12} = \dfrac{250}{12} = \$20.833 = \$20.84$

She will be charged the next higher cent in interest.

Solution

Here the principal is unknown, while the interest ($100), the rate of interest (16%), and the time (30 days) are all known. Use the formula

$$P = \frac{I}{R \times T}$$

$$I = \$100$$

$$R = 0.16$$

$$T = \frac{30}{360} = \frac{1}{12}$$

$$P = \frac{100}{0.16 \times \frac{1}{12}} = \frac{100}{0.16} \times \frac{12}{1} = \frac{1200}{0.16} = \$7500$$

Or, changing to a decimal $\left(\frac{1}{12} = 0.083 \right)$,

$$P = \frac{100}{0.16 \times 0.083} = \frac{100}{0.01328} = \$7530.12$$

The correct answer is $7500.

Remember that rounded-off decimals such as 0.083 will give you round-off errors (sometimes acceptable). Such errors can be avoided by working with fractions.

Example 6 Joe wants to borrow $1500 at 15% and is willing to pay $300 in interest. How long can he keep the money?

Solution

Here time is unknown and the principal ($1500), interest ($300), and rate (15%) are all known. Use the formula

$$T = \frac{I}{P \times R}$$

$$I = \$300$$

$$P = \$1500$$

$$R = 0.15$$

$$T = \frac{300}{1500 \times 0.15} = \frac{300}{225} = 1\frac{1}{3} \text{ years} \quad \text{or 1 year 4 months}$$

Joe can borrow the money for $1\frac{1}{3}$ years.

☐ **NOW WORK EXERCISE 3 IN THE MARGIN.**

3. What interest rate would you be paying if you borrowed $1000 for 6 months and paid $60 in interest?

Answers to Marginal Exercises

1. $I = \$480$ **2.** $I = \$112.50$ **3.** $R = 0.12 = 12\%$

NAME _____   SECTION _____   DATE _____

EXERCISES **8.1** _____

1. What interest will be earned in one year on a savings account of $800 if the bank pays 6% interest?

1. _____

2. _____

3. _____

4. _____

2. If interest is paid at 10%, what will a principal of $600 earn in one year?

3. If a principal of $900 is invested for 90 days at a rate of 14%, how much interest will be earned?

4. A loan of $5000 is made at 11% for a period of 6 months. How much interest is paid?

Answers

5. —————————

6. —————————

7. —————————

8. —————————

9. —————————

5. If you borrow $750 for 30 days at 18%, how much interest will you pay?

6. How much interest is paid on a 60-day loan of $500 at 12%?

7. Find the simple interest paid on a savings account of $1800 for 120 days at 8%.

8. A savings account of $2300 is left for 90 days, drawing interest at a rate of 7%. How much interest is earned? What is the total amount in the account at the end of 90 days?

9. Every 6 months a stock pays a 10% dividend (interest on investment). What will be the earnings of $14,600 invested for 6 months?

NAME _____ SECTION _____ DATE _____

10. One thousand dollars worth of merchandise is charged at a local department store for 60 days at 18% interest. How much is owed at the end of 60 days?

11. You buy an oven on sale marked down from $500 to $450, but you don't pay the bill for 60 days and are charged interest at a rate of 18%. How much do you pay for the oven by waiting 60 days to pay? How much did you save by buying the oven on sale?

12. A friend borrows $500 from you for a period of 8 months and pays you interest at 6%. How much interest are you paid? Suppose you ask 8% instead. Then how much interest are you paid?

13. How much would you have to invest at 8% to earn interest of $500 in 60 days?

Answers

10. _____

11. _____

12. _____

13. _____

Answers

14. _____

15. _____

16. (a) _____

    (b) _____

    (c) _____

    (d) _____

14. How many days must you leave $1000 in a savings account at $5\frac{1}{2}\%$ to earn $11.00 interest?

15. What is the rate of interest charged if a loan of $2500 for 90 days is paid off with $2562.50?

16. Determine the missing item in each row.

Principal	Rate	Time	Interest
$ 400	16%	90 days	$ (a)
$ (b)	15%	120 days	$ 5.00
$ 560	12%	(c)	$ 5.60
$2700	(d)	40 days	$25.50

NAME _____   SECTION _____   DATE _____

17. Determine the missing item in each row.

Principal	Rate	Time	Interest
$ 500	18%	30 days	$ (a)
$ 500	18%	(b)	$15.00
$ 500	(c)	90 days	$22.50
$ (d)	18%	30 days	$ 1.50

18. If you have a savings account of $25,000 drawing interest at 8%, how much interest will you earn in 6 months? How long must you leave the money in the account to earn $1500?

19. You have accumulated $50,000, and you want to live on the interest each year. If you need $800 a month to live on, what interest rate must you earn on your $50,000? If you are earning $13\frac{1}{2}\%$, payable each month, what will your monthly interest be?

Answers

17. (a) _____

(b) _____

(c) _____

(d) _____

18. _____

19. _____

20. Your \$2500 savings account draws interest at $5\frac{1}{2}\%$. How many days will it take for you to earn \$68.75? If the interest rate is then raised to 6%, what will your money earn in the next 6 months?

21. A bank decides to loan \$5 million dollars to a contractor to build some houses. How much interest will the bank earn in one year if the interest rate is $14\frac{1}{4}\%$?

22. A credit card company has \$20 million dollars loaned to its customers at 18% interest. How much interest will it earn in one month?

23. A small airline company borrowed \$7.5 million dollars to buy some new airplanes. The loan rate was 15% and the airline paid \$562,500 in interest. What was the length of time of the loan?

24. A department store keeps \$15 million dollars in merchandise in stock. If the store pays interest at 9% on a bank loan for this stock, how much interest will the store pay in 3 months' time?

8.2

COMPOUND INTEREST

> **Compound interest:** Interest paid on interest

To calculate compound interest, we calculate the simple interest over each period of time **with a new principal for each calculation.** The new principal is the previous principal plus the earned interest.

> To calculate compound interest:
>
> 1. Using $I = P \times R \times T$, calculate the simple interest where T is period of time for compounding. For example,
>
> $$T = 1 \text{ for compounding annually;}$$
>
> $$T = \frac{1}{2} \text{ for compounding semiannually;}$$
>
> $$T = \frac{1}{4} \text{ for compounding quarterly;}$$
>
> $$T = \frac{1}{12} \text{ for compounding monthly;}$$
>
> $$T = \frac{1}{360} \text{ for compounding daily.}$$
>
> 2. Add this interest to the principal to create a new value for the principal.
> 3. Continue Steps 1 and 2 until the entire interest period is covered.

> **Note:** Most savings and loan associations and banks pay interest compounded daily. A computer or a set of tables is needed to do such interest calculations over a period of time. Since the purpose here is to teach the concept of compound interest, the problems will involve compounding annually, semiannually, quarterly, or monthly, but not daily.

Example 1 You deposit $1000 in an account that pays 8% interest compounded quarterly (every 3 months). How much interest would you earn in one year?

Solution

Calculate the simple interest four times with $T = \frac{1}{4}$. Add this interest to the previous principal before each calculation. You will be calculating interest paid on interest.

$$I = P \times R \times T$$

(a) $I = \$1000 \times 0.08 \times \dfrac{1}{4}$ using $P = \$1000$

$= 80 \times \dfrac{1}{4} = \$20$ interest

(b) $I = \$1020 \times 0.08 \times \dfrac{1}{4}$ using $P = \$1000 + 20 = \$1020$

$= 81.60 \times \dfrac{1}{4} = \$20.40$ interest

(c) $I = \$1040.40 \times 0.08 \times \dfrac{1}{4}$ using $P = \$1020 + 20.40 = \$1040.40$

$= 83.23 \times \dfrac{1}{4} = \$20.81$ interest

(d) $I = \$1061.21 \times 0.08 \times \dfrac{1}{4}$ using $P = \$1040.40 + \$20.81 = \$1061.21$

$= 84.90 \times \dfrac{1}{4} = \$21.23$ interest

$$
\begin{array}{r}
\$20.00 \\
20.40 \\
20.81 \\
+\ 21.23 \\
\hline
\$82.44
\end{array}
\quad \text{total interest}
$$

Example 2   In Example 1, how much more interest did you earn by having your interest compounded quarterly for one year than if it had just been calculated as simple interest for the year?

Solution

Simple interest for one year would be

$$I = \$1000 \times 0.08 \times 1 = \$80$$

$$
\begin{array}{rl}
\$82.44 & \text{compound interest} \\
-\ 80.00 & \text{simple interest} \\
\hline
\$\ 2.44 & \text{more by compounding quarterly}
\end{array}
$$

Example 3   \$5000 is deposited in a savings account, and interest is compounded semiannually at 10%. How much interest will be earned in one year?

Solution

There are two calculations with $T = \dfrac{1}{2}$ because interest is accumulated every six months.

(a) $I = \$5000 \times 0.10 \times \dfrac{1}{2}$ using $P = \$5000$

$= \$250$

(b) $I = \$5250 \times 0.10 \times \dfrac{1}{2}$ using $P = \$5000 + \$250 = \$5250$

$\qquad = \$262.50$

$$\begin{array}{r} \$250.00 \\ +\ 262.50 \\ \hline \$512.50 \end{array}$$ interest in one year

Example 4   If an account is compounded monthly at 12%, how much interest will \$2000 earn in three months?

Solution

Use $T = \dfrac{1}{12}$ and calculate the interest three times.

(a) $I = \$2000 \times 0.12 \times \dfrac{1}{12}$ using $P = \$2000$

$\qquad = \$20$

(b) $I = \$2020 \times 0.12 \times \dfrac{1}{12}$ using $P = \$2000 + \$20 = \$2020$

$\qquad = \$20.20$

(c) $I = \$2040.20 \times 0.12 \times \dfrac{1}{12}$ using $P = \$2020 + \$20.20 = \$2040.20$

$\qquad = \$20.41$

$$\begin{array}{r} \$20.00 \\ 20.20 \\ +\ 20.41 \\ \hline \$60.61 \end{array}$$ interest in three months

Example 5   Suppose your income is \$1000 per month (or \$12,000 per year) and you will receive a cost of living raise each year. If inflation is at 9% each year, in how many years will you be making \$24,000?

Solution

Compound annually at 9% until the principal (base salary) plus interest (raise) totals $24,000.

In table form (inflation at 9%),

Year	Base	+	Raise	=	Total
1	$12,000.00		$ 0		$12,000.00
2	12,000.00		1080.00		13,080.00
3	13,080.00		1177.20		14,257.20
4	14,257.20		1283.15		15,540.35
5	15,540.35		1398.64		16,938.99
6	16,938.99		1524.51		18,463.50
7	18,463.50		1661.72		20,125.22
8	20,125.22		1811.27		21,936.49
9	21,936.49		1974.29		23,910.78

In 9 years, you would have almost doubled your salary, but your relative purchasing power would be the same as it was 9 years before.

Completion Example 6 You have invested $5000 at 12% interest to be compounded quarterly. How much interest will you earn in nine months?

Solution

Use $T = \dfrac{1}{4}$ and calculate three times to cover nine months.

(a) $I = \$5000 \times 0.12 \times \dfrac{1}{4} = $ _____

(b) $I = $ _____ $\times 0.12 \times \dfrac{1}{4} = $ _____

(c) $I = $ _____ $\times 0.12 \times \dfrac{1}{4} = $ _____

_____ is the total interest earned in nine months.

Completion Example Answer

6. (a) $I = \$5000 \times 0.12 \times \dfrac{1}{4} = \$150$

(b) $I = \$5150 \times 0.12 \times \dfrac{1}{4} = \$154.50$

(c) $I = \$5304.50 \times 0.12 \times \dfrac{1}{4} = \$159.14$

$463.64 is the total interest earned in nine months.

NAME _____   SECTION _____   DATE _____

EXERCISES **8.2** _____

Answers

1. (a) If a bank compounds interest quarterly at 12% on a certificate of deposit, what will your $13,000 deposit be worth in 6 months? (b) In one year?

1. (a) _____

   (b) _____

2. _____

2. A check for $9000 is deposited in a savings and loan account and interest is compounded monthly at 10%. What will the balance of the account be in 6 months?

3. (a) _____

   (b) _____

   (c) _____

3. (a) If an account is compounded quarterly at a rate of 6%, how much interest will be earned on $5000 in one year? (b) What will be the total amount in the account? (c) How much more interest is earned in the first year because compounding is done quarterly rather than annually?

4. (a) _____

   (b) _____

4. (a) How much interest will be earned in 4 years on a savings account of $4000 compounded semiannually at 6%? (b) What will be the balance of the account?

Answers

5. (a) _____

   (b) _____

6. _____

7. _____

8. (a) _____

   (b) _____

5. (a) How much interest will be earned on a savings account of $3000 in 2 years if interest is compounded annually at 6%? (b) If interest is compounded semiannually?

6. If interest is calculated at 10% and compounded quarterly, what will the value of $15,000 be in $\frac{3}{4}$ year?

7. You borrowed $4000 and agreed to make equal payments of $1000 each plus interest over the next 4 years. Interest was at a rate of 8% based only on what you owed. How much interest did you pay?

8. You borrow $400 from a friend and agree to pay back the principal and interest at 12% in 60 days. But you can only pay $100 plus interest. Your friend agrees to let you pay $100 plus interest every 60 days until the debt is repaid. (a) How long will it take you to repay your friend? (b) How much interest will you pay?

NAME _____ SECTION _____ DATE _____

9. Calculate the interest earned in six months on $10,000 compounded monthly at 14%.

9. _____

10. _____

11. _____

10. Calculate the interest you will pay on a loan of $6500 for one year if the interest is compounded every 3 months at 18% and you make no monthly payments.

11. If a savings account of $10,000 draws interest at a rate of 10% compounded annually, how many years will it take for the $10,000 to double in value? (Fill in the table to look like the one in Example 5 on page 388.)

Table of Interest Compounded at 10%

Year	Principal	+	Interest	=	Total

12. _____

13. (a) _____

    (b) _____

    (c) _____

    (d) _____

    (e) _____

    (f) _____

12. If a savings account of $10,000 draws interest at a rate of 12% compounded annually, how many years will it take for the $10,000 to double in value? (Fill in the table to look like the one in Example 5 on page 388.)

Table of Interest Compounded at 12%

Year	Principal	+	Interest	=	Total

13. Suppose $5000 has been invested in a savings account that is compounded annually at 8%. A table showing data for six years is given. Answer the following questions using this table.

Table of Interest Compounded at 8%

Year	Principal	+	Interest	=	Total
1	$5000.00		$400.00		$5400.00
2	5400.00		432.00		5832.00
3	5832.00		466.56		6298.56
4	6298.56		503.89		6802.45
5	6802.45		544.20		7346.65
6	7346.65		587.74		7934.39

(a) How much interest was earned during the third year?

(b) How much interest was earned during the first two years?

(c) How much interest was earned during the third and fourth years together?

(d) Why is the answer to part (c) larger than the answer to part (b)?

(e) How much interest accumulated in six years?

(f) How much interest would be earned during the seventh year?

8.3

BALANCING A CHECKING ACCOUNT

Many people do not balance their checking accounts simply because they were never told how. Others trust the bank's calculations over their own, but this is a poor practice because, for a variety of reasons, computers do make errors. Also, because the bank's statement comes only once a month, you should know your current balance so your account will not be overdrawn. Overdrawn accounts pay a penalty and can lead to a bad credit rating.

Like many applications with mathematics, balancing a checking account follows a pattern of steps. The procedure and related ideas are given in the following lists.

The bank or the savings and loan company sends you a statement of your checking account each month. This statement contains a record of

1. the beginning balance;

2. a list of all checks paid by the bank;

3. a list of deposits you made;

4. any interest paid to you; (Some checking accounts do pay interest.)

5. any service charge by the bank;

6. the closing balance.

Your checkbook register contains a record of

1. all the checks you have written;

2. the current balance.

Your current balance will not agree with the closing balance on the bank statement because

1. you have not recorded the interest;

2. you have not recorded the service charge;

3. the bank does not have a record of all the checks you have written. (Several checks will be *outstanding* because they have not been received yet for payment by the bank by the date given on the bank statement.)

To **balance your checking account** (sometimes called **reconciling** the bank statement with your checkbook register):

1. Go through your checkbook register and the bank statement and put a check mark (✔) by each check paid and deposit recorded on the bank statement.

2. In your checkbook register,
   (a) add any interest paid to your current balance;
   (b) subtract any service charge from this balance.
   This number is is your **true balance.**

3. Find the total of all the outstanding checks (no ✔ mark) in your checkbook register (checks not yet received by the bank).

4. On a reconciliation sheet,
   (a) enter the balance from the bank statement;
   (b) add any deposits you have made that are not recorded (no ✔ mark) on the bank statement;
   (c) subtract the total of your outstanding checks (found in Step 3). This number should agree with your **true balance.** Now your checking account is balanced.

If your checking account does not balance,

1. go over your arithmetic in the balancing procedure.

2. go over your arithmetic check by check in your checkbook register.

3. make an appointment with bank personnel to find the reasons for any other errors.

Example 1 The true balance is found by following the steps on the reconciliation sheet. Work through the steps yourself so you will understand the process.

Check No.	Date	Transaction Description	Payment (−)	(✓)	Deposit (+)	Balance
			Balance brought forward			505.21
152	1-3	U-Auto Lease (car lease)	198.60	✓		−198.60 / 306.61
153	1-10	ABC Power Co. (electric bill)	75.00			−75.00 / 231.61
154	1-10	Fly-By-Nite Airway (plane ticket)	112.00	✓		−112.00 / 119.61
	1-15	Deposit ( )		✓	500.00	+500.00 / 619.61
155	1-15	Safe-T Drugs (medicine)	5.60	✓		− 5.60 / 614.01
156	1-17	Milady Togs (dress)	49.80	✓		− 49.80 / 564.21
157	2-03	U-Auto Lease (car lease)	198.60			−198.60 / 365.61
	1-31	Service Charge ( )	4.00	✓		− 4.00 / 361.61
		( )				
		( )				
		True Balance				361.61

YOUR CHECKBOOK REGISTER

BANK STATEMENT
Checking Account Activity

Transaction Description	Amount	(✓)	Running Balance	Date
Beginning Balance		✓	505.21	1-01
Check #152	198.60	✓	306.61	1-06
Check #154	112.00	✓	194.61	1-14
Deposit	500.00	✓	694.61	1-15
Check #155	5.60	✓	689.01	1-18
Check #156	49.80	✓	639.21	1-20
Service Charge	4.00	✓	635.21	1-31
Ending Balance			635.21	

RECONCILIATION SHEET

A. First, mark ✓ beside each check and deposit listed in both your checkbook register and on the bank statement.

B. Second, in your checkbook register, add any interest paid and subtract any service charge listed on the bank statement.

C. Third, find the total of all outstanding checks.

Outstanding Checks	
No.	Amount
153	75.00
157	198.60
Total	273.60

Statement Balance	635.21
Add deposits not credited	+ ——
Total	635.21
Subtract total amount of checks outstanding	−273.60
True Balance	361.61

NAME _____ SECTION _____ DATE _____

EXERCISES **8.3** _____

For each of the following problems, you are given a copy of a checkbook register, the corresponding bank statement, and a reconciliation sheet. You are to find the true balance of the account on both the checkbook register and on the reconciliation sheet as shown in Example 1. Follow the directions on the reconciliation sheet.

1.

		YOUR CHECKBOOK REGISTER				
Check No.	Date	Transaction Description	Payment (−)	(✓)	Deposit (+)	Balance
		Balance brought forward				-0-
	7-15	Deposit ( )			700.00	+700.00 / 700.00
1	7-15	Quiet Town Apt. (rent/deposit)	520.00			-520.00 / 180.00
2	7-15	Pa Bell Telephone (phone installation)	32.16			-32.16 / 147.84
3	7-15	XYZ Power Co. (gas/elect. hook up)	46.49			-46.49 / 101.35
4	7-16	Foodway Stores (groceries)	51.90			-51.90 / 49.45
	7-20	Deposit ( )			350.00	+350.00 / 399.45
5	7-23	Comfy Furniture (sofa, chair)	300.50			-300.50 / 98.95
	8-1	Deposit ( )			350.00	+350.00 / 448.95
	7-31	Interest ( )				___
	7-31	Service Charge ( )				___
		True Balance				

BANK STATEMENT
Checking Account Activity

Transaction Description	Amount	(✓)	Running Balance	Date
Beginning Balance			0.00	7-15
Deposit	700.00		700.00	7-15
Check #1	520.00		180.00	7-16
Check #4	51.90		128.10	7-17
Check #2	32.16		95.94	7-18
Check #3	46.49		49.45	7-18
Deposit	350.00		399.45	7-20
Interest	2.50		401.95	7-31
Service Charge	2.00		399.95	7-31
Ending Balance			399.95	

RECONCILIATION SHEET

A. First, mark ✓ beside each check and deposit listed in both your checkbook register and on the bank statement.

B. Second, in your checkbook register, add any interest paid and subtract any service charge listed on the bank statement.

C. Third, find the total of all outstanding checks.

Outstanding Checks			
No.	Amount		
		Statement Balance	_____
		Add deposits not credited	+
		Total	_____
		Subtract total amount of checks outstanding	−
Total	_____	True Balance	_____

2.

YOUR CHECKBOOK REGISTER						
Check No.	Date	Transaction Description	Payment (−)	(✓)	Deposit (+)	Balance
					Balance brought forward	1610.39
1234	12-7	Pearl City (pearl ring)	524.00			−524.00 1086.39
1235	12-7	Comp-U-Tate (home computer)	801.60			−801.60 284.79
1236	12-8	Sportz Hutz (skis)	206.25			−206.25 78.54
1237	12-8	Guild Card Shop (christmas cards)	25.50			−25.50 53.04
	12-10	Deposit ( )			1000.00	+1000.00 1053.04
1238	12-14	Toys-R-We (stuffed panda)	80.41			−80.41
1239	12-24	Meat Markette (turkey)	18.39			−18.39
1240	12-24	Poodle Shoppe (pedigreed puppy)	300.00			−300.00
1241	12-31	Homey Sav. & Loan (mortgage payment)	600.00			−600.00
		( )	0			—
					True Balance	

BANK STATEMENT Checking Account Activity					
Transaction Description	Amount	(✓)	Running Balance	Date
Beginning Balance			1610.39	12-01
Check #1234	524.00		1086.39	12-08
Check #1236	206.25		880.14	12-09
Check #1237	25.50		854.64	12-09
Deposit	1000.00		1854.64	12-10
Check #1235	801.60		1053.04	12-11
Check #1238	80.41		972.63	12-15
Check #1239	18.39		954.24	12-27
Check #1240	300.00		654.24	12-28
Ending Balance			654.24	

RECONCILIATION SHEET

A. First, mark ✓ beside each check and deposit listed in both your checkbook register and on the bank statement.

B. Second, in your checkbook register, add any interest paid and subtract any service charge listed on the bank statement.

C. Third, find the total of all outstanding checks.

Outstanding Checks	
No.	Amount
Total	_____

Statement Balance	_____
Add deposits not credited	+ _____
Total	_____
Subtract total amount of checks outstanding	− _____
True Balance	_____

3.

		YOUR CHECKBOOK REGISTER				
Check No.	Date	Transaction Description	Payment (−)	(✔)	Deposit (+)	Balance
					Balance brought forward	756.14
271	6-15	Parts, Parts, Parts (spark plugs)	12.72			−12.72 743.42
272	6-24	Firerock Tire Co. (2 tires)	121.40			____
273	6-30	Gus' Gas Station (tune-up)	75.68			____
	7-1	Deposit ( )			250.00	____
274	7-1	Prudent Ins Co. (car insurance)	300.00			____
	6-30	Service Charge ( )				____
		( )				____
		( )				____
		( )				____
		( )				____
		True Balance				

BANK STATEMENT Checking Account Activity				
Transaction Description	Amount	(✔)	Running Balance	Date
Beginning Balance			756.14	6-01
Check #271	12.72		743.42	6-16
Check #272	121.40		622.02	6-26
Service Charge	1.00		621.02	6-30
Ending Balance			621.02	

RECONCILIATION SHEET

A. First, mark ✔ beside each check and deposit listed in both your checkbook register and on the bank statement.

B. Second, in your checkbook register, add any interest paid and subtract any service charge listed on the bank statement.

C. Third, find the total of all outstanding checks.

Outstanding Checks			
No.	Amount		
		Statement Balance	_____
		Add deposits not credited	+ _____
		Total	_____
		Subtract total amount of checks outstanding	− _____
	Total _____	True Balance	_____

4.

YOUR CHECKBOOK REGISTER						
Check No.	Date	Transaction Description	Payment (−)	(✔)	Deposit (+)	Balance
				Balance brought forward		12.14
419	1-2	Postmaster (stamps)	10.00			−10.00 2.14
	1-3	Deposit ( )			525.50	——
420	1-17	E. Hamel, DDS (dental check-up)	63.50			——
421	1-26	Cash ( )	100.00			——
422	1-31	Up-Top Apartments (rent)	350.00			——
	1-31	Service Charge ( )				——
		( )				——
		( )				——
		( )				——
		( )				——
					True Balance	

BANK STATEMENT
Checking Account Activity

Transaction Description	Amount	(✔)	Running Balance	Date
Beginning Balance			12.14	1-01
Deposit	525.50		537.64	1-03
Check #419	10.00		527.64	1-03
Check #420	63.50		464.14	1-20
Check #421	100.00		364.14	1-26
Service Charge	2.00		362.14	1-31
Ending Balance			362.14	

RECONCILIATION SHEET

A. First, mark ✔ beside each check and deposit listed in both your checkbook register and on the bank statement.

B. Second, in your checkbook register, add any interest paid and subtract any service charge listed on the bank statement.

C. Third, find the total of all outstanding checks.

Outstanding Checks			
No.	Amount		
		Statement Balance	———
		Add deposits not credited	+ ———
		Total	———
		Subtract total amount of checks outstanding	− ———
Total	———	True Balance	———

NAME _____ SECTION _____ DATE _____

5.

YOUR CHECKBOOK REGISTER						
Check No.	Date	Transaction Description	Payment (−)	(✔)	Deposit (+)	Balance
					Balance brought forward	**967.22**
772	4-13	C. P. Hay (accountant)	85.00			_____
	4-14	Deposit ( )			1200.00	_____
773	4-14	E. Z. Pharmacy (aspirin)	4.71			_____
774	4-15	I. R. S. (income tax)	2000.00			_____
775	4-30	Heavy Finance Co. (loan payment)	52.50			_____
	5-1	Deposit ( )			600.00	_____
	4-30	Interest ( )				_____
	4-30	Service Charge ( )				_____
		( )				_____
		( )				_____
					True Balance	

BANK STATEMENT Checking Account Activity				
Transaction Description	Amount	(✔)	Running Balance	Date
Beginning Balance			967.22	4-01
Deposit	1200.00		2167.22	4-14
Check #772	85.00		2082.22	4-15
Check #773	4.71		2077.51	4-15
Interest	2.82		2080.33	4-30
Service Charge	4.00		2076.33	4-30
Ending Balance			2076.33	

RECONCILIATION SHEET

A. First, mark ✔ beside each check and deposit listed in both your checkbook register and on the bank statement.

B. Second, in your checkbook register, add any interest paid and subtract any service charge listed on the bank statement.

C. Third, find the total of all outstanding checks.

Outstanding Checks			
No.	Amount		
		Statement Balance	_____
		Add deposits not credited +	_____
		Total	_____
		Subtract total amount of checks outstanding −	_____
Total	_____	True Balance	_____

6.

		YOUR CHECKBOOK REGISTER				
Check No.	Date	Transaction Description	Payment (−)	(✓)	Deposit (+)	Balance
					Balance brought forward	1403.49
86	9-1	Now Stationers (school supplies)	17.12			———
87	9-2	Young-At-Heart (clothes)	192.50			———
88	9-4	H.S.U. Bookstore (books)	56.28			———
89	9-7	Regent's Office (tuition)	380.00			———
90	9-7	Off-Campus Apts. (rent)	240.00			———
91	9-27	State Telephone Co. (phone bill)	24.62			———
92	9-30	Up-N-Up Foods (groceries)	47.80			———
	9-30	Service Charge ( )				———
		( )				———
		( )				———
					True Balance	

BANK STATEMENT Checking Account Activity				
Transaction Description	Amount	(✓)	Running Balance	Date
Beginning Balance			1403.49	9-01
Check #86	17.12		1386.37	9-02
Check #87	192.50		1193.87	9-05
Check #88	56.28		1137.59	9-05
Check #90	240.00		897.59	9-10
Service Charge	4.00		893.59	9-30
Ending Balance			893.59	

RECONCILIATION SHEET

A. First, mark ✓ beside each check and deposit listed in both your checkbook register and on the bank statement.

B. Second, in your checkbook register, add any interest paid and subtract any service charge listed on the bank statement.

C. Third, find the total of all outstanding checks.

Outstanding Checks			
No.	Amount		
		Statement Balance	———
		Add deposits not credited	+ ———
		Total	———
		Subtract total amount of checks outstanding	− ———
Total	———	True Balance	———

NAME _____ SECTION _____ DATE _____

7.

YOUR CHECKBOOK REGISTER						
Check No.	Date	Transaction Description	Payment (−)	(✓)	Deposit (+)	Balance
					Balance brought forward	602.82
14	6-20	Aisle Bridal (flowers)	402.40			——
	6-22	Deposit ( )			1000.00	——
15	6-24	Tuxedo Junction (tux)	155.65			——
16	6-28	D. Lohengrin (organist)	55.00			——
17	6-28	D-Lux Limo (limo rental)	125.00			——
18	6-30	C.C. Catering (food caterer)	700.00			——
19	7-1	Luv-Lee Stationers (thank-you cards)	35.20			——
	6-30	Service Charge ( )				——
		( )				——
		( )				——
					True Balance	

BANK STATEMENT				
Checking Account Activity				
Transaction Description	Amount	(✓)	Running Balance	Date
Beginning Balance			602.82	6-01
Deposit	1000.00		1602.82	6-22
Check #14	402.40		1200.42	6-22
Check #15	155.65		1044.77	6-26
Service Charge	1.00		1043.77	6-30
Ending Balance			1043.77	

RECONCILIATION SHEET

A. First, mark ✓ beside each check and deposit listed in both your checkbook register and on the bank statement.

B. Second, in your checkbook register, add any interest paid and subtract any service charge listed on the bank statement.

C. Third, find the total of all outstanding checks.

Outstanding Checks			
No.	Amount	Statement Balance	——
		Add deposits not credited	+
		Total	═══
		Subtract total amount of checks outstanding	−
	———		
Total	———	True Balance	═══

8.

YOUR CHECKBOOK REGISTER						
Check No.	Date	Transaction Description	Payment (−)	(✓)	Deposit (+)	Balance
			Balance brought forward			**278.32**
326	8-12	J. J. Jones (birthday check)	40.00			—
	8-15	Deposit ( )			500.00	—
327	8-15	N. O. Payne, M.D. (physical exam)	260.00			—
328	8-15	Local Waterworks (water/trash)	27.40			—
	8-22	Deposit ( )			400.12	—
329	8-27	Time-O-Life (magazine subscrip)	12.50			—
330	8-29	Biggar Mortgage (house payment)	750.00			—
	8-31	Interest ( )				—
	8-31	Service Charge ( )				—
		( )				—
				True Balance		

BANK STATEMENT Checking Account Activity					
Transaction Description	Amount	(✓)	Running Balance	Date
Beginning Balance			278.32	8-01
Deposit	500.00		778.32	8-15
Check #326	40.00		738.32	8-15
Check #327	260.00		478.32	8-17
Deposit	400.12		878.44	8-22
Check #328	27.40		851.04	8-22
Interest	1.82		852.86	8-31
Service Charge	4.00		848.86	8-31
Ending Balance			848.86	

RECONCILIATION SHEET

A. First, mark ✓ beside each check and deposit listed in both your checkbook register and on the bank statement.

B. Second, in your checkbook register, add any interest paid and subtract any service charge listed on the bank statement.

C. Third, find the total of all outstanding checks.

Outstanding Checks	
No.	Amount
Total	_____

Statement Balance	_____
Add deposits not credited	+ _____
Total	_____
Subtract total amount of checks outstanding	− _____
True Balance	_____

NAME _____ SECTION _____ DATE _____

9.

YOUR CHECKBOOK REGISTER						
Check No.	Date	Transaction Description	Payment (−)	(✔)	Deposit (+)	Balance
					Balance brought forward	147.02
203	2-3	Food Stoppe (groceries)	26.90			———
204	2-8	Ekhon Oil (gasoline bill)	71.45			———
205	2-14	Rose's Roses (flowers)	25.00			———
206	2-14	I. M. R. U. (alumni dues)	20.00			———
	2-15	Deposit ( )			600.00	———
207	2-26	SRO (theater tickets)	52.50			———
208	2-28	MPG Mtg. (house payment)	500.00			———
	2-28	Service Charge ( )				———
		( )				———
		( )				———
					True Balance	

BANK STATEMENT Checking Account Activity				
Transaction Description	Amount	(✔)	Running Balance	Date
Beginning Balance			147.02	2-01
Check #203	26.90		120.12	2-04
Check #204	71.45		48.67	2-14
Deposit	600.00		648.67	2-15
Check #205	25.00		623.67	2-15
Service Charge	3.00		620.67	2-28
Ending Balance			620.67	

RECONCILIATION SHEET

A. First, mark ✔ beside each check and deposit listed in both your checkbook register and on the bank statement.

B. Second, in your checkbook register, add any interest paid and subtract any service charge listed on the bank statement.

C. Third, find the total of all outstanding checks.

Outstanding Checks			
No.	Amount		
		Statement Balance	———
		Add deposits not credited	+ ———
		Total	———
		Subtract total amount of checks outstanding	− ———
Total	———	True Balance	———

10.

		YOUR CHECKBOOK REGISTER				
Check No.	Date	Transaction Description	Payment (−)	(✓)	Deposit (+)	Balance
			Balance brought forward			4071.82
996	10-1	Red-E Credit (loan payment)	200.75			___
997	10-10	United Ways (charity donation)	25.00			___
998	10-21	Mac-Intosh Farms (barrel of apples)	42.20			___
999	10-26	Fun Haus (costume rental)	35.00			___
1000	10-28	Yum Yum Shoppe (Halloween candy)	12.14			___
1001	10-29	B-Sharp, Inc. (piano tuners)	20.00			___
1002	10-30	Food-2-Gooo! (party platter)	78.50			___
1003	10-31	Cash ( )	300.00			___
1004	10-31	Principal S&L (house payment)	1250.60			___
	10-31	Interest ( )				___
				True Balance		

BANK STATEMENT Checking Account Activity				
Transaction Description	Amount	(✓)	Running Balance	Date
Beginning Balance			4071.82	10-01
Check #996	200.75		3871.07	10-05
Check #998	42.20		3828.87	10-23
Check #999	35.00		3793.87	10-30
Check #1003	300.00		3493.87	10-31
Interest	16.29		3510.16	10-31
Ending Balance			3510.16	

RECONCILIATION SHEET

A. First, mark ✓ beside each check and deposit listed in both your checkbook register and on the bank statement.

B. Second, in your checkbook register, add any interest paid and subtract any service charge listed on the bank statement.

C. Third, find the total of all outstanding checks.

Outstanding Checks			Statement Balance	___
No.	Amount		Add deposits not credited	+ ___
			Total	___
			Subtract total amount of checks outstanding	− ___
	Total ___		True Balance	___

8.4

BUYING AND OWNING A HOME

> EXPENSES IN BUYING A HOME
>
> **Purchase price:** The selling price (what you have agreed to pay for the home)
>
> **Down payment:** Cash you pay to the seller (usually 20% to 30% of the purchase price)
>
> **Mortgage loan (1st Trust Deed):** Loan to you by bank or savings and loan (difference between purchase price and down payment)
>
> **Mortgage fee:** Loan fee charged by lender (usually 1% to 3% of the mortgage loan)
>
> **Fire insurance:** Insurance against loss of your home by fire
>
> **Recording fees:** Local and state fees for recording you as the legal owner
>
> **Taxes:** Property taxes must be prepaid
>
> **Legal fees:** Fees charged by a lawyer or escrow company for completing all forms in a legal manner

Example 1   You buy a home for $150,000. Your down payment is 20% of the selling price and the mortgage fee is 2% of the new mortgage. You also have to pay $500 for fire insurance, $350 for taxes, $50 for recording fees, and $310 for legal fees. What is the amount of your mortgage? How much cash must you provide to complete the purchase?

Solution

(a) Find the down payment (20% of $150,000).

$$
\begin{array}{rl}
\$150,000. & \text{selling price} \\
\times \quad 0.20 & \\
\hline
\$30,000.00 & \text{down payment}
\end{array}
$$

(b) Find the mortgage and mortgage fee (2% of mortgage).

$$
\begin{array}{rl}
\$150,000 & \text{selling price} \\
- \quad 30,000 & \text{down payment} \\
\hline
\$120,000 & \text{mortgage}
\end{array}
\qquad
\begin{array}{rl}
\$120,000. & \text{mortgage} \\
\times \quad 0.02 & \\
\hline
\$2400.00 & \text{mortgage fee}
\end{array}
$$

(c) Add all cash expenses.

$$
\begin{array}{rl}
\$30,000 & \text{down payment} \\
2,400 & \text{mortgage fee (loan fee)} \\
500 & \text{fire insurance} \\
350 & \text{taxes} \\
50 & \text{recording fees} \\
+ \quad 310 & \text{legal fees} \\
\hline
\$33,610 & \text{cash to complete purchase}
\end{array}
$$

The mortgage will be $120,000 and you will need $33,610 in cash.

EXPENSES IN OWNING A HOME

Monthly mortgage payment: Payment to mortgage holder includes both principal and interest

Property taxes: May be paid monthly, semi-annually, or annually

Homeowner's insurance: Can be included with your fire insurance; includes insurance against theft and liability

Utilities: Monthly payments for gas, electricity, and water

Maintenance: Repairs, yardwork, painting, and so on

Example 2   What were your total expenses for your home the month you paid $1250 on your mortgage loan, $100 in taxes, $50 for fire insurance, $200 for all utilities, and $75 for maintenance? If your salary was $2500, what percentage of your salary was spent on your home?

Solution

(a) Find your total expenses.

$1250	loan payment
100	taxes
50	fire insurance
200	utilities
+ 75	maintenance
$1675	total spent on home

(b) Calculate the percent of your salary spent on your home.

$$\frac{1675}{2500} = 0.67 = 67\%$$

Note: This percentage is very high. You still have to buy food and clothes and pay income tax. Possibly you should get a raise.

NAME _____ SECTION _____ DATE _____

EXERCISES **8.4** _____

Answers

1. A home is sold for $162,500. The buyer has to make a down payment of 25% of the selling price, pay a loan fee of 2% of the mortgage, $200 for fire insurance, $50 for recording fees, $580 for taxes, and $570 for legal fees. (a) What is the amount of the down payment? (b) What is the amount of the mortgage? (c) How much cash does the buyer need to complete the purchase?

1. (a) _____

   (b) _____

   (c) _____

2. (a) _____

   (b) _____

2. The purchase price on a home is $98,000 and the buyer makes a down payment of $9,800 (10% of the selling price) and pays a loan fee of $2\frac{1}{2}\%$ of the new mortgage. He also pays $250 for legal fees, $320 for taxes, and $425 for fire insurance. (The seller agreed to pay all recording fees.) (a) What is the amount of the loan fee? (b) How much does the buyer owe to complete the purchase?

Answers

3. (a) _____

   (b) _____

4. (a) _____

   (b) _____

   (c) _____

   (d) _____

5. (a) _____

   (b) _____

   (c) _____

   (d) _____

3. You bought a home for $85,000 and made a down payment of $17,000. If you paid a mortgage fee (loan fee) of $1360, (a) what percent of the first trust deed was this fee? (b) You also paid $250 for fire insurance, $35 for recording fees, $170 for taxes, and $195 for legal fees. How much cash did you need to complete the purchase?

4. A house is sold for $125,000 and the buyer makes a 30% down payment and is charged a 1% loan fee on the first trust deed. The buyer is also charged $220 for recording fees, $345 for legal fees, $450 for taxes, and $520 for fire insurance. (a) How much is the down payment? (b) How much is the first trust deed? (c) How much is the loan fee? (d) What amount of cash does the buyer need?

5. A condominium was sold for $96,000. The buyer made a down payment of 25% of the selling price and paid a loan fee of $1\frac{1}{2}$% of the amount of the mortgage. She also paid $420 for fire insurance, $35 for recording fees, and $380 for taxes. (The seller paid all legal fees.) (a) What was the amount of the down payment? (b) What was the amount of the mortgage? (c) How much was the loan fee? (d) How much cash did she need to complete the purchase?

6. In March, Ms. Smith made a mortgage payment of $625, and paid $25 for taxes, $10 for water, $35 for electricity, $40 for gas, and $55 for a plumber's bill. (a) What were her home expenses in March? (b) If her income was $1975, what percent of her income did she spend on her home?

Answers

6. (a) _____

(b) _____

7. (a) _____

(b) _____

7. During July, the Johnsons made a loan payment of $875, paid taxes of $90, $85 for utilities, $70 for fire insurance, and $285 for a painter and other repairs. (a) How much did the Johnsons spend on their home in July? (b) If the combined income of Mr. and Mrs. Johnson was $4620, what percent of their income did they spend on their home?

8. (a) _____

    (b) _____

9. (a) _____

    (b) _____

10. _____

8. In one month Sam paid $150 for home repairs, $390 for his home loan, $60 for utilities, $20 for fire insurance, and $40 for taxes. (a) What were his home expenses? (b) What percent of his $2000 income did he spend on his home?

9. Your income for one month was $1800. (a) If you paid a mortgage payment of $578, taxes of $25, a water bill of $15, an electric bill of $35, a gas bill of $45, and a fire insurance premium of $40, how much did you spend on your home? (b) What percent of your income was this?

10. Your income was $2800 per month and you figured you could afford to spend 30% of this each month on a home. What mortgage payment could you make if you estimated taxes at $60 per month, utilities at $80 per month, fire insurance at $65 per month, and repairs at 7% of your income per month?

8.5
BUYING AND OWNING A CAR

EXPENSES IN BUYING A CAR

Purchase price: The selling price agreed on by the owner and the buyer

Sales tax: A fixed percent (approximately 6%) that varies from state to state

License fee: Fixed by the state, often based on the value and type of the car

Example 1   You are going to buy a new car for $8500. The bank will loan you 70% of the purchase price. But you must pay a 6% sales tax and a $150 license fee. How much cash do you need to buy the car?

Solution

(a) Calculate the down payment (30% of $8500 since the bank will loan 70%).

$$\begin{array}{r} \$8500 \\ \times \quad 0.30 \\ \hline \$2550.00 \end{array} \quad \text{down payment}$$

(b) Calculate the sales tax (6% of $8500).

$$\begin{array}{r} \$8500 \\ \times \quad 0.06 \\ \hline \$510.00 \end{array} \quad \text{sales tax}$$

(c) Add all the cash expenses.

$$\begin{array}{r} \$2550.00 \quad \text{down payment} \\ 510.00 \quad \text{sales tax} \\ + \quad 150.00 \quad \text{license fee} \\ \hline \$3210.00 \quad \text{total cash needed} \end{array}$$

You need $3210 cash to buy the car.

EXPENSES IN OWNING A CAR

Monthly payments: Payments made if you borrowed money to buy the car

Insurance: Auto insurance can cover a variety of situations (liability, collision, towing, theft, and so on)

Operating costs: Basic items such as gasoline, oil, tires, tune-ups

Repairs: Replacing worn or damaged parts

Example 2 In one month, Joe's car expenses were as follows: loan payment, $350; insurance, $80; gasoline, $100; oil and filter, $22; new headlight, $32. What was the total of his car expenses that month? What percent of his expenses was the loan payment?

Solution

(a) Find his total expenses.

$$
\begin{array}{rl}
\$350 & \text{loan payment} \\
80 & \text{insurance} \\
100 & \text{gasoline} \\
22 & \text{oil and filter} \\
+\ \ 32 & \text{headlight} \\
\hline
\$584 & \text{total expenses}
\end{array}
$$

(b) Find the percent spent on the loan payment.

$$\frac{350}{584} \approx 0.5993 \approx 60\%$$

Joe spent $584 on his car and the loan payment was approximately 60% of his expenses.

EXERCISES **8.5** _____

1. To buy a used car for $6800, you must pay a 6% sales tax, a 15% down payment, and a license fee of $120. How much cash do you need to buy the car?

1. _____

2. _____

3. _____

2. John wants to buy a new convertible for $25,000. His credit union will loan him 70% of the selling price. The license fee is $250. If he has to pay a 6% sales tax, what amount of cash does he need to buy the car?

4. _____

5. _____

3. How much cash would you need to buy a car for $18,000 if the sales tax is calculated at 6%, the license fee is $200, and the loan company will let you borrow 75% of the selling price?

4. A used car is priced at $4500. Your old car is worth $800 on a trade-in. The sales tax is figured at 6% of the selling price and the license fee is $80. If the savings and loan company will loan you $3000, how much cash do you need?

5. Your old car is worth $1500 if you trade it in for a new car priced at $11,200. Sales tax is 6% and the license fee is $225. If the bank will loan you 80% of the purchase price, how much cash do you need to buy the new car?

Answers

6. (a) _____

(b) _____

7. _____

8. (a) _____

(b) _____

9. _____

10. (a) _____

(b) _____

(c) _____

6. (a) If your car expenses in one month were $325 for the loan payment, $35 for insurance, $120 for gasoline, and $145 for two new tires, what were your total car expenses? (b) If your income was $2000, what percent of your income was used for car expenses?

7. Sue decided her old car needed painting. If the paint job was priced at $650, including repairing some dents, and she figured that driving her car cost an average of 24¢ per mile, including gas, oil, and insurance, what were her car expenses that month if she drove 800 miles?

8. (a) If you own your car, but it needs a new transmission for $1200 (installed), what are your car expenses that month if you also spent $65 for insurance, $75 for gas, $15 for oil and filter, and $65 for a tune-up? (b) If you took $1000 from your savings account to help pay for the transmission, what percent of your income of $1600 was used for the remainder of your car expenses?

9. Betty decided she would like to have a new car but she could not afford the car if expenses averaged more than 20% of her monthly income. If she figured the expenses would average $300 for a loan payment, $70 for insurance, $65 for gas, $8 for oil, $10 for tire wear, and $40 for general repairs, could she afford the car with a monthly income of $2100?

10. Suppose you owned your car, your monthly income was $1800, and you figured you could spend 15% of this to operate a car. (a) What would you be able to spend on gas if insurance cost $85, oil and filter $20, and you estimated $60 per month for other expenses? (b) How many gallons of gas could you buy if gas cost $1.25 per gallon? (c) How many miles could you drive if your car averaged 19 miles per gallon?

NAME _____ SECTION _____ DATE _____

CHAPTER **8** TEST _____

1. How much interest will be earned in one year on a savings account of $1500 if the bank pays $6\frac{1}{2}$% interest?

1. _____

2. _____

3. _____

2. If the interest earned on a savings account in 9 months at 7% interest is $42, what is the original amount in the savings account?

3. If you have $3000 in your savings account earning 5.5% interest, how long will it take for you to earn $330 interest?

4. If an investment of $7500 earns $131.25 in 45 days, what is the rate of earnings?

5. $1000 is deposited in a certificate account earning 12% compounded quarterly. How much will be in the account at the end of the first year (to the nearest dollar)?

6. Referring to Problem 5, how much less interest would be earned on the same amount in the same time at 12% simple interest?

7. Jeremy is buying a car for $7500. Sales tax is 6% and his license and transfer fees total $213. If he receives a $400 trade-in allowance and a $200 factory rebate, how much will his new car cost?

NAME _____ SECTION _____ DATE _____

8. Referring to Problem 7, if Jeremy finances 80% of the price of his car, what will the down payment be?

8. _____

9. _____

10. _____

9. The purchase of a home is $120,000. The buyer makes a down payment of 20% of the purchase price and pays a loan fee of 1% of the mortgage. (a) What is the amount of the down payment? (b) What is the amount of the loan fee?

10. Referring to Problem 9, if the buyer pays $50 for recording fees, $450 for taxes, and $350 for legal fees, how much cash does the buyer need to complete the purchase?

11. Find the true balance of the account on both the checkbook register and on the reconciliation sheet.

Check No.	Date	Transaction Description	Payment (−)	(✔)	Deposit (+)	Balance
			Balance brought forward			1201.40
203	2-3	Food Mart (groceries)	49.12			———
204	2-8	Enco Oil (gasoline bill)	82.50			———
205	2-14	Top Drawer (blouse)	25.00			———
206	2-14	I.O. Dyne (medicine)	20.00			———
	2-15	Deposit ( )			800.00	———
207	2-26	Stop 'n Shop (groceries)	52.50			———
208	2-28	Mayne Mtg. (house payment)	900.00			———
	2-28	Service Charge ( )				———
		( )				———
		( )				———
					True Balance	

YOUR CHECKBOOK REGISTER

BANK STATEMENT
Checking Account Activity

Transaction Description	Amount	(✔)	Running Balance	Date
Beginning Balance			1201.40	2-01
Check #203	49.12		1152.28	2-04
Check #204	82.50		1069.78	2-14
Deposit	800.00		1869.78	2-15
Check #205	25.00		1844.78	2-15
Service Charge	3.00		1841.78	2-28
Ending Balance			1841.78	

RECONCILIATION SHEET

A. First, mark ✔ beside each check and deposit listed in both your checkbook register and on the bank statement.

B. Second, in your checkbook register, add any interest paid and subtract any service charge listed on the bank statement.

C. Third, find the total of all outstanding checks.

Outstanding Checks			
No.	Amount		
		Statement Balance	———
		Add deposits not credited	+
		Total	———
		Subtract total amount of checks outstanding	− ———
Total ———		True Balance	———

NAME _____ SECTION _____ DATE _____

12. In one month, Jan paid $200 for repairs, $665 for her home loan, $80 for utilities, $60 for fire insurance, and $85 for taxes. (a) What were her home expenses? (b) What percent of her $3000 monthly income did she spend on her home?

Answers

12. (a) _____

    (b) _____

13. _____

13. A home was sold for $150,000. The buyer made a down payment of 20% of the selling price and paid a loan fee of 2% of the amount of the mortgage. He also paid $520 for fire insurance, $800 for taxes, and $350 for other fees. How much cash did he need to complete the purchase?

Statistics

9.1
MEAN, MEDIAN, AND RANGE

The study of statistics is basically the study of numerical characteristics (or measures) indicating some type of information about large amounts of numerical values. In this section, we will discuss only three such measures (or statistics): **mean, median,** and **range.** Other measures that you might study or read about are:

standard deviation (indicates how "spread out" data is)

z-scores (used to compare relative positions on different tests)

correlation coefficients (used, for example, to measure any correlation between the amount of schooling you have and the amount of your lifetime savings).

An understanding of the following terms is necessary for working with the problems in this section.

> TERMS USED IN THE STUDY OF STATISTICS
>
> **Data:** Value(s) measuring some characteristic of interest (We will consider only numerical data.)
>
> **Statistic:** A single number describing some characteristic of the data
>
> **Mean:** The arithmetic average of the data (Add all the data and divide by the number of data items.)
>
> **Median:** The middle data item (Arrange the data in numerical order and pick out the middle item.)
>
> **Range:** The difference between the largest and smallest data items

In Examples 1–3, the questions are related to the following two groups (or sets) of data.

GROUP A: Annual Income for 8 Families

$18,000; $12,000; $15,000; $17,000; $35,000; $70,000; $15,000; $20,000

GROUP B: Grade Point Average (GPA) for 11 Students

2.0; 2.0; 1.9; 3.1; 3.5; 2.9; 2.5; 3.6; 2.0; 2.4; 3.4

Example 1   Find the mean income for the families in Group A.

Solution

Find the sum of the 8 incomes and divide by 8.

```
  $ 18,000              $25,250
    12,000          8)202,000
    15,000             16
    17,000             42
    35,000             40
    70,000              2 0
    15,000              1 6
  +  20,000             40
   $202,000             40
                        00
                         0
                         0
```

You may want to use a calculator to do this arithmetic. The mean annual income is $25,250.

To find the median:

1. Arrange the data in numerical order.
2. If there is an **even** number of items, the median is the average of the two middle items.
3. If there is an **odd** number of items, the median is the middle item.

Example 2 Find the median income for Group A and the median GPA for Group B.

Solution

Arrange both sets of data in order.

GROUP A

$12,000; $15,000; $15,000; $17,000; $18,000; $20,000; $35,000; $70,000

GROUP B

1.9; 2.0; 2.0; 2.0; 2.4; 2.5; 2.9; 3.1; 3.4; 3.5; 3.6

For Group A, the median is the average of the 4th and 5th items because there are 8 (an even number) of items.

$$\text{median} = \frac{\$17,000 + \$18,000}{2} = \frac{35,000}{2} = \$17,500$$

For Group B, the median is the 6th item because, with 11 (an odd number) of items, the sixth item is the middle item.

$$\text{median} = 2.5$$

Example 3 Find the range for both Group A and Group B.

Solution

The range is the difference between the largest and smallest items of data:

Group A: range = $70,000 − $12,000 = $58,000

Group B: range = 3.6 − 1.9 = 1.7

Commentary

The mean and median are commonly used statistics. Many people feel that the mean (or arithmetic average) is relied on too much in reporting central tendencies for data such as income, housing costs, and taxes where a few very high items can **distort** the picture of a central tendency. As you can see in the Group A data, the median of $17,500 is probably more representative of the data than the mean of $25,250. This is because the one high income of $70,000 raises the mean considerably.

When you read an article in a magazine or newspaper that reports means or medians, you should now have a better understanding of the implications of these statistics.

Example 4   On an English exam, two students scored 95 points each, five students scored 86 points each, one student scored 82 points, another student scored 78 points, and six students scored 75 points each. What is the class mean?

Solution

There are 15 students in the class. Rather than add all 15 scores directly, we do some multiplying first.

$$
\begin{array}{ccccc}
95 & 86 & 82 & 78 & 75 \\
\times\ 2 & \times\ 5 & \times\ 1 & \times\ 1 & \times\ 6 \\
\hline
190 & 430 & 82 & 78 & 450
\end{array}
$$

Now we add these five products and divide the sum by 15 (not by 5) because the sum represents 15 scores.

$$
\begin{array}{r}
190 \\
430 \\
82 \\
78 \\
+\ 450 \\
\hline
1230
\end{array}
\qquad
\begin{array}{r}
82 \quad \text{class mean} \\
15\overline{)1230} \\
\underline{120} \\
30 \\
\underline{30} \\
0
\end{array}
$$

The class mean is 82 points.

Completion Example 5 If you had scores of 82, 75, and 72 on three mathematics tests, what is the lowest score you could get on the final exam and still get a "B" grade? (To get a "B" you must average between 80 and 89.)

Solution

The lowest average you could have and still get a "B" for the course is 80.

(a) If four exams are to average 80, then the total number of points on these exams must be

$$
\begin{array}{r}
80 \\
\times\ 4 \\
\hline
\end{array}
$$
_____ total points

(b) On the first three exams you have accumulated

$$
\begin{array}{r}
82 \\
75 \\
+72 \\
\hline
\end{array}
$$
_____ points

(c) To average 80 on your four exams, you need

$$
\begin{array}{r}
320 \\
- \\
\hline
\end{array}
$$
_____ points on the final exam

(d) Thus, you need at least a score of _____ on your final exam to get a "B" grade for the course.

Completion Example Answer

5. (a) 320   (b) 229   (c) $\begin{array}{r} 320 \\ -229 \\ \hline 91 \end{array}$ (d) 91

NAME _____ SECTION _____ DATE _____

EXERCISES **9.1** _____

For Exercises 1–12, find (a) the mean, (b) the median, and (c) the range of the given data.

1. Ten mathematics students had the following scores on a final exam:

   75, 83, 93, 65, 85,
   85, 88, 90, 55, 71

2. Joe did the following number of sit-ups each morning for a week:

   25, 52, 48, 42, 38, 58, 52

3. Fifteen college students reported the following hours of sleep the night before an exam:

   4, 6, 6, 7, 6.5, 6.5, 7.5, 8.5
   5, 6, 4.5, 5.5, 9, 3, 8

1. (a) _____

   (b) _____

   (c) _____

2. (a) _____

   (b) _____

   (c) _____

3. (a) _____

   (b) _____

   (c) _____

Answers

4. (a) _____

    (b) _____

    (c) _____

5. (a) _____

    (b) _____

    (c) _____

6. (a) _____

    (b) _____

    (c) _____

7. (a) _____

    (b) _____

    (c) _____

4. The local high school basketball team scored the following points per game during their 20-game season:

85, 60, 62, 70, 75, 52, 88, 50, 80, 72,

90, 85, 85, 93, 70, 75, 68, 73, 65, 82

5. Stacey went to six different repair shops to get the following estimates to repair her car. (The accident was not her fault; her car was parked at the time.)

$425, $525, $325, $300, $500, $325

6. Mike kept track of his golf scores for twelve rounds of eighteen holes each. His scores were:

85, 90, 82, 85, 87, 80,

78, 82, 88, 82, 86, 81

7. The local weather station recorded the following daily high temperatures for one month:

75, 76, 76, 78, 85, 82, 85, 88, 90, 90,

88, 95, 96, 92, 88, 88, 80, 80, 78, 80,

78, 76, 77, 75, 75, 74, 70, 70, 72, 73

8. The Big City fire department reported the following mileage for tires used on their nine fire trucks:

14,000; 14,000; 11,000; 15,000;
9,000; 14,000; 12,000; 10,000; 9,000

Answers

8. (a) _____

(b) _____

(c) _____

9. The city planning department issued the following numbers of building permits over a three-week period (fifteen working days):

17, 19, 18, 35, 30, 29, 23, 14,
18, 16, 20, 18, 18, 25, 30

9. (a) _____

(b) _____

(c) _____

10. Police radar measured the following speeds of 35 cars on one street:

28, 24, 22, 38, 40, 25, 24, 35, 25,
23, 22, 50, 31, 37, 45, 28, 30, 30,
30, 25, 35, 32, 45, 52, 24, 26, 18,
20, 30, 32, 33, 48, 58, 30, 25

10. (a) _____

(b) _____

(c) _____

11. (a) _____

(b) _____

11. On a one-day fishing trip, Mr. and Mrs. Johnson recorded the following lengths of fish they caught (measured in inches):

14.3; 13.6; 10.5; 15.5; 20.1;
10.9; 12.4; 25.0; 30.2; 32.5

(c) _____

Answers

12. (a) _____

    (b) _____

    (c) _____

13. _____

14. _____

15. _____

12. A machine puts out parts measured in thickness to the nearest hundredth of an inch. One hundred parts were measured and the results are tallied in the following chart:

Thickness measured	0.80	0.83	0.84	0.85	0.87
Number of parts	22	41	14	20	3

13. On an exam in history, two students scored 95 each, six students scored 90 each, three students scored 80 each, and one student scored 50. What was the class average?

14. Nick bought shares in the stock market from two companies. He paid $450 for nine shares in one company and $690 for eleven shares in a second company. What did he pay as an average price per share for the twenty shares?

15. Three families, each with two children, had incomes of $18,942 each. Two families, each with four children, had incomes of $20,512 each. Four families, each with two children, had incomes of $31,111 each. One family had no children and an income of $22,026. What was the average income per family?

9.2
CIRCLE GRAPHS

Graphs are pictures of numerical information. Graphs appear almost daily in newspapers and magazines and frequently in textbooks and corporate reports. Communication with well-drawn graphs can be accurate, effective, and fast. Most computers can be programmed to draw graphs, and anyone whose work involves a computer in any way will probably be expected to understand graphs.

> All graphs should:
>
> 1. Be clearly labeled.   2. Be easy to read.   3. Have titles.

In this section we will discuss circle graphs and their uses.

> **Circle graph:** A circle separated into pie-shaped regions called **sectors,** each sector representing a portion (or percent) of the whole

Circle graphs are particularly helpful in illustrating and understanding percents and relative amounts (or ratios).

Example 1 The circle graph shown in Figure 9.1 represents the various sources of income for a city government with a total income of $100,000.000. (a) What is the city's largest source of income? (b) What percent of income comes from Goods and Services? (c) What is the ratio of income from Taxes to the total income?

SOURCES OF CITY REVENUES

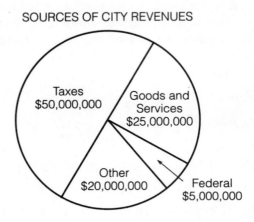

FIGURE 9.1 Circle Graph

Solution

(a) The largest source of income is Taxes.

(b) To find percent, we set up a ratio, reduce and then divïde:

$$\frac{25,000,000 \text{ Goods and Services}}{100,000,000 \text{ Total Income}} = \frac{1}{4}$$

$$\frac{1}{4} = 25\%$$

25% of the city's revenue comes from Goods and Services.

(c) The ratio of income from Taxes to the total revenue is:

$$\frac{50,000,000 \text{ Taxes}}{100,000,000 \text{ Total}} = \frac{1}{2}$$

Completion Example 2 Figure 9.2 shows percents budgeted for various items in a home for one year. Suppose a single person has an annual income of $15,000. Using Figure 9.2, calculate how much will be allocated to each item indicated in the graph.

Solution

ITEM		AMOUNT
Housing	$0.25 \times \$15,000 =$	$ 3750.00
Food	$0.20 \times \$15,000 =$	3000.00
Taxes	$0.05 \times \$15,000 =$	750.00
Clothing	$0.07 \times \$15,000 =$	1050.00
Savings	$0.10 \times \$15,000 =$	_____
Education	$0.15 \times \$15,000 =$	_____
Entertainment	_____ =	_____
Transportation and Maintenance	_____ =	_____

What is the total of all the amounts? _____

HOME BUDGET FOR 1 YEAR

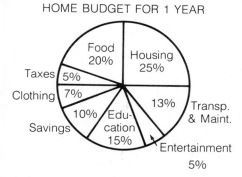

FIGURE 9.2 Circle Graph

Completion Example Answer

2. Savings $1500.00
   Education $2250.00
   Entertainment $0.05 \times 15,000 = \$750.00$
   Transportation and Maintenance $0.13 \times 15,000 = \$1950.00$
   Total of all amounts is 100% of $15,000 which is $15,000.

NAME _____ SECTION _____ DATE _____

EXERCISES **9.2** _____

Answers

1. The school budget shown is based on a total budget of $12,500,000. (a) What amount will be spent on each category? (b) How much more will be spent on teachers' salaries than on administration salaries? (c) What percent will be spent on items other than salaries? (d) How much will be spent on items other than salaries?

1. (a) _____

_____

_____

_____

_____

_____

_____

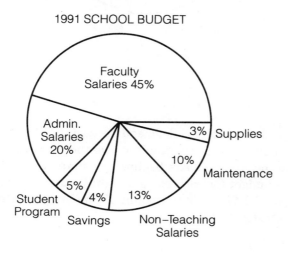

1991 SCHOOL BUDGET

(b) _____

(c) _____

(d) _____

2. Station XYZ is off the air from 2 A.M. to 6 A.M., so there are only 20 hours of daily programming. Sports are not shown here because they are considered special events. In the 20-hour period shown, how much time (in minutes) is devoted to each category?

2. (a) _____

(b) _____

(c) _____

(d) _____

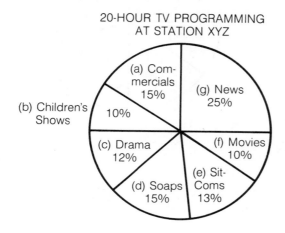

20-HOUR TV PROGRAMMING
AT STATION XYZ

(e) _____

(f) _____

(g) _____

Answers

3. (a) _____

   (b) _____

   (c) _____

4. (a) _____

   (b) _____

   (c) _____

3. Sally's monthly car expenses are shown in the circle graph. (a) What were her total car expenses for the month? (b) What percent of her expenses did she spend on each category? (c) What was the ratio of her expenses on insurance to expenses on gas?

MONTHLY CAR EXPENSES

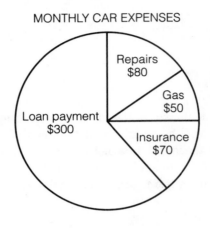

4. Mike just graduated from college and decided that he should try to live within a budget. The circle graph shows the categories he chose and the percents he allowed. His beginning salary was $24,000. (a) How much did he budget for each category? (b) What category was smallest in his budget? (c) What total amount did he budget for personal items (savings, clothing, and fun)?

MIKE'S BUDGET

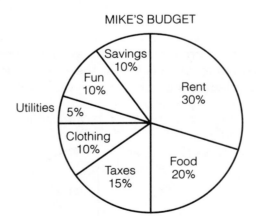

9.3

BAR GRAPHS AND LINE GRAPHS _____

Bar graphs and line graphs are particularly useful in indicating tendencies (or trends) over a period of time. In this section, we will be concerned with learning to read graphs of these types and doing some related calculations.

> All graphs should:
> 1. Be clearly labeled.   2. Be easy to read.   3. Have titles.

Completion Example 1   Figure 9.3 shows a bar graph. Note that the scale on the left and the data on the bottom (months) are clearly labeled and the graph itself has a title. The following questions can be easily answered by looking at the graph.

(a) What were the sales in January? $100,000

(b) During what month were sales lowest? <u>March</u>

(c) During what months were sales highest? <u>February and June</u>

(d) What were the sales during each of the highest sales months?

_____

(e) What were the sales in April? _____

The following questions will take some calculations after reading the graph.

(f) What was the amount of decrease in sales between February and March?

$$
\begin{array}{ll}
\$150,000 & \text{February sales} \\
-\ \ \ 50,000 & \text{March sales} \\
\hline
\$100,000 & \text{decrease in sales}
\end{array}
$$

(g) What was the percent of decrease?

$$
\frac{100,000}{150,000} \begin{array}{l} \text{decrease} \\ \text{February sales} \end{array} = \frac{2}{3} = 0.666
$$

$$
= 67\% \text{ decrease (approximately)}
$$

FIGURE 9.3 Bar Graph

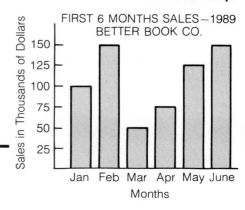

Completion Example Answer

1. (d) $150,000
   (e) $75,000

Completion Example 2 Figure 9.4 is a line graph that shows the relationships between daily high and low temperatures. From the graph you can see that temperatures tended to rise during the week but fell sharply on Saturday.

What was the lowest high temperature? _____66°_____

On what day did this occur? _____Sunday_____

What was the highest low temperature? _____70°_____

On what day did this occur? _____Friday_____

Find the average difference between the daily high and low temperatures for the week shown.

Solution

First find the differences, then average these differences.

Sunday   66 − 60 = _____6°_____

Monday   70 − 62 = _____8°_____

Tuesday   76 − 66 = _____10°_____

Wednesday 72 − 66 = _____

Thursday   80 − 68 = _____

Friday   80 − 70 = _____

Saturday   74 − 62 = _____

                                          _____ total of differences

                                          _____ average difference (approximately)
                                        7)‾‾‾‾‾‾‾‾

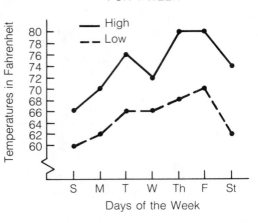

HIGH & LOW TEMPERATURES
FOR 1 WEEK

— High
-- Low

Temperatures in Fahrenheit

80
78
76
74
72
70
68
66
64
62
60

S M T W Th F St

Days of the Week

(Note: The jagged line ⟩ on the left-hand scale indicates the numbers 0–59 are missing.)

FIGURE 9.4 Line Graph

Completion Example Answer

2. Sunday   6°
   Monday   8°
   Tuesday   10°
   Wednesday   6°
   Thursday   12°
   Friday   10°
   Saturday   12°
              ‾‾‾‾
              64°   total of differences

        9.1°   average difference
      7)64.0
        63
        ‾‾‾
         1 0
           9
         ‾‾‾
           1

NAME _____ SECTION _____ DATE _____

EXERCISES **9.3** _____

Answer the questions related to each of the graphs in Exercises 1–8. Some questions can be answered directly from the graphs, while others may require some calculations.

1. (a) Which field of study has the largest number of declared majors?
(b) Which field of study has the smallest number of declared majors?
(c) How many declared majors are indicated in the entire graph?
(d) What percent are computer science majors?

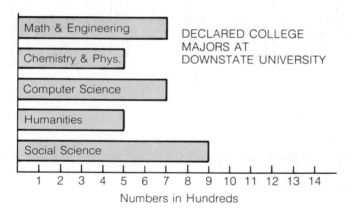

2. (a) What was the total number of vehicles that crossed this intersection in the 2 weeks? (b) Which day averaged the highest number? (c) What was the average on that day? (d) What percent of the total intersection traffic was counted on Sundays? (e) On Mondays?

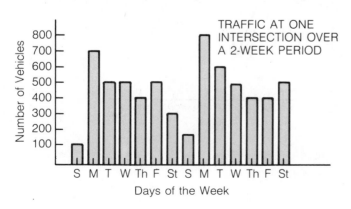

Answers

1. (a) _____

   (b) _____

   (c) _____

   (d) _____

2. (a) _____

   (b) _____

   (c) _____

   (d) _____

   (e) _____

3. Assume that all five students (Al, Bob, Ron, Sue, Ann) had graduated with comparable grades from the same high school. (a) What would be a major difficulty in putting the two graphs shown here together on one graph? (b) Who worked the most hours per week? (c) Who had the lowest college GPA? (d) If Bob spent 30 hours per week studying for his classes, what percent of his total work week (part-time work plus study time) did he spend studying? (e) Ann also spent 30 hours per week studying. What percent of her work week did she spend studying? (f) Does there seem to be any relationship between college GPA and job hours?

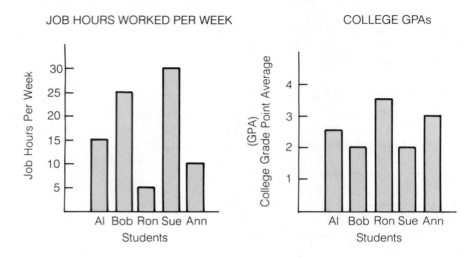

4. (a) In what year was the rainfall least? (b) What was the most rainfall in a year? (c) In what year did the most rainfall occur? (d) What was the average rainfall over the six-year period?

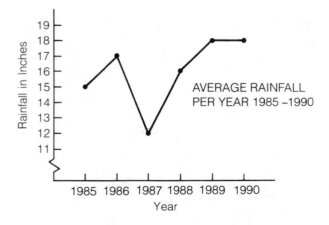

NAME ———————————————— SECTION ————————— DATE ————————————

5. (a) If on Monday morning you had 100 shares in each of the three stocks shown (oil, steel, wheat), and you held the stock all week, in which stock would you have lost money? (b) How much? (c) In which stock would you have gained money? (d) How much? (e) In which stock could you have made the most money if you had sold at the best time? (f) How much? (g) In which stock could you have lost the most money if you had sold at the worst time? (h) How much?

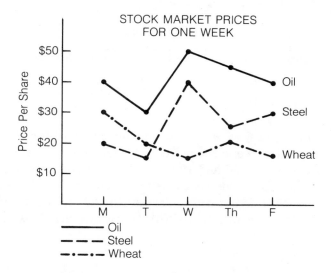

Answers

5. (a) ——————————————

(b) ——————————————

(c) ——————————————

(d) ——————————————

(e) ——————————————

(f) ——————————————

(g) ——————————————

(h) ——————————————

6. (a) During what month or months in 1990 were interest rates highest? (b) When were interest rates lowest? (c) What was the average of the interest rates over the entire year?

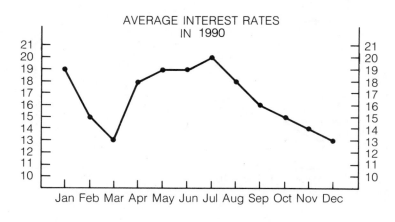

6. (a) ——————————————

(b) ——————————————

(c) ——————————————

Answers

7. (a) _____

  (b) _____

  (c) _____

  (d) _____

  (e) _____

  (f) _____

8. (a) _____

  (b) _____

  (c) _____

  (d) _____

  (e) _____

  (f) _____

  (g) _____

  (h) _____

7. (a) During which month were domestic sales highest? (b) How much higher were they than for the lowest month? (c) What was the difference in import sales for the same two months? (d) What was the difference between domestic and import sales in March? (e) What percent of sales were imports in December 1989? (f) In December 1990? (Answers will be approximate.)

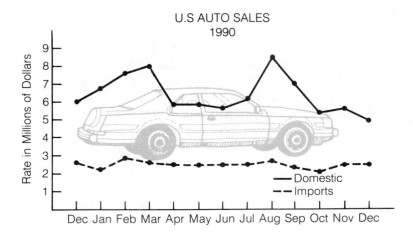

8. (a) What percent of workers were in each of the four areas in 1860? (b) In 1980? (c) Which area of work seems to have had the most stable percent of workers between 1860 and 1980? (d) What is the difference between the highest and lowest percents for this area? (e) Which area has had the most growth? (f) What was its lowest percent and when? (g) What was its highest percent and when? (h) Which area has had the most decline?

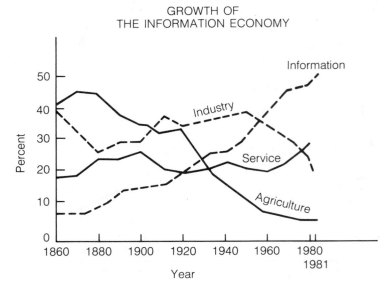

Source: Marc Porat, *The Information Economy: Definition and Measurement,* Office of Telecommunications Special Publication 77-12 (1), May 1977.

9.4

HISTOGRAMS AND FREQUENCY POLYGONS _____

For the bar graphs in Section 9.3 we labeled the base line for each bar with individual names for people, months, days of the week, or other categories. In this section, we will study a special type of bar graph called a **histogram** and related line graphs called **frequency polygons**. In a histogram, the base line is labeled with numbers that indicate the boundaries of a range of numbers called a **class**. The bars are placed next to each other with no spaces between them.

> **Histogram:** A special type of bar graph with intervals of numbers marked on the base line
>
> **Class:** A range (or interval) of numbers that contains data items
>
> **Lower class limit:** The smallest number that belongs to a class
>
> **Upper class limit:** The largest number that belongs to a class
>
> **Class boundaries:** Numbers that are halfway between the upper limit of one class and the lower limit of the next class
>
> **Class width:** The difference between the class boundaries of a class (the width of each bar)
>
> **Frequency:** The number of data items in a class

Example 1 Twenty students took a mathematics exam and the teacher grouped the scores as shown in Table 9.1, and then graphed the related histogram to get a picture of the distribution of the scores. (Note that in the histogram, the class boundaries are used in marking the base line. No data item (or score) is equal to a class boundary. This eliminates any ambiguity as to which class a particular data item belongs in.)

TABLE 9.1 A Frequency Distribution of Mathematics Exam Scores

Class Number	Class Limits (Range of exam scores)	Frequency (Number of students in each class)
1	50–59	1
2	60–69	4
3	70–79	8
4	80–89	5
5	90–99	$\underline{2}$
		20 students

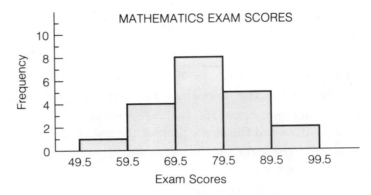

FIGURE 9.5 Histogram for Table 9.1

The following statements describe various characteristics and facts related to the histogram.

(a) There are 5 classes represented.

(b) For the first class, 59 is the upper class limit and 50 is the lower class limit.

(c) There are 8 scores in the third class. That is, 8 students had scores in the class 70 to 79.

(d) The class boundaries for the second class are 59.5 and 69.5.

(e) The width of each class is 10 (69.5 − 59.5 = 10). (Note that this is **not** the difference in the class limits since 69 − 60 = 9.)

(f) The percent of scores in the fifth class is found by dividing the frequency of the class (2) by the total number of scores (20):

$$\frac{2}{20} = \frac{1}{10} = 0.1 = 10\%$$

Completion Example 2 Figure 9.6 shows a histogram that summarizes the scores of 50 students on an English placement test. Answer the following questions by referring to the graph.

(a) How many classes are represented?   6

(b) What are the class limits of the first class?   201 and 250

(c) What are the class boundaries of the second class?   250.5 and 300.5

(d) What is the width of each class? _____

(e) Which class has the greatest frequency? _____

(f) What is this frequency? _____

(g) What percent of the scores are between 200.5 and 250.5? _____

(h) What percent of the scores are above 400? _____

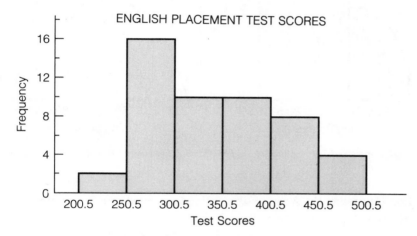

FIGURE 9.6 Histogram of Test Scores

Completion Example Answer

2. (d) 50

(e) second class

(f) 16

(g) 4%

(h) 24%

A variation of a histogram can be created by connecting the midpoints of the tops of the rectangles with straight line segments. The resulting line graph is called a **frequency polygon.** Two additional line segments can be drawn connecting the points at each end to 0. This is done to indicate the fact that the frequency is 0 outside the classes indicated by the histogram.

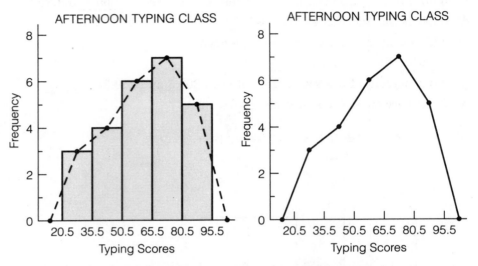

FIGURE 9.7 Histogram of Typing Scores   **FIGURE 9.8 Frequency Polygon of Typing Scores**

Completion Example 3   Answer the following questions by referring to the frequency polygon in Figure 9.8.

(a) How many students scored in the highest class?   5

(b) What were their typing scores?   between 80.5 and 95.5

(c) What was the class width?   _____

(d) How many students took the typing test?   _____

(e) If a score of 51 or higher is passing, what percent of the scores were passing?   _____

Completion Example Answer

3. (c) 15

(d) 25

(e) 72%

NAME _____ SECTION _____ DATE _____

EXERCISES **9.4** _____

Answer the questions by referring to the graphs given in each of the following exercises.

1.

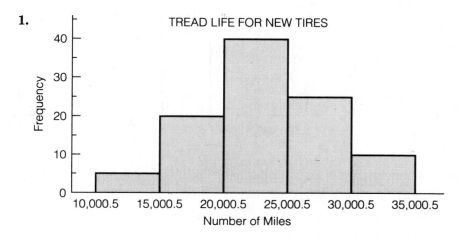

TREAD LIFE FOR NEW TIRES

(a) How many classes are represented?

(b) What is the width of each class?

(c) Which class has the highest frequency?

(d) What is this frequency?

(e) What are the class boundaries of the second class?

(f) How many tires were tested?

(g) What percent of the tires were in the first class?

(h) What percent of the tires lasted more than 25,000 miles?

Answers

1. (a) _____

   (b) _____

   (c) _____

   (d) _____

   (e) _____

   (f) _____

   (g) _____

   (h) _____

Answers

2. (a) ——————————

   (b) ——————————

   (c) ——————————

   (d) ——————————

   (e) ——————————

   (f) ——————————

   (g) ——————————

   (h) ——————————

2.

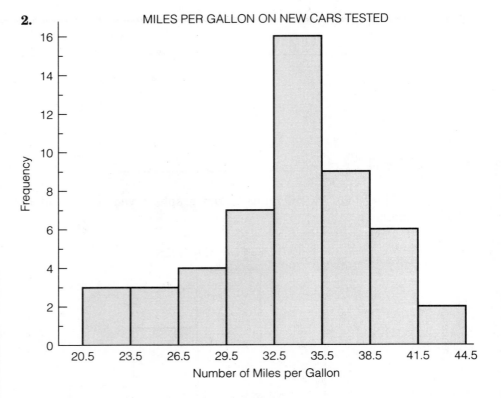

MILES PER GALLON ON NEW CARS TESTED

(a) How many classes are represented?

(b) What is the class width?

(c) Which class has the smallest frequency?

(d) What is this frequency?

(e) What are the class limits for the third class?

(f) How many cars were tested?

(g) How many cars tested below 30 miles per gallon?

(h) What percent of the cars tested above 38 miles per gallon?

NAME _____ SECTION _____ DATE _____

3.

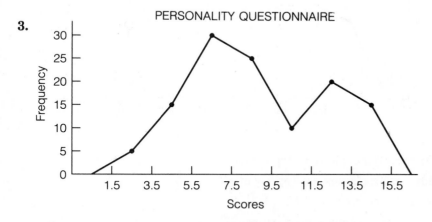

PERSONALITY QUESTIONNAIRE

(a) Which class contains the highest number of test scores?

(b) How many scores are between 3.5 and 5.5?

(c) How many students took the test?

(d) How many students scored less than 7.5?

(e) How many students scored more than 11?

(f) Approximately what percent of the students scored between 5.5 and 11.5?

Answers

3. (a) _____

   (b) _____

   (c) _____

   (d) _____

   (e) _____

   (f) _____

Answers

4. (a) _____

   (b) _____

   (c) _____

   (d) _____

   (e) _____

   (f) _____

4.

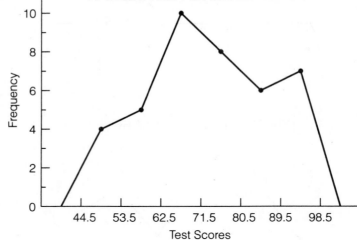

APTITUDE TEST FOR SOCIAL WORKERS

(a) How many people took the aptitude test?

(b) How many classes are represented?

(c) What is the class width?

(d) How many people scored below 63?

(e) What percent of the scores were in the first class?

(f) If a score of 72 or higher is considered acceptable for employment as a social worker, what percent of the test takers are employable?

NAME _____ SECTION _____ DATE _____

CHAPTER **9** TEST _____

Find (a) the mean, (b) the median, and (c) the range for the data given in Exercises 1 and 2.

1. The number of hours of television viewing per day for one person for eleven days:

   3, 2, 2, 0, 1, 4, 1, 2, 3, 2, 2

2. The number of inches of rain per month for six months in a particular region:

   2.54, 10.16, 7.62, 20.32, 12.70, 7.62

Answers

1. (a) _____

   (b) _____

   (c) _____

2. (a) _____

   (b) _____

   (c) _____

Answers

3. _____

4. _____

5. _____

In Problems 3–5, use the circle graph below.

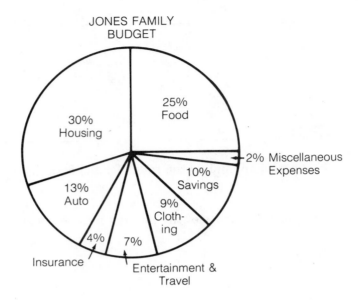

JONES FAMILY BUDGET

3. If the combined Jones family income is $3300 per month, how much do they spend for food in one year (to the nearest dollar)?

4. How much more do they save in one month than they spend for clothing (to the nearest dollar)?

5. The Jones family would like to take a $3000 tour of Europe. Will one year's entertainment and travel budget accommodate the trip? If so, what will they have left over? If not, how short will they be?

NAME _____ SECTION _____ DATE _____

In Problems 6–8, refer to the graph below.

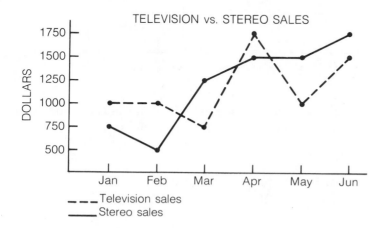

TELEVISION vs. STEREO SALES

- - - Television sales
——— Stereo sales

6. _____

7. _____

8. _____

6. What were average monthly television sales during the six months (to the nearest dollar)?

7. What percent of the total March sales were stereos (to the nearest tenth of a percent)?

8. What percent of the total six months' sales were televisions (to the nearest tenth of a percent)?

Answers

9. (a) _____

(b) _____

(c) _____

10. _____

11. _____

In Problems 9–11, refer to the histogram below.

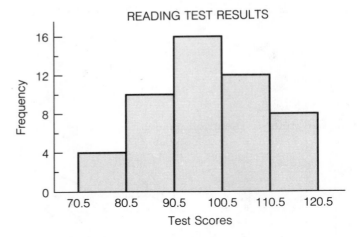

READING TEST RESULTS

9. (a) How many classes are represented? (b) What is the width of each class? (c) What is the frequency of the fourth class?

10. What percent of the scores were between 80.5 and 100.5?

11. What percent of the scores were below 101?

Measurement

10

10.1

METRIC SYSTEM: LENGTH AND AREA

About 90% of the people in the world use the metric system of measurement. The United States is the only major industrialized country still committed to the US Customary System (formerly called the English system). Even in the United States, the metric system has been used for years in such fields as medicine, science, and military activities.

Length

The **meter** is the basic unit of length in the metric system. Smaller and larger units are named by putting a prefix in front of the basic unit, for example **centi**meter and **kilo**meter. The prefixes* we will use are, from largest to smallest unit size,

kilo	hecto	deka	deci	centi	milli
(thousand)	(hundred)	(ten)	(tenth)	(hundredth)	(thousandth)

These prefixes should be memorized, and memorized in order.

---

* Other prefixes that indicate extremely small units are micro, nano, pico, femto, and atto. Prefixes that indicate extremely large units are mega, giga, and tera. These prefixes will not be used in this text.

The basic unit of length is the **meter**.
Here are some examples of metric lengths:

one millimeter (mm): About the width of the wire in a paper clip

1 mm →

one centimeter (cm): About the width of a paper clip

1 cm →

one meter: Just over 39 inches, or slightly longer than a yard
one kilometer (km): About 0.62 of a mile

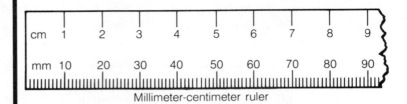

Millimeter-centimeter ruler

The first table below shows the various prefixes and their values along with their abbreviations. The other table shows measures of length and their abbreviations.

Metric Prefixes and Their Values

Prefix	Value
milli	0.001—thousandths
centi	0.01—hundredths
deci	0.1—tenths
basic unit	1—ones
deka	10—tens
hecto	100—hundreds
kilo	1000—thousands

Measures of Length

1 **milli**meter	(mm)	= 0.001 meter
1 **centi**meter	(cm)	= 0.01 meter
1 **deci**meter	(dm)	= 0.1 meter
1 meter	(m)	= 1.0 meter
1 **deka**meter	(dam)	= 10 meters
1 **hecto**meter	(hm)	= 100 meters
1 **kilo**meter	(km)	= 1000 meters

A chart can be used to change from one unit to another.

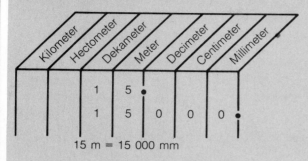

15 m = 15 000 mm

1. List each unit across the top. Memorize the unit prefixes in order.
2. Enter the given number so that each digit is in one column and the decimal point is on the given unit line.
3. Move the decimal point to the desired unit line.
4. Fill in all spaces with 0's.

In the metric system,

1. A 0 is written to the left of the decimal point if there is no whole number part. (0.25 m)
2. No commas are used in writing numbers. If a number has more than four digits (left or right of the decimal point), the digits are grouped in threes from the decimal point with a space between the groups. (14 000 m)

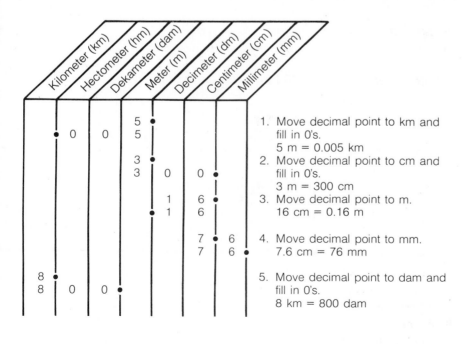

1. Move decimal point to km and fill in 0's.
   5 m = 0.005 km
2. Move decimal point to cm and fill in 0's.
   3 m = 300 cm
3. Move decimal point to m.
   16 cm = 0.16 m
4. Move decimal point to mm.
   7.6 cm = 76 mm
5. Move decimal point to dam and fill in 0's.
   8 km = 800 dam

Using the chart, fill in the blanks in Examples 1–5.

Example 1   5 m = _____ km

Example 2   3 m = _____ cm

Example 3   16 cm = _____ m

Example 4   7.6 cm = _____ mm

Example 5   8 km = _____ dam

Using the chart, change the units as indicated in Examples 6–10.

Completion Example 6   4 m = _____ mm

Completion Example 7   3.1 cm = _____ mm

Completion Example 8   50 cm = _____ m

Completion Example 9   18 km = _____ m

Completion Example 10   25 km = _____ m

Completion Example Answers

6. 4 m = 4000 mm
7. 3.1 cm = 31 mm
8. 50 cm = 0.5 m
9. 18 km = 18 000 m
10. 25 km = 25 000 m

☐ **NOW WORK EXERCISES 1–5 IN THE MARGIN.**

CHANGING METRIC MEASURES WITHOUT A CHART

1. To change to a measure that is
   one unit smaller, multiply by 10   3 cm = 30 mm
   two units smaller, multiply by 100   5 m = 500 cm
   three units smaller, multiply by 1000   14 m = 14 000 mm

   and so on.

2. To change to a measure that is
   one unit larger, divide by 10   50 cm = 5 dm
   two units larger, divide by 100   50 cm = 0.5 m
   three units larger, divide by 1000   13 mm = 0.013 m

   and so on.

Completion Example 11   42 m = 420 dm = 4200 cm = _____ mm

Completion Example 12   17 m = _____ dm = _____ cm = _____ mm

Completion Example 13   6 m = 0.6 dam = 0.06 hm = _____ km

Completion Example 14   112 m = _____ dam = _____ hm = _____ km

☐ **NOW WORK EXERCISES 6–10 IN THE MARGIN.**

Completion Example Answers

11. 42 m = 420 dm = 4200 cm = 42 000 mm
12. 17 m = 170 dm = 1700 cm = 17 000 mm
13. 6 m = 0.6 dam = 0.06 hm = 0.006 km
14. 112 m = 11.2 dam = 1.12 hm = 0.112 km

Make your own chart on a piece of paper and then use the chart to change the following units as indicated.

1. 5 m = _____ mm

2. 1.6 m = _____ cm

3. 83 m = _____ mm

4. 7 cm = _____ m

5. 32 mm = _____ m

Without a chart, change the units as indicated.

6. 35 m = _____ cm

7. 35 m = _____ mm

8. 6.4 cm = _____ mm

9. 5.9 cm = _____ m

10. 8 m = _____ km

Area

Area is a measure of the interior, or enclosure, of a surface. For example, the two rectangles in Figure 10.1 have different areas because they have different amounts of interior space, or different amounts of space are enclosed by the sides of the figures.

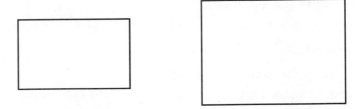

FIGURE 10.1 Two Rectangles with Different Areas

Area: A measure of the interior or enclosure of a surface

Area is measured in square units. A square that is 1 centimeter long on each side is said to have an area of 1 square centimeter, or 1 cm^2.

1 cm

1 cm

Area = 1 cm^2

A rectangle that is 7 cm on one side and 4 cm on the other side encloses 28 squares that each have an area of 1 cm^2. So the rectangle is said to have an area of 28 square centimeters, or 28 cm^2.

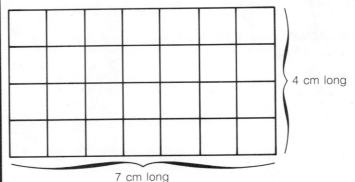

4 cm long

7 cm long

Area = 7 cm × 4 cm = 28 cm^2

There are 28 squares that are each 1 cm^2 in the large rectangle

Measures of Small Area

$$1 \text{ cm}^2 = 100 \text{ mm}^2$$
$$1 \text{ dm}^2 = 100 \text{ cm}^2 = 10\ 000 \text{ mm}^2$$
$$1 \text{ m}^2 = 100 \text{ dm}^2 = 10\ 000 \text{ cm}^2 = 1\ 000\ 000 \text{ mm}^2$$

Note: Each unit of area is equal to 100 times the next smaller unit of area—**not** just 10 times.

Example 15 A square 1 centimeter on a side encloses 100 square millimeters.

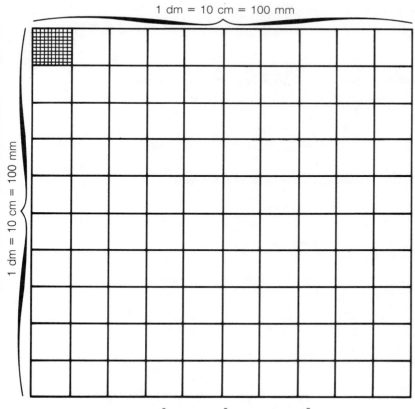

1 cm
Area = 1 cm²

10 mm
Area = 1 cm² = 100 mm²

Example 16 A square 1 decimeter (10 cm) on a side encloses 1 square decimeter. ($1 \text{ dm}^2 = 100 \text{ cm}^2 = 10\ 000 \text{ mm}^2$)

1 dm = 10 cm = 100 mm

1 dm = 10 cm = 100 mm

$$1 \text{ dm}^2 = 100 \text{ cm}^2 = 10\ 000 \text{ mm}^2$$

Change the units as indicated.

11. 22 cm² = _____ mm²

12. 500 mm² = _____ cm²

13. 3.7 dm² = _____ cm²

= _____ mm²

14. 5200 mm² = _____ cm²

= _____ dm²

A chart similar to that used earlier in this section can be used to change measures of area. The key difference is that there must be two digits in each column. This corresponds to multiplication of the previous unit of area by 100.

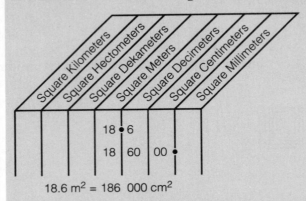

A chart can be used to change from one unit of area to another.

1. List each area unit across the top. (Abbreviations will do.)
2. Enter the given number so that there are **two** digits in each column with the decimal point on the given unit line.
3. Move the decimal point to the desired unit line.
4. Fill in the spaces with 0's using two digits per column.

Completion Example 17 Change the units as indicated.

(a) 5 cm² = _____ mm²
(b) 3 dm² = _____ cm² = _____ mm²
(c) 1.4 m² = _____ dm² = _____ cm² = _____ mm²

☐ **NOW WORK EXERCISES 11–14 IN THE MARGIN.**

Completion Example Answers

17. (a) 5 cm² = 500 mm²
(b) 3 dm² = 300 cm² = 30 000 mm²
(c) 1.4 m² = 140 dm² = 14 000 cm² = 1 400 000 mm²

Are (a): A square with each side 10 m long; encloses an area of 1 are (**Are** is pronounced "air.")

Hectare (ha): 100 ares

Ares and hectares are used to measure land area.

Measures of Land Area

$$1\ a = 100\ m^2$$
$$1\ ha = 100a = 10\ 000\ m^2$$

Example 18   A square 10 m on each side encloses 100 m² or 1 are.

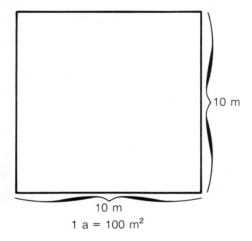

10 m

10 m

1 a = 100 m²

Completion Example 19   Change the units as indicated

(a) 3.2 a = _____ m²

(b) 7.63 ha = _____ a = m²

Completion Example Answer

19. (a) 3.2 a = 320 m²

   (b) 7.63 ha = 763 a = 76 300 m²

Change the units as indicated.

15. 3.6 a = _____ m²

16. 0.73 ha = _____ a

= _____ m²

17. 5.4 ha = _____ a

= _____ m²

18. 49 500 m² = _____ a

= _____ ha

Example 20   How many ares are in 1 km²? (**Note:** One km is about 0.6 mile, so 1 km² is about 0.6 × 0.6 = 0.36 square mile.)

Remember that 1 km = 1000 m, so

$$1 \text{ km}^2 = (1000 \text{ m}) \times (1000 \text{ m})$$

$$= 1\ 000\ 000 \text{ m}^2$$

$$= 10\ 000 \text{ a} \quad \text{Divide m}^2 \text{ by 100 to get}$$
ares because every 100 m²
is equal to 1 are.

☐ **NOW WORK EXERCISES 15–18 IN THE MARGIN.**

Answers to Marginal Exercises

1. 5000 mm   **2.** 160 cm   **3.** 83 000 mm   **4.** 0.07 m   **5.** 0.032 m   **6.** 3500 cm
7. 35 000 mm   **8.** 64 mm   **9.** 0.059 m   **10.** 0.008 km   **11.** 2200 mm²
12. 5 cm²   **13.** 370 cm² = 37 000 mm²   **14.** 52 cm² = 0.52 dm²   **15.** 360 m²
16. 73 a = 7300 m²   **17.** 540 a = 54 000 m²   **18.** 495 a = 4.95 ha

NAME _____ SECTION _____ DATE _____

EXERCISES **10.1** _____

Answers

1. Write the six metric prefixes discussed in this section in order from smallest to largest unit size.

1. _____

2. _____

3. _____

Change the following units as indicated.

4. _____

2. 1 m = _____ cm

3. 5 m = _____ cm

5. _____

4. 12 m = _____ cm

5. 6 m = _____ cm

6. _____

6. 2 m = _____ mm

7. 0.3 m = _____ mm

7. _____

8. _____

8. 0.7 m = _____ mm

9. 1.4 m = _____ mm

9. _____

10. _____

10. 1.6 cm = _____ mm

11. 1.8 cm = _____ mm

11. _____

12. _____

12. 25 cm = _____ mm

13. 35 cm = _____ mm

13. _____

14. 4 m = _____ dm

15. 16 m = _____ dm

14. _____

15. _____

16. 7 dm = _____ cm

17. 2.1 dm = _____ cm

16. _____

17. _____

Answers

18. _____

19. _____

20. _____

21. _____

22. _____

23. _____

24. _____

25. _____

26. _____

27. _____

28. _____

29. _____

30. _____

31. _____

32. _____

33. _____

34. _____

35. _____

36. _____

37. _____

18. 3 km = _____ m

19. 5 km = _____ m

20. 5.28 km = _____ m

21. 6.4 km = _____ m

22. 11 mm = _____ cm

23. 26 mm = _____ cm

24. 72 mm = _____ cm

25. 48 mm = _____ cm

26. 6 mm = _____ dm

27. 12 mm = _____ dm

28. 20 mm = _____ m

29. 30 mm = _____ m

30. 145 mm = _____ m

31. 256 mm = _____ m

32. 25 cm = _____ m

33. 32 cm = _____ m

34. 150 cm = _____ m

35. 170 cm = _____ m

36. 3000 m = _____ km

37. 2400 m = _____ km

NAME _____ SECTION _____ DATE _____

38. 500 m = _____ km

39. 400 m = _____ km

40. 3.45 m = _____ cm

41. 4.62 m = _____ cm

42. 6.3 cm = _____ m

43. 5.2 cm = _____ m

44. 3.25 m = _____ mm

45. 6.41 m = _____ mm

46. 3 mm = _____ cm

47. 5 mm = _____ cm

48. 32 mm = _____ m

49. 57 mm = _____ m

50. 20 000 m = _____ km

51. 35 000 m = _____ km

Answers

38. _____

39. _____

40. _____

41. _____

42. _____

43. _____

44. _____

45. _____

46. _____

47. _____

48. _____

49. _____

50. _____

51. _____

Answers

52. _____

53. _____

54. _____

55. _____

56. _____

57. _____

58. _____

59. _____

60. _____

61. _____

62. _____

63. _____

64. _____

65. _____

66. _____

67. _____

52. 1.5 km = _____ m

53. 2.3 km = _____ m

54. 0.5 m = _____ km

Change the following units as indicated.

55. 3 cm^2 = _____ mm^2

56. 5.6 cm^2 = _____ mm^2

57. 8.7 cm^2 = _____ mm^2

58. 3.61 cm^2 = _____ mm^2

59. 600 mm^2 = _____ cm^2

60. 28 mm^2 = _____ cm^2

61. 1400 mm^2 = _____ cm^2

62. 20 000 mm^2 = _____ cm^2

63. 4 dm^2 = _____ cm^2 = _____ mm^2

64. 7.3 dm^2 = _____ cm^2 = _____ mm^2

65. 57 dm^2 = _____ cm^2 = _____ mm^2

66. 0.6 dm^2 = _____ cm^2 = _____ mm^2

67. 17 m^2 = _____ dm^2 = _____ cm^2 = _____ mm^2

NAME _____ SECTION _____ DATE _____

68. $2.9 \text{ m}^2 =$ _____ $\text{dm}^2 =$ _____ $\text{cm}^2 =$ _____ mm^2

69. $0.03 \text{ m}^2 =$ _____ $\text{dm}^2 =$ _____ $\text{cm}^2 =$ _____ mm^2

70. $0.5 \text{ m}^2 =$ _____ $\text{dm}^2 =$ _____ $\text{cm}^2 =$ _____ mm^2

71. $142 \text{ mm}^2 =$ _____ cm^2 **72.** $5800 \text{ mm}^2 =$ _____ cm^2

73. $200 \text{ dm}^2 =$ _____ m^2 **74.** $35 \text{ dm}^2 =$ _____ m^2

75. $7.8 \text{ a} =$ _____ m^2 **76.** $300 \text{ a} =$ _____ m^2

77. $0.04 \text{ a} =$ _____ m^2 **78.** $0.53 \text{ a} =$ _____ m^2

79. $8.69 \text{ ha} =$ _____ $\text{a} =$ _____ m^2

80. $7.81 \text{ ha} =$ _____ $\text{a} =$ _____ m^2

81. $0.16 \text{ ha} =$ _____ $\text{a} =$ _____ m^2

Answers

68. _____

69. _____

70. _____

71. _____

72. _____

73. _____

74. _____

75. _____

76. _____

77. _____

78. _____

79. _____

80. _____

81. _____

Answers

82. _____

83. _____

84. _____

85. _____

86. _____

82. 1 a = _____ ha

83. 15 a = _____ ha

84. 5 km^2 = _____ a = _____ ha

85. 4.76 km^2 = _____ a = _____ ha

86. 0.3 km^2 = _____ a = _____ ha

10.2
METRIC SYSTEM: WEIGHT AND VOLUME

Mass (Weight)

The basic unit of mass in the metric system is the kilogram, about 2.2 pounds. In some fields, such as medicine, the gram (about the mass of a paper clip) is more convenient as a basic unit than the kilogram.

> **Mass:** The amount of material in an object
>
> **Weight:** A measure of the Earth's gravitational pull on an object (At the Earth's surface, mass and weight have the same measure. The further an object is from the Earth, the less its weight, but its mass remains constant.)
>
> Here are some examples of metric measures of mass:
>
> **one gram:** The mass (or weight) of a paper clip
>
> **one kilogram:** About 2.2 pounds, a little less than the weight of this text
>
> **one ton:** About 2200 pounds;* a truck would carry several tons
>
> In this text, mass and weight will be used interchangeably. A **mass** of 20 kilograms will be said to **weigh** 20 kilograms.
>
>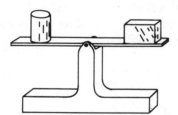
>
> The two objects have the same *mass* and balance on an equal arm balance, regardless of their location in space.

---

*This is called a **metric** ton. It is about 200 pounds heavier than the US ton. A US ton is 2000 pounds.

Measures of Mass

1 **milli**gram (mg) = 0.001 gram
1 **centi**gram (cg) = 0.01 gram
1 **deci**gram   (dg) = 0.1 gram
1 gram   (g) = 1.0 gram
1 **deka**gram (dag) = 10 grams
1 **hecto**gram (hg) = 100 grams
1 **kilo**gram   (kg) = 1000 grams
1 metric ton   (t) = 1000 kilograms

Relationships Between Smaller
and Larger Measures of Mass

1000 mg = 1 g   0.001 g = 1 mg
 1000 g = 1 kg   0.001 kg = 1 g
1000 kg = 1 t   0.001 t = 1 kg
1 t = 1000 kg = 1 000 000 g = 1 000 000 000 mg

The centrigram, decigram, dekagram, and hectogram have little practical use and are not included in the exercises.

A chart may also be used to change units of mass. The headings are parts of a gram. Enter one digit per column in Examples 1–5 and move the decimal point to the new unit line. Fill in any spaces with 0's.

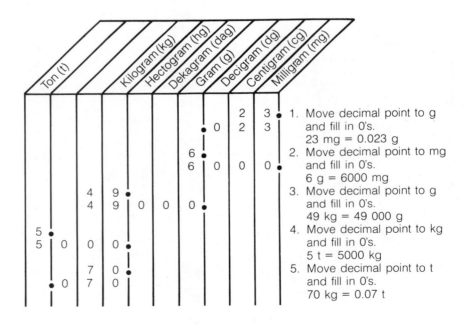

1. Move decimal point to g and fill in 0's.
   23 mg = 0.023 g
2. Move decimal point to mg and fill in 0's.
   6 g = 6000 mg
3. Move decimal point to g and fill in 0's.
   49 kg = 49 000 g
4. Move decimal point to kg and fill in 0's.
   5 t = 5000 kg
5. Move decimal point to t and fill in 0's.
   70 kg = 0.07 t

Using the chart, fill in the blanks in Examples 1–5.

Example 1   23 mg = _____ g

Example 2   6 g = _____ mg

Example 3   49 kg = _____ g

Example 4   5 t = _____ kg

Example 5   70 kg = _____ t

In Examples 6–10, change the units as indicated.

Completion Example 6   60 mg = 0.06 g = _____ kg

Completion Example 7   135 mg = _____ g = _____ kg

Completion Example 8   5 700 000 g = 5700 kg = _____ t

Completion Example 9   100 g = _____ kg

Completion Example 10   78 g = _____ mg

☐ **NOW WORK EXERCISES 1–5 IN THE MARGIN.**

Change the units as indicated.

1. 43 g = _____ mg

2. 7.8 g = _____ mg

3. 350 mg = _____ g

4. 75 kg = _____ g

5. 3940 kg = _____ t

Completion Example Answers
6. 60 mg = 0.06 g = 0.00006 kg
7. 135 mg = 0.135 g = 0.000135 kg
8. 5 700 000 g = 5700 kg = 5.7 t
9. 100 g = 0.1 kg
10. 78 g = 78 000 mg

Volume

Volume: A measure of the space enclosed by a three-dimensional figure

Volume is measured in cubic units. The volume or space contained within a cube that is 1 cm on each edge is 1 cubic centimeter, or 1 cm^3 (about the size of a sugar cube).

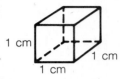

1 cm

1 cm

1 cm

Volume = 1 cm^3

A rectangular solid that has edges of 3 cm, 2 cm, and 5 cm has a volume of 3 cm $\times$ 2 cm $\times$ 5 cm = 30 cm^3.

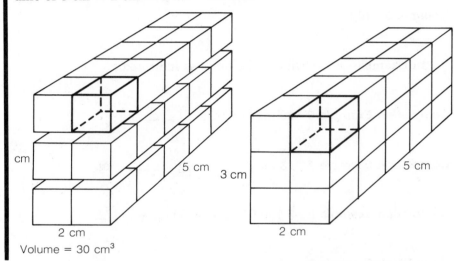

cm

5 cm

3 cm

5 cm

2 cm

2 cm

Volume = 30 cm^3

If a cube is 1 decimeter along each edge, then the volume of the cube is 1 cubic decimeter (or 1 dm^3). In terms of centimeters, this same cube has volume.

$$10 \text{ cm} \times 10 \text{ cm} \times 10 \text{ cm} = 1000 \text{ cm}^3$$

That is, as shown in Figure 10.2, 1 dm^3 = 1000 cm^3.

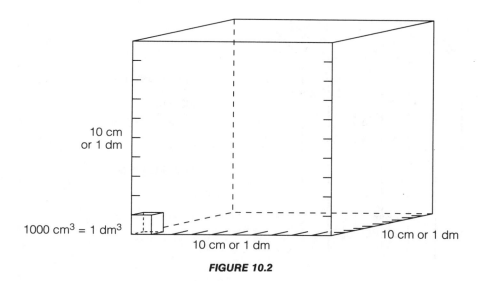

10 cm
or 1 dm

1000 cm³ = 1 dm³

10 cm or 1 dm

10 cm or 1 dm

FIGURE 10.2

This relationship is true of cubic units in the metric system: equivalent cubic units can be found by multiplying the previous unit by 1000. Again, we can use a chart; however, this time **there must be three digits in each column.**

A chart can be used to change from one unit of volume to another.

Cubic Kilometers (km³)
Cubic Hectometers (hm³)
Cubic Dekameters (dam³)
Cubic Meters (m³)
Cubic Decimeters (dm³)
Cubic Centimeters (cm³)
Cubic Millimeters (mm³)

156•32

156 | 320 |000•

156.32 m³ = 156 320 000 cm³

1. List each volume unit across the top. (Abbreviations will do.)

2. Enter the given number so that there are **three** digits in each column with the decimal point on the given unit line.

3. Move the decimal point to the desired unit line.

4. Fill in the spaces with 0's using three digits per column.

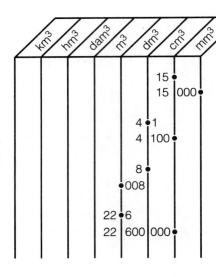

1. Move decimal point to mm³ and fill in 0's.
   15 cm³ = 15 000 mm³

2. Move decimal point to cm³ and fill in 0's.
   4.1 dm³ = 4100 cm³

3. Move decimal point to m³ and fill in 0's.
   8 dm³ = 0.008 m³

4. Move decimal point to cm³ and fill in 0's.
   22.6 m³ = 22 600 000 cm³

Using the chart, fill in the blanks in Examples 11–14.

Example 11   15 cm³ = _____ mm³

Example 12   4.1 dm³ = _____ cm³

Example 13   8 dm³ = _____ m³

Example 14   22.6 m³ = _____ m³

Using the chart, change the units as indicated in Examples 15–17.

Completion Example 15   3.7 dm³ = _____ cm³

Completion Example 16   0.8 m³ = _____ dam³

Completion Example 17   4m³ = _____ dm³ = _____ cm³

Completion Example Answers

15. 3700 cm³   **16.** 0.0008 dam³   **17.** 4000 dm³ = 4 000 000 cm³

Liquid Volume

Liquid volume is measured in **liters** (abbreviated L). You are probably familiar with 1L and 2L bottles of soda on your grocer's shelf.

> **Liter (L):** The volume enclosed in a cube that is 10 cm on each edge
>
> $$1 \text{ liter} = 10 \text{ cm} \times 10 \text{ cm} \times 10 \text{ cm} = 1000 \text{ cm}^3$$
> $$= 1 \text{ dm} \times 1 \text{ dm} \times 1 \text{ dm} = 1 \text{ dm}^3$$
> $$1 \text{ L} = 1 \text{ dm}^3 = 1000 \text{ cm}^3$$

Measures of Liquid Volume		
1 **milli**liter	(mL)	= 0.001 liter
1 liter	(L)	= 1.0 liter
1 **hecto**liter	(hL)	= 100 liters
1 **kilo**liter	(kL)	= 1000 liters

Equivalent Measures of Volume		
1000 mL = 1 L	1 mL = 1 cm^3	
1000 L = 1 kL	1 L = 1 dm^3	
10 hL = 1 kL	1kL = 1m^3	

The following chart shows several unit changes. The prefixes kilo, hecto, deka, deci, centi, and milli all indicate the same parts of a liter as they do of the meter and the gram. However, the centiliter, deciliter, and dekaliter are not commonly used and are not included in any exercises.

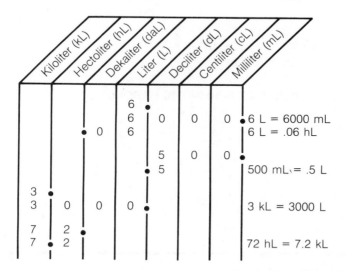

Using the chart, change the units in Completion Examples 18–21 as indicated.

Change the units as indicated.

6. 2 mL = _____ L

7. 3.6 kL = _____ L

8. 500 mL = _____ L

9. 500 mL = _____ cm³

10. 42 hL = _____ kL

Completion Example 18 5000 mL = _____ L

Completion Example 19 3.2 L = _____ mL

Completion Example 20 60 hL = _____ kL

Completion Example 21 637 mL = _____ L

There is an interesting "crossover" relationship between liquid volume measures and cubic volume measures. Since

$$1 \text{ L} = 1000 \text{ mL} \quad \text{and} \quad 1 \text{ L} = 1000 \text{ cm}^3$$

we have

$$\mathbf{1 \text{ mL} = 1 \text{ cm}^3}$$

Also,

$$1 \text{ kL} = 1000 \text{ L} = 1\,000\,000 \text{ cm}^3 \quad \text{and} \quad 1\,000\,000 \text{ cm}^3 = 1 \text{ m}^3.$$

This gives

$$\mathbf{1 \text{ kL} = 1000 \text{ L} = 1 \text{ m}^3.}$$

Example 22 70 mL = _____ cm³
70 mL = 70 cm³

Example 23 3.8 kL = _____ m³
3.8 kL = 3.8 m³

☐ **NOW WORK EXERCISES 6–10 IN THE MARGIN.**

Completion Example Answers
18. 5000 mL = 5 L   **19.** 3.2 L = 3200 mL   **20.** 60 hL = 6 kL
21. 637 mL = 0.637 L

Answers to Marginal Exercises
1. 43 000   **2.** 7800 mg   **3.** 0.35 g   **4.** 75 000 g   **5.** 3.94 t   **6.** 0.002 L
7. 3600 L   **8.** 0.5 L   **9.** 500 cm³   **10.** 4.2 kL

NAME _____ SECTION _____ DATE _____

EXERCISES **10.2** _____

Change the following units as indicated.

1. 7g = _____ mg

2. 2 kg = _____ g

3. 34.5 mg = _____ g

4. 3700 kg = _____ t

5. 4000 kg = _____ t

6. 5600 g = _____ kg

7. 73 kg = _____ mg

8. 91 kg = _____ t

9. 0.54 g = _____ mg

10. 0.7 g = _____ mg

11. 5 t = _____ kg

12. 17 t = _____ kg

Answers

13. _____

14. _____

15. _____

16. _____

17. _____

18. _____

19. _____

20. _____

21. _____

22. _____

23. _____

24. _____

25. _____

26. _____

27. _____

28. _____

29. _____

30. _____

13. 2 t = _____ kg

14. 896 mg = _____ g

15. 896 g = _____ mg

16. 342 kg = _____ g

17. 75 000 g = _____ kg

18. 3000 mg = _____ g

19. 7 t = _____ g

20. 0.4 t = _____ g

21. 0.34 g = _____ kg

22. 0.78 g = _____ mg

23. 16 mg = _____ g

24. 2.5 g = _____ mg

25. 3.94 g = _____ mg

26. 92.3 g = _____ kg

27. 5.6 t = _____ kg

28. 7.58 t = _____ kg

29. 3547 kg = _____ t

30. 2963 kg = _____ t

NAME _____ SECTION _____ DATE _____

Complete the following tables.

31. ━━━━━━━━━━━━━

$1\ cm^3$ = _____ mm^3

$1\ dm^3$ = _____ cm^3

$1\ m^3$ = _____ dm^3

$1\ km^3$ = _____ m^3

32. ━━━━━━━━━━━━━

$1\ dm$ = _____ cm

$1\ dm$ = _____ mm

$1\ dm^2$ = _____ cm^2

$1\ dm^2$ = _____ mm^2

$1\ dm^3$ = _____ cm^3

$1\ dm^3$ = _____ mm^3

31. _____

32. _____

33. _____

33. ━━━━━━━━━━━━━

$1\ m$ = _____ dm

$1\ m$ = _____ cm

$1\ m^2$ = _____ dm^2

$1\ m^2$ = _____ cm^2

$1\ m^3$ = _____ dm^3

$1\ m^3$ = _____ cm^3

34. ━━━━━━━━━━━━━

$1\ km$ = _____ m

$1\ km^2$ = _____ m^2

$1\ km^3$ = _____ m^3

$1\ km$ = _____ dm

$1\ km^2$ = _____ ha

$1\ km^3$ = _____ kL

34. _____

35. _____

36. _____

Change the following units as indicated.

35. 73 kL = _____ L

36. 0.9 kL = _____ L

37. _____

38. _____

37. 400 mL = _____ L

38. 525 mL = _____ L

39. _____

39. 63 L = _____ mL

40. 8.7 L = _____ mL

40. _____

Answers

41. _____

42. _____

43. _____

44. _____

45. _____

46. _____

47. _____

48. _____

49. _____

50. _____

51. _____

52. _____

53. _____

54. _____

55. _____

41. 5 hL = _____ kL

42. 69 hL = _____ kL

43. 19 mL = _____ cm^3

44. 5 cm^3 = _____ mm^3

45. 2 dm^3 = _____ cm^3

46. 76.4 mL = _____ L

47. 5.3 L = _____ mL

48. 30 cm^3 = _____ mL

49. 30 cm^3 = _____ L

50. 5.3 mL = _____ L

51. 48 kL = _____ L

52. 72 000 L = _____ kL

53. 32 L = _____ hL

54. 80 L = _____ mL

55. 290 L = _____ kL

10.3

US CUSTOMARY MEASUREMENTS AND METRIC EQUIVALENTS

Temperature: US Customary measure is in **degrees** Fahrenheit (°F)

Metric measure is in **degrees** Celsius (°C)

The two scales are shown here on thermometers. Approximate conversions can be found by reading along a ruler or the edge of a piece of paper held horizontally across the page.

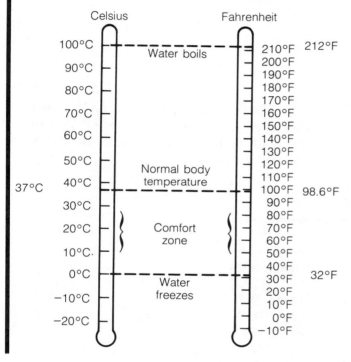

Example 1 Hold a straight edge horizontally across the two thermometers and you will read:

$$100° \text{ C} = 212° \text{ F} \quad \text{water boils at sea level}$$

$$40°\text{C} = 104° \text{ F} \quad \text{hot day in the desert}$$

$$20° \text{ C} = 68° \text{ F} \quad \text{comfortable room temperature}$$

Two formulas that give exact conversions are given here.

F = Fahrenheit temperature and C = Celsius temperature.

$$C = \frac{5(F - 32)}{9} \qquad F = \frac{9 \cdot C}{5} + 32$$

A calculator will give answers accurate to 8 digits. Answers that are not exact may be rounded off to whatever place of accuracy you choose.

Example 2 Let F = 86° and find the equivalent measure in Celsius.

Solution

$$C = \frac{5(86 - 32)}{9} = \frac{5(54)}{9} = 30$$

86° F = 30° C.

Example 3 Let C = 40° and convert this to degrees Fahrenheit.

Solution

$$F = \frac{9 \cdot 40}{5} + 32 = 72 + 32 = 104$$

40° C = 104° F.

In the tables of Length Equivalents, Area Equivalents, and Volume Equivalents that follow, the equivalent measures are rounded off. Any calculations with these measures (with or without a calculator) cannot be any more accurate than the measure in the table.

Length Equivalents

US to Metric	Metric to US
1 in = 2.54 cm (exact)	1 cm = 0.394 in.
1 ft = 0.305 m	1 m = 3.28 ft
1 yd = 0.914 m	1 m = 1.09 yd
1 mi = 1.61 km	1 km = 0.62 mi

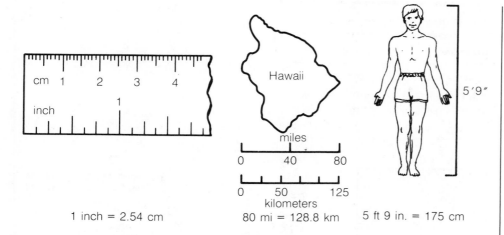

1 inch = 2.54 cm   80 mi = 128.8 km   5 ft 9 in. = 175 cm

Use the Length Equivalents table to convert measurements as indicated.

1. 20 in = _____ cm

2. 100 yd = _____ m

3. 10 m = _____ ft

4. 20 km = _____ mi

In Examples 4–7, use the Length Equivalents table to convert measurements as indicated.

Example 4   6 ft = _____ cm

Solution

6 ft = 72 in. = 72(2.54 cm) = 183 cm (rounded off)

or

6 ft = 6(0.305 m) = 1.83 m = 183 cm

Example 5   25 mi = _____ km

Solution

25 mi = 25(1.61 km) = 40.25 km

Example 6   30 m = _____ ft

Solution

30 m = 30(3.28 ft) = 98.4 ft

Example 7   10 km = _____ mi

Solution

10 km = 10(0.62 mi) = 6.2 mi

☐ **NOW WORK EXERCISES 1–4 IN THE MARGIN.**

Use the Area Equivalents table to convert the measures as indicated.

5. $6 \text{ in.}^2 =$ _____ cm^2

6. $625 \text{ ft}^2 =$ _____ m^2

7. $50 \text{ m}^2 =$ _____ ft^2

8. $3 \text{ ha} =$ _____ acres

Area Equivalents

US to Metric	Metric to US
$1 \text{ in.}^2 = 6.45 \text{ cm}^2$	$1 \text{ cm}^2 = 0.155 \text{ in.}^2$
$1 \text{ ft}^2 = 0.093 \text{ m}^2$	$1 \text{ m}^2 = 10.764 \text{ ft}^2$
$1 \text{ yd}^2 = 0.836 \text{ m}^2$	$1 \text{ m}^2 = 1.196 \text{ yd}^2$
$1 \text{ acre} = 0.405 \text{ ha}$	$1 \text{ ha} = 2.47 \text{ acres}$

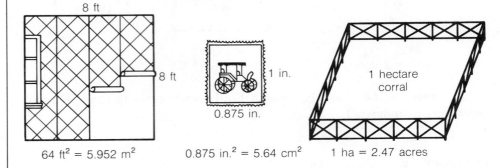

$64 \text{ ft}^2 = 5.952 \text{ m}^2$ $0.875 \text{ in.}^2 = 5.64 \text{ cm}^2$ $1 \text{ ha} = 2.47 \text{ acres}$

In Examples 8–11, use the Area Equivalents table to convert the measures as indicated.

Example 8 $40 \text{ yd}^2 =$ _____ m^2

Solution

$40 \text{ yd}^2 = 40(0.836 \text{ m}^2) = 33.44 \text{ m}^2$

Example 9 $5 \text{ acres} =$ _____ ha

Solution

$5 \text{ acres} = 5(0.405 \text{ ha}) = 2.025 \text{ ha}$

Example 10 $5 \text{ ha} =$ _____ acres

Solution

$5 \text{ ha} = 5(2.47 \text{ acres}) = 12.35 \text{ acres}$

Example 11 $100 \text{ cm}^2 =$ _____ in.^2

Solution

$100 \text{ cm}^2 = 100(0.155 \text{ in.}^2) = 15.5 \text{ in.}^2$

☐ **NOW WORK EXERCISES 5–8 IN THE MARGIN.**

Volume Equivalents

US to Metric	Metric to US
1 in.^{3} = 16.387 cm^3	1 cm^3 = 0.06 in.^{3}
1 ft^3 = 0.028 m^3	1 m^3 = 35.315 ft^3
1 qt = 0.946 L	1 L = 1.06 qt
1 gal = 3.785 L	1 L = 0.264 gal

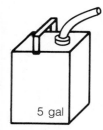

1 in.^{3} ice

5 gal = 18.925 L   1 L = 1.06 qt   3 in.^{3} = 49.161 cm^3

In Examples 12–15, use the Volume Equivalents table to convert the measures as indicated.

Example 12 20 gal = _____ L

Solution

20 gal = 20(3.785 L) = 75.7 L

Example 13 42 L = _____ gal

Solution

42 L = 42(0.264 gal) = 11.088 gal

or

42 L = 11.1 gal (rounded off)

Example 14 6 qt = _____ L

Solution

6 qt = 6(0.946 L) = 5.676 L

or

6 qt = 5.7 L (rounded off)

Example 15 10 cm^3 = _____ in.^{3}

Solution

10 cm^3 = 10(0.06 in.^{3}) = 0.6 in.^{3}

☐ **NOW WORK EXERCISES 9–12 IN THE MARGIN.**

Use the Volume Equivalents table to convert the measures as indicated.

9. 3 qt = _____ L

10. 5 gal = _____ L

11. 100 cm^3 = _____ in.^{3}

12. 5 m^3 = _____ ft^3

Mass Equivalents

US to Metric	Metric to US
1 oz = 28.35 g	1g = 0.035 oz
1 lb = 0.454 kg	1 kg = 2.205 lb

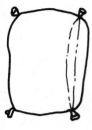

25 lb = 11.35 kg

9 kg = 19.85 lb

In Examples 16 and 17, use the Mass Equivalents table to convert the measures as indicated.

Example 16   5 lb = _____ kg

Solution

5 lb = 5(0.454 kg) = 2.27 kg

Example 17   15 kg = _____ lb

Solution

15 kg = 15(2.205 lb) = 33.075 lb

Answers to Marginal Exercises

1. 50.8 cm   **2.** 91.5 m   **3.** 32.8 ft   **4.** 12.4 mi   **5.** 38.7 cm²   **6.** 58.125 m²
7. 538.2 ft²   **8.** 7.41 acres   **9.** 2.838 L   **10.** 18.925 L   **11.** 6 in.³
12. 176.575 ft³

NAME _____ SECTION _____ DATE _____

EXERCISES **10.3** _____

Answers

Use the appropriate formula to convert the degrees as indicated.

1. 25° C = _____° F

2. 80° C = _____° F

3. 10° C = _____° F

4. 35° C = _____° F

5. 50° F = _____° C

6. 100° F = _____° C

7. 32° F = _____° C

8. 41° F = _____° C

Use the appropriate table to convert the following measures as indicated.

9. 5 ft 2 in. = _____ cm

10. 6 ft 3 in. = _____ cm

11. 3 yd = _____ m

12. 5 yd = _____ m

1. _____

2. _____

3. _____

4. _____

5. _____

6. _____

7. _____

8. _____

9. _____

10. _____

11. _____

12. _____

Answers

13. _____

14. _____

15. _____

16. _____

17. _____

18. _____

19. _____

20. _____

21. _____

22. _____

23. _____

24. _____

25. _____

26. _____

13. 60 mi = _____ km

14. 100 mi = _____ km

15. 400 mi = _____ km

16. 350 mi = _____ km

17. 200 km = _____ mi

18. 65 km = _____ mi

19. 35 km = _____ mi

20. 450 km = _____ mi

21. 50 cm = _____ in.

22. 100 cm = _____ in.

23. 3 in.^{2} = _____ cm^2

24. 16 in.^{2} = _____ cm^2

25. 600 ft^2 = _____ m^2

26. 300 ft^2 = _____ m^2

NAME _____ SECTION _____ DATE _____

27. 100 yd^2 = _____ m^2

28. 250 yd^2 = _____ m^2

29. 1000 acres = _____ ha

30. 250 acres = _____ ha

31. 300 ha = _____ acres

32. 400 ha = _____ acres

33. 5 m^2 = _____ ft^2

34. 10 m^2 = _____ yd^2

35. 30 cm^2 = _____ in.^{2}

36. 50 cm^2 = _____ in.^{2}

37. 10 qt = _____ L

38. 20 qt = _____ L

Answers

27. _____

28. _____

29. _____

30. _____

31. _____

32. _____

33. _____

34. _____

35. _____

36. _____

37. _____

38. _____

Answers

39. _____

40. _____

41. _____

42. _____

43. _____

44. _____

45. _____

46. _____

47. _____

48. _____

49. _____

50. _____

39. 25 gal = _____ L

40. 18 gal = _____ L

41. 10 L = _____ qt

42. 25 L = _____ qt

43. 42 L = _____ gal

44. 50 L = _____ gal

45. 200 in.^{3} = _____ cm^3

46. 10 m^3 = _____ ft^3

47. 10 lb = _____ kg

48. 16 oz = _____ g

49. 500 kg = _____ lb

50. 100 g = _____ oz

10.4

APPLICATIONS IN MEDICINE

This section is designed to provide an introduction to techniques (proportions) used in medicine for changing units, and formulas related to dosages of medicine.

Household fluid measure: Measures based on the use of household cooking and eating utensils

The household system is **not** accurate, but it is useful.

Table of Abbreviations

Household Fluid Measure	Apothecaries' Fluid Measure
gtt = 1 drop	$\mathfrak{m}$ = 1 minim (about 1 drop of water)
tsp = 1 teaspoonful	$f\mathfrak{z}$ = 1 fluidram (60 minims)
tbsp = 1 tablespoonful	$f\mathfrak{z}$ = 1 fluidounce (480 minims)

Fluid Measure Equivalents (Approximate)

Household		Metric		Apothecaries'*
1 gtt	=	0.06 mL	=	$1\,\mathfrak{m}$
15 gtt	=	1.0 mL	=	$15\,\mathfrak{m}$
75 gtt	=	5.0 mL	=	$75\,\mathfrak{m}$
1 tsp	=	5.0 mL	=	$1\frac{1}{3}\,f\mathfrak{z}$
3 tsp	=	15.0 mL	=	$4\,f\mathfrak{z}$
1 tbsp	=	15.0 mL	=	$4\,f\mathfrak{z} = \frac{1}{2}\,f\mathfrak{z}$
1 glassful	=	240.0 mL	=	$8\,f\mathfrak{z}$

---

*Nursing students will learn to use the Roman numeral system with Apothecaries' measures.

1 drop (gtt) = 1 minim (♏)
= 0.06 mL

1 teaspoonful (tsp) = 75 drops (gtt)
= 5.0 mL

1 tablespoonful (tbsp)
= 3 teaspoonfuls (tsp)
= 15.0 mL

1 fluidram ($f\!\!\!\mathfrak{Z}$) = 60 minims (♏)
(about $\frac{4}{5}$ teaspoonful)

1 fluidounce ($f\!\!\!\mathfrak{Z}$) = 8 fluidrams ($f\!\!\!\mathfrak{Z}$)

To use proportions to change from one system to another,

1. find the corresponding equivalent units in the table of Fluid Measure Equivalents;
2. write x for the unknown quantity;
3. write a proportion using the ratio of equivalent measures and the ratio of x to the known measure being changed;
4. solve the proportion.

Example 1   How many milliliters (mL) are in 6 drops (gtt)?

Solution

From the table of Fluid Measure Equivalents,
1 gtt = 0.06 mL.
Let x = unknown number of milliliters.
Set up the proportion:

$$\frac{x \text{ mL}}{6 \text{ gtt}} = \frac{0.06 \text{ mL}}{1 \text{ gtt}}$$

$$1 \cdot x = 6(0.06)$$

$$x = 0.36 \text{ mL}$$

So, 6 drops (gtt) = 0.36 milliliters (mL) or there are 0.36 milliliters in 6 drops.

Example 2   How many tablespoonfuls (tbsp) are in 60 milliliters (mL)?

Solution

From the table of Fluid Measure Equivalents,
1 tbsp = 15.0 mL.
Let x = unknown number of tablespoonfuls.
Set up the proportion:

$$\frac{x \text{ tbsp}}{60 \text{ mL}} = \frac{1 \text{ tbsp}}{15.0 \text{ mL}}$$

$$15 \cdot x = 60 \cdot 1$$

$$\frac{\cancel{15}x}{\cancel{15}} = \frac{60}{15}$$

$$x = 4 \text{ tbsp}$$

So, 60 mL = 4 tbsp or 4 tablespoonfuls are 60 milliliters.

Example 3   How many teaspoonfuls are in 8 fluidrams?

Solution

From the table of Fluid Measure Equivalents,

$1 \text{ tsp} = 1\frac{1}{3} f\mathfrak{z}$.

Let x = unknown number of teaspoonfuls.
Set up the proportion:

$$\frac{x \text{ tsp}}{8 f\mathfrak{z}} = \frac{1 \text{ tsp}}{1\frac{1}{3} f\mathfrak{z}}$$

$$\frac{4}{3} \cdot x = 8 \cdot 1$$

$$\frac{\cancel{3}}{\cancel{4}} \cdot \frac{\cancel{4}}{\cancel{3}} \cdot x = \frac{3}{4} \cdot 8$$

$$x = 6 \text{ tsp}$$

There are 6 teaspoonfuls in 8 fluidrams or 6 tsp = $8 f\mathfrak{z}$.

Example 4   How many glassfuls are in 16 fluidounces?

Solution

From the table of Fluid Measure Equivalents,
1 glassful = $8 f\mathfrak{z}$.
Let x = unknown number of glassfuls.
Set up the proportion:

$$\frac{x \text{ glassfuls}}{16 f\mathfrak{z}} = \frac{1 \text{ glassful}}{8 f\mathfrak{z}}$$

$$8 \cdot x = 16 \cdot 1$$

$$\frac{\cancel{8} \cdot x}{\cancel{8}} = \frac{16}{8}$$

$$x = 2 \text{ glassfuls}$$

There are 2 glassfuls in 16 fluidounces or 2 glassfuls = $16 f\mathfrak{z}$.

Example 5   How many glassfuls are in 480 milliliters?

Solution

From the table of Fluid Measure Equivalents,
1 glassful = 240.0 mL.

Let x = unknown number of glassfuls.

Set up the proportion:

$$\frac{x \text{ glassfuls}}{480 \text{ mL}} = \frac{1 \text{ glassful}}{240 \text{ mL}}$$

$$240 \cdot x = 480 \cdot 1$$

$$\frac{\cancel{240} \cdot x}{\cancel{240}} = \frac{480}{240}$$

$$x = 2 \text{ glassfuls}$$

Thus, 2 glassfuls = 480 mL or there are 2 glassfuls in 480 milliliters.
From Example 4, we can also tell that 2 glassfuls = 480 mL = 16 f℥ .

THREE FORMULAS USED TO CALCULATE DOSAGES OF MEDICINE FOR CHILDREN*

1. **Fried's Rule** (Time of birth to age 2)

$$\text{Child's dose} = \frac{\text{age in months} \times \text{adult dose}}{150}$$

2. **Young's Rule** (ages 1 to 12)

$$\text{Child's dose} = \frac{\text{age in years} \times \text{adult dose}}{\text{age in years} + 12}$$

3. **Clark's Rule** (ages 2 and over)

$$\text{Child's dose} = \frac{\text{weight in lbs} \times \text{adult dose}}{150}$$

---

*These three formulas are commonly studied in nursing. However, there are other formulas also studied in nursing considered more accurate that do not depend on the child's age but more on factors such as height and weight.

Example 6   Use Fried's Rule to find a child's dose, given:

$$\text{Child's age} = 12 \text{ months}$$

$$\text{Adult dose} = 5 \text{ mL of paregoric}$$

Solution

Substituting into the formula gives:

$$\text{Child's dose} = \frac{12 \text{ months} \times 5 \text{ mL}}{150}$$

$$= \frac{60}{150}$$

$$= 0.4 \text{ mL}$$

The child's dose would be 0.4 mL of paregoric.

Example 7   Use Young's Rule to find a child's dose, given:

$$\text{Child's age} = 30 \text{ months (2.5 years)}$$

$$\text{Adult dose} = 15 \text{ mL of castor oil}$$

Solution

Substituting into the formula gives

$$\text{Child's dose} = \frac{2.5 \text{ yrs} \times 15 \text{ mL}}{2.5 + 12}$$

$$= \frac{37.5}{14.5}$$

$$= 2.6 \text{ mL (rounded off)}$$

According to Young's Rule, the child's dose would be 2.6 mL of castor oil.

NAME _____ SECTION _____ DATE _____

EXERCISES **10.4** _____

Use the table of Fluid Measure Equivalents and proportions to find the
following equivalent measures.

1. 5 gtt = _____ mL

2. 7 mL = _____ gtt

3. 4 tbsp = _____ mL

4. 3 tbsp = _____ f̵3

5. 6 f̵3 = _____ tsp

6. 10 mL = _____ tsp

7. 0.18 mL = _____ gtt

8. $2\frac{1}{2}$ glassfuls = _____ mL

Answers

1. _____

2. _____

3. _____

4. _____

5. _____

6. _____

7. _____

8. _____

Answers

9. _____

10. _____

11. _____

12. _____

13. _____

14. _____

15. _____

16. _____

17. _____

18. _____

9. 5 tbsp = _____ $f\!\!\!3$

10. 4 tbsp = _____ $f\!\!\!3$

11. 360 mL = _____ glassfuls

12. 4 gtt = _____ $\mathfrak{m}$

13. 5 $\mathfrak{m}$ = _____ gtt

14. 30 gtt = _____ mL

15. 30 mL = _____ tsp

16. 6 $f\!\!\!3$ = _____ glassfuls

17. 45 mL = _____ tbsp

18. 5 tsp = _____ $f\!\!\!3$

NAME _____ SECTION _____ DATE _____

19. 150 ℳ = _____ gtt

20. 6 glassfuls = _____ mL

19. _____

20. _____

21. _____

22. _____

Use Fried's Rule to find the child's dose from the given information.

23. _____

21. Child's age: 9 months
Adult dose: 250 mL liquid
Ampicillin

22. Child's age: 5 months
Adult dose: 30 mL milk of
magnesia

23. Child's age: 20 months
Adult dose: 15 mL castor oil

Answers

24. _____

25. _____

26. _____

27. _____

28. _____

29. _____

Use Young's Rule to find the child's dose from the given information.

24. Child's age: 5 years
Adult dose: 250 mL liquid
Ampicillin

25. Child's age: 8 years
Adult dose: 10 mL
Robitussin®

26. Child's age: 6 years
Adult dose: 15 mL castor oil

Use Clark's Rule to find the child's dose from the given information.

27. Child's weight: 20 lb
Adult dose: 5 mL Polara-
mine® syrup

28. Child's weight: 60 lb
Adult dose: 250 mL
penicillin

29. Child's weight: 75 lb
Adult dose: 15 mL castor oil

NAME _____ SECTION _____ DATE _____

CHAPTER 10 TEST _____

In Problems 1–30, change the units as indicated.

1. 7 cm = _____ m

2. 15 m = _____ cm

3. 173 mm = _____ m

4. 0.3 m = _____ mm

5. 17 g = _____ kg

6. 103 kg = _____ g

7. 42 kg = _____ t

8. 6 t = _____ kg

9. 4 m^2 = _____ cm^2

10. 16 cm^2 = _____ m^2

11. 45 gtt = _____ mL

12. 12 tsp = _____ f℥

13. 2 L = _____ cm^3

14. 175 mL = _____ L

15. 2.75 ft = _____ in.

16. $6\frac{1}{2}$ yd = _____ ft

Answers

1. _____

2. _____

3. _____

4. _____

5. _____

6. _____

7. _____

8. _____

9. _____

10. _____

11. _____

12. _____

13. _____

14. _____

15. _____

16. _____

Answers

17. _____

18. _____

19. _____

20. _____

21. _____

22. _____

23. _____

24. _____

25. _____

26. _____

27. _____

28. _____

29. _____

30. _____

31. _____

17. 74 in. = _____ ft

18. 2 ft = _____ yd

19. $4\frac{3}{4}$ lb = _____ oz

20. 1.3 t = _____ lb

21. 54 oz = _____ lb

22. 800 lb = _____ t

23. 3.2 hr = _____ min

24. $2\frac{5}{12}$ days = _____ hr

25. 168 sec = _____ min

26. 84 hr = _____ days

27. 1.6 pt = _____ fl oz

28. $2\frac{3}{8}$ gal = _____ qt

29. $3\frac{1}{4}$ qt = _____ gal

30. 22 fl oz _____ pt

31. Use the appropriate rule for calculating the child's dose from the following information:

> Child's weight: 45 lb
> Adult dose: 250 mL liquid Ampicillin

Geometry

11.1

LENGTH AND PERIMETER

Introduction to Geometry

Plane geometry is the study of the properties of figures in a plane. The most basic ideas or concepts are **point, line,** and **plane.** These terms name ideas that cannot be formally defined. They are called **undefined terms.**

UNDEFINED TERM	REPRESENTATION	DISCUSSION
1. **Point**	$A \bullet$ point A	A dot represents a point. Points are labeled with capital letters.
2. **Line**	line ℓ or line $\overleftrightarrow{AB}$	A line has no beginning or end. Lines are labeled with small letters or by two points on the line.
3. **Plane**	plane P	Flat surfaces, such as a table top or wall, represent planes. Planes are labeled with capital letters.

In a more formal approach to plane geometry than we will use in this text, these undefined terms provide the basis for defining other terms such as line segment, ray, angle, triangle, and so on. Then, with these terms and certain assumptions known as axioms, other statements about geometry, called theorems, are proven. This approach to the study of geometry is credited to Euclid (about 300 BC) which is why the plane geometry courses generally given in high school are known as Euclidean geometry.

In this chapter, we will discuss formulas and properties related to several plane geometric figures, including rectangles, triangles, and circles, and some three-dimensional figures such as rectangular solids and spheres. The formulas do not depend on any particular measurement system, and both the metric system and US customary system of measure will be used in the exercises.

Be sure to label all answers with the correct unit of measure indicating length, area, or volume.

Length and Perimeter

From the metric system, some of the units of length are meter (m), decimeter (dm), centimeter (cm), and millimeter (mm). From the US customary system, some of the units of length are foot (ft), inch (in.), and yard (yd). The following geometric figures and the related formulas illustrate applications of measures of length. The formulas are independent of the measurement system used.

> **Perimeter:** Total distance around a plane geometric figure
>
> **Circumference:** Perimeter of a circle
>
> **Radius:** The distance from the center of a circle to a point on the circle
>
> **Diameter:** Distance from one point on a circle to another point on the circle measured through the center
>
> Six geometric figures and the formulas for finding their perimeters are shown here.
>
>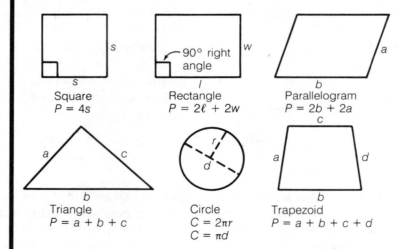
>
> Square   Rectangle   Parallelogram
> $P = 4s$ $P = 2\ell + 2w$ $P = 2b + 2a$
>
> Triangle   Circle   Trapezoid
> $P = a + b + c$ $C = 2\pi r$ $P = a + b + c + d$
> $C = \pi d$
>
> **Note:** π is the symbol used for the constant number 3.1415926535 . . . . This constant is an infinite decimal with no pattern to its digits. For our purposes, we will use $\pi = 3.14$, but you should be aware that 3.14 is only an approximation to π.

Example 1   Find the perimeter of a rectangle with length 20 in. and width 15 in.

Solution

Sketch the figure first.

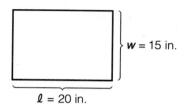

$w = 15$ in.

$\ell = 20$ in.

$P = 2\ell + 2w$

$P = 2 \cdot 20 + 2 \cdot 15$

$\quad = 40 + 30 = 70$ in.

The perimeter is 70 inches.

Example 2   Find the circumference of a circle with diameter 3 m.

Solution

Sketch the figure first.

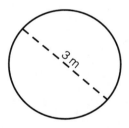

3 m

$C = \pi d$

$C = 3.14(3) = 9.42$ m

The circumference is 9.42 m. (**Note:** Remember that 9.42 m is only an approximate answer because 3.14 is an approximation of π.)

1. Find the circumference of a circle with radius 4 in.

Example 3 Find the perimeter of a triangle with sides of 4 cm, 0.7 dm, and 80 mm. Write your answer in millimeters.

Solution

Sketch the figure and change all the units to millimeters.

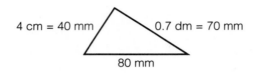

4 cm = 40 mm   0.7 dm = 70 mm

80 mm

$P = a + b + c$

$P = 40 + 70 + 80$

$\quad = 190$ mm

The perimeter is 190 mm.

Completion Example 4 Find the circumference of a circle with radius 6 ft.

Solution

Sketch the figure first.

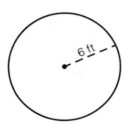

6 ft

$C = 2\pi r$

$C = \underline{\quad} \cdot \underline{\quad} \cdot \underline{\quad}$

$\quad = \underline{\quad}$ ft

The circumference is _____ ft.

2. Find the perimeter of a square with sides of 3.6 cm.

☐ **NOW WORK EXERCISES 1 AND 2 IN THE MARGIN.**

Completion Example Answer

4. $C = 2 \cdot 3.14 \cdot 6$
$\quad = 37.68$ ft
The circumference is 37.68 ft.

Answers to Marginal Exercises

1. 25.12 in.   **2.** 14.4 cm

NAME _____ SECTION _____ DATE _____

EXERCISES **11.1** _____

Answers

1. Match each formula for perimeter to its corresponding geometric figure.

    ____ (a) square   (1) $P = 2l + 2w$

    ____ (b) parallelogram   (2) $P = 4s$

    ____ (c) circle   (3) $P = 2b + 2a$

    ____ (d) rectangle   (4) $C = 2\pi r$

    ____ (e) trapezoid   (5) $P = a + b + c$

    ____ (f) triangle   (6) $P = a + b + c + d$

1. (a) _____

    (b) _____

    (c) _____

    (d) _____

    (e) _____

    (f) _____

2. Find the perimeter of a triangle with sides of 4 cm, 8.3 cm, and 6.1 cm.

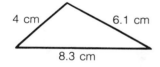

2. _____

3. _____

3. Find the perimeter of a rectangle with length 35 mm and width 17 mm.

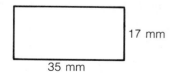

Answers

4. _____

5. _____

6. _____

7. _____

4. Find the perimeter of a square with sides of 13.3 m.

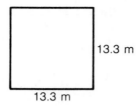

13.3 m

13.3 m

5. Find the circumference of a circle with radius 5 ft. (Use $\pi = 3.14$.)

5 ft

6. Find the perimeter of a parallelogram with one side 43 cm and another side 20 mm. Write your answer in millimeters.

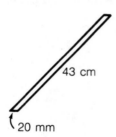

43 cm

20 mm

7. Find the circumference of a circle with diameter 6.2 yd. (Use $\pi = 3.14$.)

6.2 yd

NAME _____ SECTION _____ DATE _____

8. Find the perimeter of a rectangle with length 50 m and width 50 dm. Write your answer in meters.

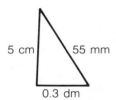

50 dm

50 m

9. Find the perimeter of a triangle with sides of 5 cm, 55 mm, and 0.3 dm. Write your answers in centimeters.

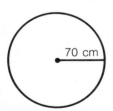

5 cm   55 mm

0.3 dm

10. Find the circumference of a circle in meters if its radius is 70 cm. (Use $\pi = 3.14$.)

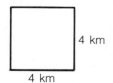

70 cm

11. Find the perimeter of a square in meters if one side is 4 km long.

4 km

4 km

Answers

8. _____

9. _____

10. _____

11. _____

Answers

12. _____

13. _____

14. _____

15. _____

12. Find the perimeter of a triangle with sides of 1 yd, 4 ft, and 5 ft. Write your answer in feet.

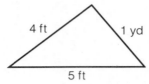

13. Find the perimeter of the trapezoid shown here.

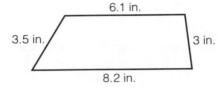

14. Find the perimeter of a parallelogram with sides of $3\frac{1}{2}$ ft and 14 inches. Write your answer in both inches and feet.

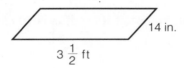

15. Find the circumference of a circle with diameter 1.5 in. Write your answer in centimeters.

11.2
AREA

Area is the measure of the interior, or enclosure, of a surface and is measured in **square units.** From the metric system, some of the units of area are square meters (m^2), square decimeters (dm^2), square centimeters (cm^2), and square millimeters (mm^2). From the US customary system, some of the units of area are square feet (ft^2), square inches ($in.^2$), and square yards (yd^2). The following formulas for the areas of the geometric figures shown are valid regardless of the system of measurement used.

Be sure to label your answers with the correct units.

Six geometric figures and the formulas for finding their **areas** are shown here.

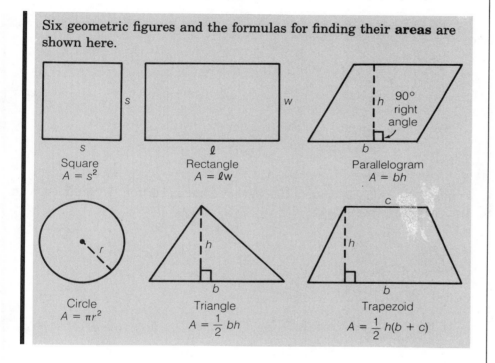

Square
$A = s^2$

Rectangle
$A = \ell w$

Parallelogram
$A = bh$

Circle
$A = \pi r^2$

Triangle
$A = \frac{1}{2} bh$

Trapezoid
$A = \frac{1}{2} h(b + c)$

Example 1 Find the area of the figure shown here with the indicated dimensions.

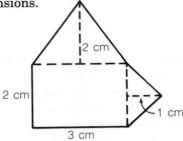

Solution

There are two triangles and one rectangle.

RECTANGLE	LARGER TRIANGLE	SMALLER TRIANGLE
$A = lw$	$A = \dfrac{1}{2}bh$	$A = \dfrac{1}{2}bh$
$A = 2 \cdot 3 = 6 \text{ cm}^2$	$A = \dfrac{1}{2} \cdot 3 \cdot 2 = 3 \text{ cm}^2$	$A = \dfrac{1}{2} \cdot 2 \cdot 1 = 1 \text{ cm}^2$

$$\text{Total area} = 6 \text{ cm}^2 + 3 \text{ cm}^2 + 1 \text{ cm}^2$$
$$= 10 \text{ cm}^2$$

Example 2 Find the area of the washer (shaded portion) with dimensions as shown. (Use $\pi = 3.14$.)

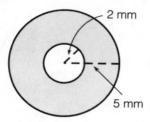

Solution

Subtract the area of the inside (smaller) circle from the area of the outside (larger) circle.

LARGER CIRCLE	SMALLER CIRCLE
$A = \pi r^2$	$A = \pi r^2$
$A = 3.14(5^2)$	$A = 3.14(2^2)$
$= 3.14(25)$	$= 3.14(4)$
$= 78.50 \text{ mm}^2$	$= 12.56 \text{ mm}^2$

WASHER

$$\begin{array}{r} 78.50 \text{ mm}^2 \\ -12.56 \text{ mm}^2 \\ \hline 65.94 \text{ mm}^2 \end{array} \quad \text{area of washer}$$

Example 3 Find the area of a trapezoid with altitude 6 in. and parallel sides of length 1 ft and 2 ft.

Solution

First draw a figure and label the lengths of the known parts. The units of length must be the same throughout, either all in inches or all in feet.

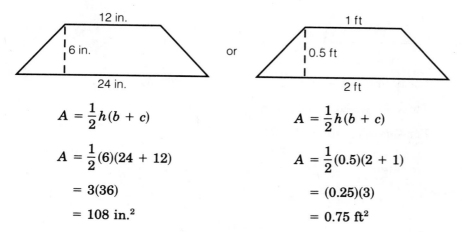

$$A = \frac{1}{2}h(b + c) \qquad\qquad A = \frac{1}{2}h(b + c)$$

$$A = \frac{1}{2}(6)(24 + 12) \qquad\qquad A = \frac{1}{2}(0.5)(2 + 1)$$

$$= 3(36) \qquad\qquad\qquad = (0.25)(3)$$

$$= 108 \text{ in.}^2 \qquad\qquad\qquad = 0.75 \text{ ft}^2$$

Two correct areas are 108 in.² and 0.75 ft². (Note that, since 1 ft² = (12 in.)(12 in.) = 144 in.², we have 0.75 ft² = 0.75(144) in.² = 108 in.² The two answers are equivalent. They are simply in different units of area.)

1. Find the area of a circle with diameter 6 in.

Completion Example 4 Find the area enclosed by a semicircle and a diameter if the diameter is 10 in.

Solution

First sketch the figure. (A semicircle is half of a circle.)

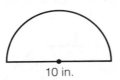

10 in.

For a circle, $A = \pi r^2$.

Thus, for a semicircle, $A = \dfrac{1}{2}\pi r^2$.

For this semicircle, $d = 10$, so $r = 5$.

$$A = \frac{1}{2}(\underline{\quad}) \cdot (\underline{\quad})^2$$

$$= \underline{\quad} \cdot \underline{\quad}$$

$$= \underline{\quad} \text{ in.}^2$$

The area enclosed by the semicircle and a diameter is $\underline{\quad}$ in.^{2}

☐ **NOW WORK EXERCISES 1 AND 2 IN THE MARGIN.**

2. Find the area of a triangle with base 10 cm and height 5 cm.

Completion Example Answer

4. $A = \dfrac{1}{2}(3.14)(5)^2$
   $= 1.57 \cdot 25$
   $= 39.25$ in.^{2}

   The area enclosed by the semicircle and a diameter is 39.25 in.^{2}

Answers to Marginal Exercises

1. 28.26 in.^{2} 2. 25 cm^2

NAME _____ SECTION _____ DATE _____

EXERCISES **11.2** _____

Answers

1. Match each formula for area to its corresponding geometric figure.

_____ (a) square

_____ (b) parallelogram

_____ (c) circle

_____ (d) rectangle

_____ (e) trapezoid

_____ (f) triangle

(1) $A = lw$

(2) $A = bh$

(3) $A = s^2$

(4) $A = \pi r^2$

(5) $A = \frac{1}{2}bh$

(6) $A = \frac{1}{2}h(b + c)$

First sketch the figure and label the given parts; then find the area of each of the following figures.

2. A rectangle 35 in. long and 25 in. wide

3. A triangle with base 2 cm and altitude 6 cm

4. A triangle with base 5 mm and altitude 8 mm

1. (a) _____

(b) _____

(c) _____

(d) _____

(e) _____

(f) _____

2. _____

3. _____

4. _____

5. _____

6. _____

7. _____

8. _____

9. _____

5. A circle of radius 5 yd (Use $\pi = 3.14$.)

6. A circle of radius 1.5 ft (Use $\pi = 3.14$.)

7. A trapezoid with parallel sides of 3 in. and 10 in., and altitude of 35 in.

8. A trapezoid with parallel sides of 3.5 mm and 4.2 mm, and altitude of 1 cm

9. A parallelogram with altitude 10 cm to a base of 5 mm

NAME _____ SECTION _____ DATE _____

Find the area of each of the following figures with the indicated dimensions. (Use $\pi = 3.14$.)

10.

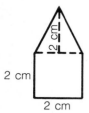

11.

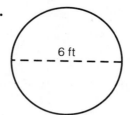

12.

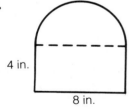

13.

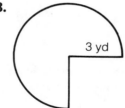

14.

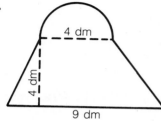

15.

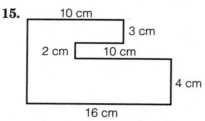

10. _____

11. _____

12. _____

13. _____

14. _____

15. _____

16. _____

17. _____

18. _____

19. _____

20. _____

21. _____

22. _____

Find the areas of the shaded portions in ares. (Use $\pi = 3.14$.)

16.

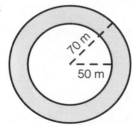

70 m
50 m

17.

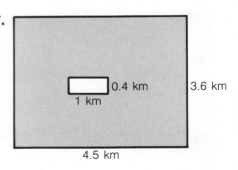

0.4 km
1 km
3.6 km
4.5 km

18. Find the area of a circle of radius 1 ft.

19. Find the area of a circle with diameter 1 ft.

20. Find the area of a rectangle 2 m long and 60 cm wide. Write your answer in both cm^2 and m^2.

21. Find the area of a square with sides of 50 cm. Write your answer in both cm^2 and m^2.

22. Find the area of a rectangle 0.5 ft long and 35 in. wide. Write your answer in both in.^{2} and ft^2.

11.3
VOLUME

Volume is a measure of the space enclosed by a three-dimensional figure and is measured in **cubic units.** Cubic meters (m^3), cubic decimeters (dm^3), cubic centimeters (cm^3), and cubic millimeters (mm^3) are from the metric system. Cubic feet (ft^3), cubic inches (in.^{3}), and cubic yards (yd^3) are from the US customary system. The following formulas for the volumes of the geometric solids shown do not depend on the measurement system used.

Five geometric solids and the formulas for their volumes are shown here.

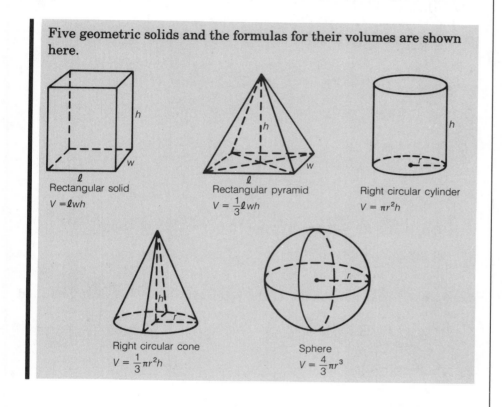

Rectangular solid
$V = \ell w h$

Rectangular pyramid
$V = \frac{1}{3}\ell w h$

Right circular cylinder
$V = \pi r^2 h$

Right circular cone
$V = \frac{1}{3}\pi r^2 h$

Sphere
$V = \frac{4}{3}\pi r^3$

Example 1 Find the volume of the rectangular solid with length 8 in., width 4 in., and height 1 ft. Write your answer in cubic inches and in cubic feet.

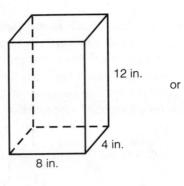 or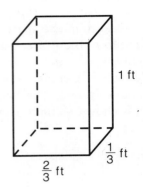

Solution

$$V = \ell wh$$

$$V = 8 \cdot 4 \cdot 12$$

$$= 384 \text{ in.}^3$$

$$V = \ell wh$$

$$V = \frac{2}{3} \cdot \frac{1}{2} \cdot 1$$

$$= \frac{2}{9} \text{ ft}^3$$

(Note that 1 ft^3 = (12 in.)(12 in.)(12 in.) = 1728 in.^{3} and $\frac{2}{9}$(1728 in.^{3}) = 384 in.^{3})

Example 2 Find the volume of the solid with the dimensions indicated. (Use $\pi = 3.14$.)

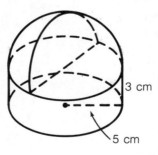

Solution

On top of the cylinder is a hemisphere (half a sphere). Find the volume of the cylinder and the hemisphere and add the results.

CYLINDER   HEMISPHERE

$V = \pi r^2 h$ $V = \dfrac{1}{2} \cdot \dfrac{4}{3} \pi r^3$

$V = 3.14(5^2)(3)$ $V = \dfrac{2}{3}(3.14)(5^3)$

$\quad = 235.5 \text{ cm}^3$ $\quad = 261.67 \text{ cm}^3$

TOTAL VOLUME

$\quad 235.50 \text{ cm}^3$
$\underline{+261.67 \text{ cm}^3}$
$\quad 497.17 \text{ cm}^3$ (or 497.17 mL)   total volume

☐ **NOW WORK EXERCISE 1 IN THE MARGIN.**

1. Find the volume of a sphere with diameter 8 ft.

Answer to Marginal Exercise

1. 267.95 ft^3

NAME _____   SECTION _____   DATE _____

EXERCISES **11.3** _____

1. Match each formula for volume to its corresponding geometric figure.

    ____ (a) rectangular solid   (1) $V = \dfrac{4}{3}\pi r^3$

    ____ (b) rectangular pyramid   (2) $V = \dfrac{1}{3}\pi r^2 h$

    ____ (c) right circular cylinder   (3) $V = lwh$

    ____ (d) right circular cone   (4) $V = \pi r^2 h$

    ____ (e) sphere   (5) $V = \dfrac{1}{3} lwh$

Find the volume of each of the following solids in a convenient unit. (Use $\pi = 3.14$.)

2. A rectangular solid with length 5 in., width 2 in., and height 7 in.

3. A right circular cylinder 15 in. high and 1 ft in diameter.

Answers

1. (a) _____

   (b) _____

   (c) _____

   (d) _____

   (e) _____

2. _____

3. _____

Answers

4. _____

5. _____

6. _____

7. _____

4. A sphere with radius 4.5 cm.

5. A sphere with diameter 12 ft.

6. A right circular cone 3 dm high with a 2-dm radius.

7. A rectangular pyramid with length 8 cm, width 10 mm, and height 3 dm.

NAME _____ SECTION _____ DATE _____

Find the volume of each of the following solids with the dimensions indicated.

8.

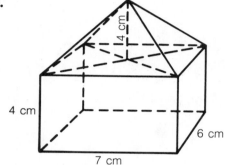

9.

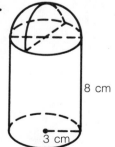

10.

Answers
8. _____
9. _____
10. _____

Answers

11. _____

12. _____

13. _____

11.

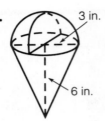

3 in.

6 in.

12.

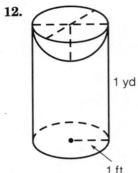

1 yd

1 ft

13.

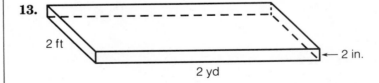

2 ft

2 yd

2 in.

11.4
ANGLES

We begin the discussion of angles with two definitions using the undefined terms, **point** and **line**.

TERM	DEFINITION	ILLUSTRATION
1. **Ray**	A **ray** consists of a point (called the **endpoint**) and all the points on a line on one side of that point.	ray $\overrightarrow{PQ}$ with endpoint P
2. **Angle**	An **angle** consists of two rays with a common endpoint (called a **vertex**).	$\angle AOB$ with vertex O

In an angle, the two rays are called the **sides** of the angle.

Every angle has a **measurement** or **measure** associated with it. Suppose that a circle is divided into 360 equal arcs. If two rays are drawn from the center of the circle through two consecutive points of division on the circle, then that angle is said to **measure one degree** (symbolized 1°). For example, in Figure 11.1, a device called a protractor shows that the measure of $\angle AOB$ is 60 degrees. (We write m$\angle AOB = 60°$.)

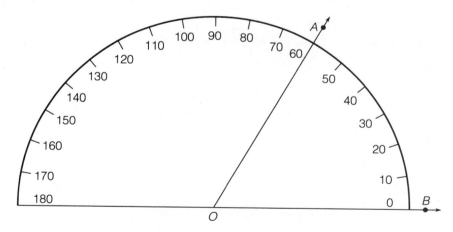

FIGURE 11.1 The Protractor Shows m$\angle AOB = 60°$

To measure an angle with a protractor, lay the bottom edge of the protractor along one side of the angle with the vertex at the marked center point. Then read the measure from the protractor where the other side of the angle crosses it. (See Figure 11.2.)

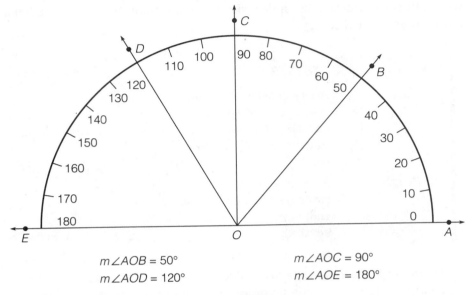

$m \angle AOB = 50°$ $m \angle AOC = 90°$

$m \angle AOD = 120°$ $m \angle AOE = 180°$

FIGURE 11.2

Three common ways of labeling angles (Figure 11.3) are:

1. Using three capital letters with the vertex as the middle letter.
2. Using single numbers such as 1, 2, 3.
3. Using the single capital letter at the vertex when the meaning is clear.

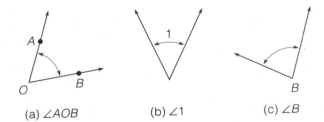

(a) ∠AOB   (b) ∠1   (c) ∠B

FIGURE 11.3 Three Ways of Labeling Angles

Angles can be classified (or named) according to their measures.

Note: We will use the two inequality symbols

$<$ (read "is less than")

$>$ (read "is greater than")

to indicate the relative sizes of the measures of angles.

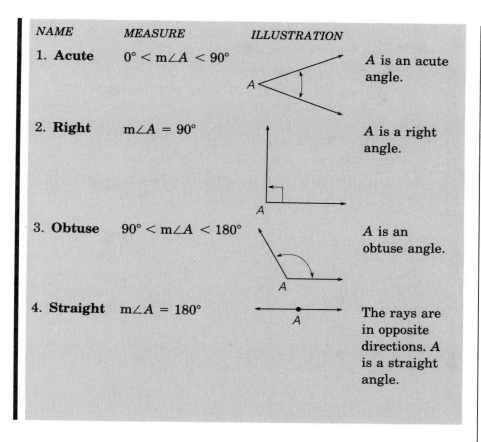

NAME	MEASURE	ILLUSTRATION	
1. **Acute**	$0° < m\angle A < 90°$		A is an acute angle.
2. **Right**	$m\angle A = 90°$		A is a right angle.
3. **Obtuse**	$90° < m\angle A < 180°$		A is an obtuse angle.
4. **Straight**	$m\angle A = 180°$		The rays are in opposite directions. A is a straight angle.

The following figure is used for Examples 1–4.

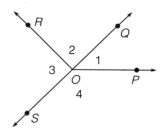

Example 1 Use a protractor to check the measures of the following angles. (a) $m\angle 1 = 45°$ (b) $m\angle 2 = 90°$ (c) $m\angle 3 = 90°$ (d) $m\angle 4 = 135°$

Example 2 Tell whether each of the following angles is acute, right, obtuse, or straight.

(a) $\angle 1$

(b) $\angle 2$

(c) $\angle POR$

Solution

(a) $\angle 1$ is acute since $0° < m\angle 1 < 90°$.

(b) $\angle 2$ is a right angle since $m\angle 2 = 90°$.

(c) $\angle POR$ is obtuse since $m\angle POR = 45° + 90° = 135° > 90°$.

> **Two angles are:**
>
>   1. **Complementary** if the sum of their measures is 90°.
>   2. **Supplementary** if the sum of their measures is 180°.
>   3. **Equal** if they have the same measure.

Example 3 In the figure, (a) $\angle 1$ and $\angle 2$ are **complementary** since $m\angle 1 + m\angle 2 = 90°$. (b) $\angle COD$ and $\angle COA$ are **supplementary** since $m\angle COD + m\angle COA = 70° + 100° = 180°$. (c) $\angle AOD$ is a straight angle since $m\angle AOD = 180°$.

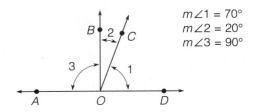

$m\angle 1 = 70°$
$m\angle 2 = 20°$
$m\angle 3 = 90°$

Example 4 In the figure, $\overrightarrow{PS}$ is a straight line and $m\angle QOP = 30°$. Find the measures of (a) $\angle QOS$ and (b) $\angle SOP$. (c) Are any pairs supplementary?

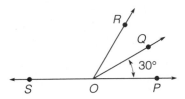

Solution

(a) $m\angle QOS = 150°$.

(b) $m\angle SOP = 180°$.

(c) Yes. $\angle QOP$ and $\angle QOS$ are supplementary and $\angle ROP$ and $\angle ROS$ are supplementary.

If two lines intersect, then two pairs of vertical angles are formed. **Vertical angles** are also called **opposite angles.** (See Figure 11.4.)

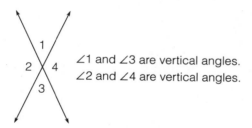

∠1 and ∠3 are vertical angles.
∠2 and ∠4 are vertical angles.

FIGURE 11.4

Vertical angles are equal. That is, **vertical angles have the same measure.** (In Figure 11.4, m∠1 = m∠3 and m∠2 = m∠4.)

Two angles are **adjacent** if they have a common side. (See Figure 11.5.)

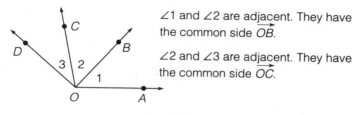

∠1 and ∠2 are adjacent. They have the common side $\overrightarrow{OB}$.

∠2 and ∠3 are adjacent. They have the common side $\overrightarrow{OC}$.

FIGURE 11.5

Example 5 In the figure, $\overleftrightarrow{AC}$ and $\overleftrightarrow{BD}$ are straight lines. (a) Name an angle adjacent to ∠EOD. (b) What is m∠AOD?

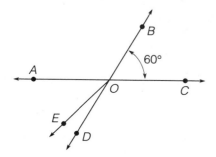

Solution

(a) There are three angles adjacent to ∠EOD: ∠AOE, ∠BOE, ∠COD.

(b) Since ∠BOC and ∠AOD are vertical angles, they have the same measure. So, m∠AOD = 60°.

NAME _____ SECTION _____ DATE _____

EXERCISES **11.4** _____

Answers

1. If ∠1 and ∠2 are complementary, and (a) m∠1 = 15°, what is m∠2? (b) m∠1 = 3°, what is m∠2? (c) m∠1 = 45°, what is m∠2? (d) m∠1 = 75°, what is m∠2?

1. (a) _____

(b) _____

(c) _____

2. If ∠3 and ∠4 are supplementary and (a) m∠3 = 45°, what is m∠4? (b) m∠3 = 90°, what is m∠4? (c) m∠3 = 110°, what is m∠4? (d) m∠3 = 135°, what is m∠4?

(d) _____

2. (a) _____

(b) _____

3. The supplement of an acute angle is an obtuse angle. (a) What is the supplement of a right angle? (b) What is the supplement of an obtuse angle? (c) What is the complement of an acute angle?

(c) _____

(d) _____

3. (a) _____

(b) _____

4. In the figure shown, $\overleftrightarrow{DC}$ is a straight line and m∠BOA = 90°. (a) What type of angle is ∠AOC? (b) What type of angle is ∠BOC? (c) What type of angle is ∠BOA? (d) Name a pair of complementary angles. (e) Name two pairs of supplementary angles.

(c) _____

4. (a) _____

(b) _____

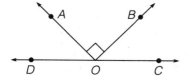

(c) _____

(d) _____

(e) _____

5. An **angle bisector** is a ray that divides an angle into two angles with equal measures. If $\overrightarrow{OX}$ bisects $\angle COD$ and $m\angle COD = 50°$, what is the measure of each of the equal angles formed?

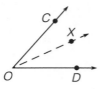

Given $m\angle AOB = 30°$ and $m\angle BOC = 80°$ and $\overrightarrow{OX}$ and $\overrightarrow{OY}$ are angle bisectors, find the measures of the following angles.

6. $\angle AOX$　　　　　　7. $\angle BOY$

8. $\angle COX$　　　　　　9. $\angle YOX$

10. $\angle AOY$　　　　　11. $\angle AOC$

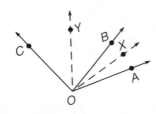

12. In the figure shown, (a) name all the pairs of supplementary angles. (b) Name all the pairs of complementary angles.

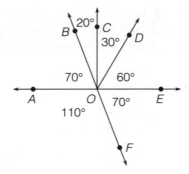

NAME _____ SECTION _____ DATE _____

Use a protractor to measure all the angles in each figure. Each line
segment may be extended as a ray to form the side of an angle.

Answers

13.

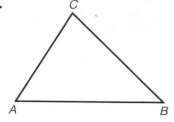

14.

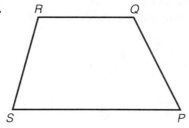

13. _____

_____

14. _____

_____

15. _____

_____

15.

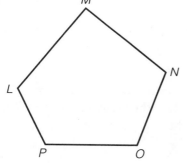

16. F

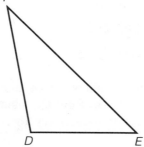

16. _____

17. (a) _____

(b) _____

(c) _____

17. Name the type of angle formed by the hands on a clock (a) at six
o'clock (b) at three o'clock (c) at one o'clock (d) at five o'clock.

(d) _____

18. (a) _____

(b) _____

18. What is the measure of each angle formed by the hands of the clock
in Exercise 17?

(c) _____

(d) _____

19. The figure shows two intersecting lines. (a) If m∠1 = 30°, what is m∠2? (b) Is m∠3 = 30°? Give a reason for your answer other than the fact that ∠1 and ∠3 are vertical angles. (c) Name four pairs of adjacent angles.

20. In the figure shown, $\overrightarrow{AB}$ is a straight line. (a) Name two pairs of adjacent angles. (b) Name two vertical angles if there are any.

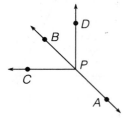

21. Given that m∠1 = 30° in the figure shown, find the measures of the other three angles.

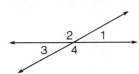

22. In the figure shown, ℓ, m, and n are straight lines with m∠1 = 20° and m∠6 = 90°. (a) Find the measures of the other four angles. (b) Which angle is supplementary to ∠6? (c) Which angles are complementary to ∠1?

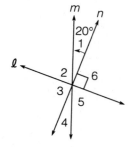

23. In the figure shown, m∠2 = m∠3 = 40°. Find all other pairs of angles that have equal measures.

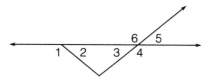

11.5
TRIANGLES _____

A **triangle** consists of the three line segments that join three points that do not lie on a straight line. The line segments are called the **sides** of the triangle, and the points are called the **vertices** of the triangle. If the points are labeled A, B, and C, the triangle is symbolized $\triangle ABC$. (See Figure 11.6.)

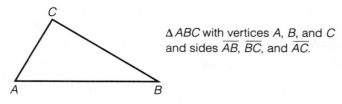

$\triangle ABC$ with vertices A, B, and C
and sides $\overline{AB}$, $\overline{BC}$, and $\overline{AC}$.

FIGURE 11.6

The sides of a triangle are said to determine three angles, and these angles are labeled by the vertices. Thus, the angles of $\triangle ABC$ are $\angle A$, $\angle B$, and $\angle C$. (Since the definition of angle involves rays, we can think of the sides of the triangle extended as rays to form these angles.)

Triangles are classified in two ways: (1): according to lengths of their sides, and (2) according to the measures of their angles. The corresponding names and properties are listed in the following tables.

(**Note:** We will indicate the length of a line segment, such as $\overline{AB}$, by writing only the letters, such as AB.)

TRIANGLES CLASSIFIED BY SIDES

NAME	PROPERTY	ILLUSTRATION
1. **Scalene**	No two sides are equal.	$\triangle ABC$ is scalene since no two sides are equal.
2. **Isosceles**	At least two sides are equal.	$\triangle PQR$ is isosceles since $PR = QR$.
3. **Equilateral**	All three sides are equal.	$\triangle XYZ$ is equilateral since $XY = XZ = YZ$.

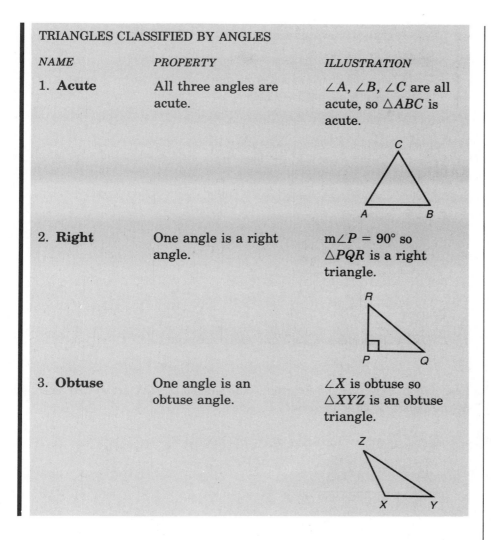

TRIANGLES CLASSIFIED BY ANGLES

NAME	PROPERTY	ILLUSTRATION
1. **Acute**	All three angles are acute.	$\angle A$, $\angle B$, $\angle C$ are all acute, so $\triangle ABC$ is acute.
2. **Right**	One angle is a right angle.	$m\angle P = 90°$ so $\triangle PQR$ is a right triangle.
3. **Obtuse**	One angle is an obtuse angle.	$\angle X$ is obtuse so $\triangle XYZ$ is an obtuse triangle.

Every triangle is said to have six parts, namely three angles and three sides. Two sides of a triangle are said to **include** the angle at their common endpoint or vertex. The third side is said to be **opposite** this angle.

The sides in a right triangle have special names. The longest side, opposite the right angle, is called the **hypotenuse** and the other two sides are called **legs.** (See Figure 11.7.)

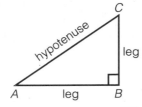

$\triangle ABC$ is a right triangle. $m\angle B = 90°$. $\overline{AC}$ is opposite $\angle B$.

FIGURE 11.7

> **Two important statements can be made about any triangle:**
>
> 1. The sum of the measures of the angles is 180°.
> 2. The sum of the lengths of any two sides must be greater than the length of the third side.

Example 1   In the figure of △ABC, AB = AC. What kind of triangle is △ABC?

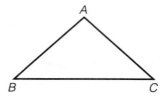

Solution

△ABC is isosceles because two sides are equal.

Example 2   Suppose the lengths of the sides of △PQR are stated as shown in the figure. Is this possible?

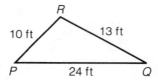

Solution

This is not possible because PR + QR = 10 ft + 13 ft = 23 ft and PQ = 24 ft, which is longer than the sum of the other two sides. In a triangle, the sum of the lengths of any two sides must be greater than the length of the third side.

Example 3   In the figure △BOR, m∠B = 50° and m∠O = 70°. (a) What is m∠R? (b) What kind of triangle is △BOR? (c) Which side is opposite ∠R? (d) Which sides include ∠R? (e) Is △BOR a right triangle? Why?

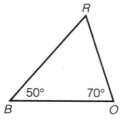

Solution

(a) The sum of the measures of the angles must be 180°. Since 50° + 70° = 120°,

$$m\angle R = 180° - 120° = 60°$$

(b) $\triangle BOR$ is an acute triangle since all the angles are acute. Also, $\triangle BOR$ is scalene because no two sides are equal.

(c) $\overline{BO}$ is opposite $\angle R$.

(d) $\overline{RB}$ and $\overline{RO}$ include $\angle R$.

(e) $\triangle BOR$ is not a right triangle because none of the angles is a right angle.

On an intuitive level, two triangles are said to be **similar** if they have the same "shape." They may or may not have the same "size." More formally (and more usefully), similar triangles have the following two important properties.

Two triangles are similar if:

1. Their **corresponding angles are equal.**

2. Their **corresponding sides are proportional.**
   (See Figure 11.8.)

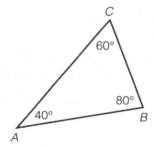

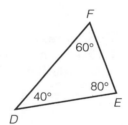

∠A corresponds to ∠D.
∠B corresponds to ∠E.
∠C corresponds to ∠F.

The corresponding sides are proportional.

We write $\triangle ABC \sim \triangle DEF$. ($\sim$ is read "is similar to.")

FIGURE 11.8

Example 4 Consider the two triangles △*ABC* and △*AXY* as shown with
m∠*ABC* = m∠*AXY* = 90°.

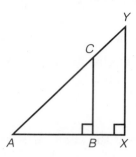

Solution

We can show that the corresponding angles are equal as follows:
m∠*CAB* = m∠*YAX* because they are the same angle.
m∠*CBA* = m∠*YXA* because both are right angles (90°).
m∠*BCA* = m∠*XYA* because the sum of the measures of the angles in
each triangle must be 180°.
Therefore, the corresponding angles are equal, and the triangles are
similar.

Example 5 Refer to the figure used in Example 4. If *AB* = 4 cm,
AX = 8 cm, and *BC* = 3 cm, find *XY*.

Solution

Since $\overline{AB}$ and $\overline{AX}$ are corresponding sides (they are opposite equal angles) and $\overline{BC}$ and $\overline{XY}$ are corresponding sides (they are opposite equal
angles), we have the proportion:

$$\frac{AB}{AX} = \frac{BC}{XY}$$

Thus,

$$\frac{4 \text{ cm}}{8 \text{ cm}} = \frac{3 \text{ cm}}{XY}$$

$$4 \cdot XY = 3 \cdot 8$$

$$\frac{\cancel{4} \cdot XY}{\cancel{4}} = \frac{24}{4}$$

$$XY = 6 \text{ cm}$$

NAME _____ SECTION _____ DATE _____

EXERCISES **11.5** _____

Name each of the following triangles with the indicated measures of angles and lengths of sides.

1.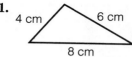

2.

3.

1. _____

2. _____

3. _____

4. _____

5. _____

4.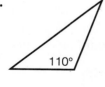

5.

6.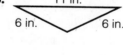

6. _____

7. _____

8. _____

9. _____

7.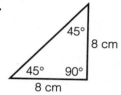

8.

9.

Answers

10. _____

11. _____

12. _____

13. _____

14. _____

10. The Pythagorean Theorem states that, in a right triangle, the square of the hypotenuse is equal to the sum of the squares of the two legs. See the following figures. Are the triangles △ABC and △DEF right triangles? If so, which sides are the hypotenuses? Are the triangles similar? If so, which angles are equal?

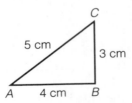

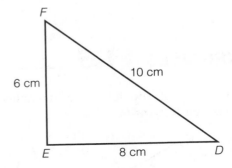

11. In the figure, m∠Q, = m∠T. Is △PQR ~ △PST? Give reasons for your answer.

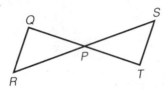

12. Suppose △XYZ ~ △UVW, m∠Z = 30°, and m∠W = 30°. If both triangles are isosceles, what are the measures of the other four angles? (In an isosceles triangle, the angles opposite the equal sides must be equal.)

13. Can you form a △STV if ST = 12 cm, TV = 9 cm, and SV = 15 cm? If so, what kind of triangle is △STV?

14. Can you form a triangle with the lengths of the sides 4 ft, 6 ft, and 10 ft? Why or why not?

11.6

SQUARE ROOTS AND THE PYTHAGOREAN THEOREM

A number is squared when it is multiplied by itself. If a whole number is squared, the result is called a **perfect square.** For example, squaring 7 gives $7^2 = 49$, and 49 is a perfect square. Table 11.1 shows the perfect square numbers found by squaring the whole numbers from 1 to 20. A more complete table is located in the appendix.

TABLE 11.1 Squares of Whole Numbers from 1 to 20

$1^2 = 1$	$6^2 = 36$	$11^2 = 121$	$16^2 = 256$
$2^2 = 4$	$7^2 = 49$	$12^2 = 144$	$17^2 = 289$
$3^2 = 9$	$8^2 = 64$	$13^2 = 169$	$18^2 = 324$
$4^2 = 16$	$9^2 = 81$	$14^2 = 196$	$19^2 = 361$
$5^2 = 25$	$10^2 = 100$	$15^2 = 225$	$20^2 = 400$

Since $5^2 = 25$, the **square root** of 25 is 5. The symbol for square root is $\sqrt{}$; the symbol is called a **square root sign** or **radical sign.** Thus,

$$\sqrt{25} = 5 \quad \text{since} \quad 5^2 = 25$$

$$\sqrt{49} = 7 \quad \text{since} \quad 7^2 = 49$$

$$\sqrt{64} = 8 \quad \text{since} \quad 8^2 = 64$$

Table 11.2 contains the square roots of the perfect square numbers from 1 to 400. Both Table 11.1 and Table 11.2 should be memorized.

TABLE 11.2 Square Roots to Perfect Squares from 1 to 400

$\sqrt{1} = 1$	$\sqrt{36} = 6$	$\sqrt{121} = 11$	$\sqrt{256} = 16$
$\sqrt{4} = 2$	$\sqrt{49} = 7$	$\sqrt{144} = 12$	$\sqrt{289} = 17$
$\sqrt{9} = 3$	$\sqrt{64} = 8$	$\sqrt{169} = 13$	$\sqrt{324} = 18$
$\sqrt{16} = 4$	$\sqrt{81} = 9$	$\sqrt{196} = 14$	$\sqrt{361} = 19$
$\sqrt{25} = 5$	$\sqrt{100} = 10$	$\sqrt{225} = 15$	$\sqrt{400} = 20$

The square roots of some numbers are not as easily found as those in the tables. In fact, most square roots are **irrational numbers (nonrepeating infinite decimals);** that is, most square roots can only be approximated with decimals.

Decimal approximations to $\sqrt{2}$ are shown here so that you will understand that there is no finite decimal number whose square is 2.

```
    1.4          1.414          1.41421          1.41422
  × 1.4        ×  1.414       ×  1.41421       ×  1.41422
    56           5656           141421           282844
  1 4            1414           282842           282844
  1.96           5656           565684           565688
               1 414            141421           141422
               1.999396         565684           565688
                              1 41421          1 41422
                              1.9999899241     2.0000182084
```

So, $\sqrt{2}$ is between 1.41421 and 1.41422. Using a calculator, we find

$$\sqrt{2} = 1.414213562$$

rounded off to nine decimal places.

> **Radical:** An expression using the radical sign, such as $\sqrt{a}$
>
> **Radicand:** The number under the radical sign in a radical

To simplify radicals in general, we need the following property.

> **PROPERTY OF SQUARE ROOTS***
>
> For numbers a and b,
>
> $$\sqrt{ab} = \sqrt{a}\,\sqrt{b}$$

With this property, we can look for a perfect square factor and write

$$\sqrt{18} = \sqrt{9 \cdot 2} = \sqrt{9} \cdot \sqrt{2} = 3\sqrt{2} \quad \text{(9 is a square factor.)}$$

The expression $3\sqrt{2}$ is simplified because the radicand, 2, has no perfect square factor. That is, **a radical is considered to be in simplest form when the radicand has no square number factor.**

---

* This property is valid only if a and b are positive numbers or zero. Positive and negative numbers are discussed in Chapter 12.

Simplify the following radicals.

Simplify the following radical expressions.

Example 1 $\sqrt{50}$

Solution

$$\sqrt{50} = \sqrt{25 \cdot 2} = \sqrt{25} \cdot \sqrt{2} = 5\sqrt{2} \quad \text{(25 is a square factor.)}$$

1. $\sqrt{8}$

Example 2 $\sqrt{75}$

Solution

$$\sqrt{75} = \sqrt{25 \cdot 3} = \sqrt{25} \cdot \sqrt{3} = 5\sqrt{3}$$

2. $\sqrt{20}$

Example 3 $\sqrt{450}$

Solution

$$\sqrt{450} = \sqrt{9 \cdot 50} = \sqrt{9 \cdot 25 \cdot 2} = \sqrt{9} \cdot \sqrt{25} \cdot \sqrt{2} = 3 \cdot 5\sqrt{2} = 15\sqrt{2}$$

or

$$\sqrt{450} = \sqrt{225 \cdot 2} = \sqrt{225} \cdot \sqrt{2} = 15\sqrt{2}$$

3. $\sqrt{135}$

☐ **NOW WORK EXERCISES 1–3 IN THE MARGIN.**

The following discussion involving right triangles serves as an application using squares and square roots.

> **Right Triangle:** A triangle with a right angle (90°)
>
> **Hypotenuse:** The longest side (opposite the 90° angle) of a right triangle
>
> **Leg:** Each of the other two sides of a right triangle (not the hypotenuse)

Pythagoras, a famous Greek mathematician, is given credit for discovering the following theorem.

PYTHAGOREAN THEOREM

In a right triangle the square of the hypotenuse is equal to the sum of the squares of the two legs:

$$c^2 + a^2 + b^2$$

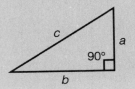

The following examples illustrate the Pythagorean Theorem.

Example 4

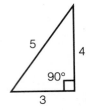

$$5^2 = 3^2 + 4^2$$
$$25 = 9 + 16$$

Example 5

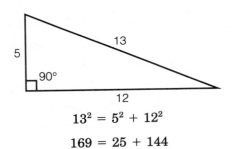

$$13^2 = 5^2 + 12^2$$
$$169 = 25 + 144$$

In most right triangles, all three legs do not have integer values. Examples 6 and 7 illustrate irrational numbers.

Example 6

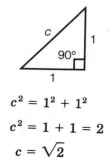

$$c^2 = 1^2 + 1^2$$
$$c^2 = 1 + 1 = 2$$
$$c = \sqrt{2}$$

That is, the length of the hypotenuse is $\sqrt{2}$.

Example 7

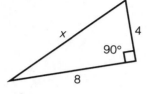

$$x^2 = 8^2 + 4^2$$
$$x^2 = 64 + 16 = 80$$
$$x = \sqrt{80} = \sqrt{16 \cdot 5} = \sqrt{16} \cdot \sqrt{5} = 4\sqrt{5}$$

The length of the hypotenuse is $4\sqrt{5}$.

Answers to Marginal Exercises

1. $2\sqrt{2}$ **2.** $2\sqrt{5}$ **3.** $3\sqrt{15}$

EXERCISES **11.6** _____

State whether or not each number is a perfect square in Exercises 1–10.

1. 144   **2.** 169   **3.** 81   **4.** 16   **5.** 400

6. 225   **7.** 242   **8.** 48   **9.** 45   **10.** 40

11. Show by squaring that $\sqrt{3}$ is between 1.732 and 1.733.

12. Show by squaring that $\sqrt{5}$ is between 2.236 and 2.237.

Simplify the radical expressions in Exercises 13–36.

13. $\sqrt{12}$ **14.** $\sqrt{28}$ **15.** $\sqrt{24}$

Answers

1. _____

2. _____

3. _____

4. _____

5. _____

6. _____

7. _____

8. _____

9. _____

10. _____

11. _____

12. _____

13. _____

14. _____

15. _____

Answers

16. _____

17. _____

18. _____

19. _____

20. _____

21. _____

22. _____

23. _____

24. _____

25. _____

26. _____

27. _____

28. _____

29. _____

30. _____

16. $\sqrt{32}$ **17.** $\sqrt{48}$ **18.** $\sqrt{288}$

19. $\sqrt{363}$ **20.** $\sqrt{242}$ **21.** $\sqrt{500}$

22. $\sqrt{300}$ **23.** $\sqrt{128}$ **24.** $\sqrt{125}$

25. $\sqrt{72}$ **26.** $\sqrt{98}$ **27.** $\sqrt{605}$

28. $\sqrt{150}$ **29.** $\sqrt{169}$ **30.** $\sqrt{196}$

31. $\sqrt{800}$ **32.** $\sqrt{80}$ **33.** $\sqrt{90}$

Answers

31. _____

32. _____

33. _____

34. _____

34. $\sqrt{40}$ **35.** $\sqrt{256}$ **36.** $\sqrt{361}$

35. _____

36. _____

37. _____

38. _____

Use the Pythagorean Theorem to determine whether or not each of the triangles in Exercises 37 and 38 is a right triangle.

37.

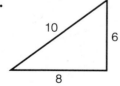

38.

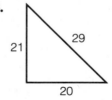

Answers

39. _____

40. _____

41. _____

42. _____

43. _____

44. _____

Find the length of the hypotenuse of each of the following right triangles.

39.

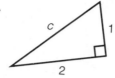

40.

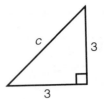

41.

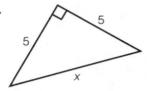

42.

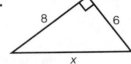

43.

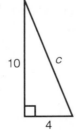

44.

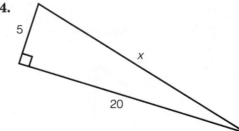

NAME _____ SECTION _____ DATE _____

CHAPTER **11** TEST _____

Answers

1. Find the circumference of the following circle. (Use $\pi = 3.14$.)

1. _____

2. _____

3. _____

4. _____

2. Find the area of the circle in Problem 1.

3. Find the perimeter of the following figure.

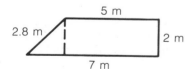

4. Find the area of the figure in Problem 3.

Answers

5. _____

6. _____

7. (a) _____

   (b) _____

   (c) _____

   (d) _____

   (e) _____

8. (a) _____

   (b) _____

   (c) _____

In Problems 5 and 6, find the volume of each solid in liters. (Use $\pi \approx 3.14$.)

5.
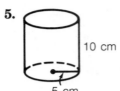
10 cm

5 cm

6.
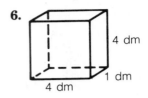
4 dm

1 dm

4 dm

7. In the figure shown, $\overleftrightarrow{AD}$ and $\overleftrightarrow{BE}$ are straight lines. (a) What type of angle is $\angle BOC$? (b) What type of angle is $\angle AOE$? (c) Name a pair of vertical angles. (d) Name a pair of complementary angles. (e) Name three pairs of supplementary angles.

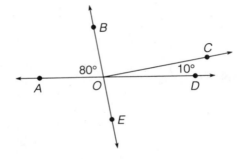

8. Name each of the following triangles based on the measures shown.

(a)
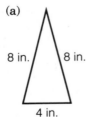
8 in.   8 in.

4 in.

(b)
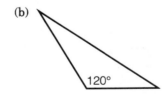
120°

(c)
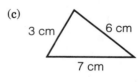
3 cm   6 cm

7 cm

NAME _____ SECTION _____ DATE _____

9. Find the value of x given that $\triangle ABC \sim \triangle ADE$.

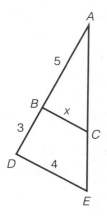

Simplify the following radicals.

10. $\sqrt{28}$

11. $\sqrt{45}$

12. $\sqrt{48}$

13. $\sqrt{300}$

14. $\sqrt{90}$

15. $\sqrt{338}$

Answers

9. _____

10. _____

11. _____

12. _____

13. _____

14. _____

15. _____

Answers

16. _____

17. _____

16. Use the Pythagorean Theorem to determine whether or not the triangle is a right triangle.

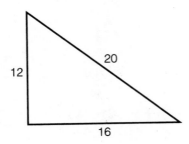

17. Use the Pythagorean Theorem to find the length of the hypotenuse.

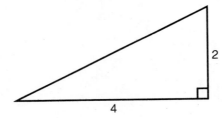

Introduction to Algebra

12.1
NUMBER LINES AND ABSOLUTE VALUE (INTEGERS)

> **Number line:** A line with a scale of points marked on it
>
>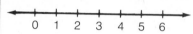
>
> **Graph of a number (a point):** A shaded dot at the corresponding point on a number line
>
> **Coordinate of a point:** The number that corresponds to a point or shaded dot
>
> The terms **point** and **number** will be used interchangeably.

Example 1 Graph the set of numbers {0, 1, 5} on a number line.

Solution

Example 2 Graph the set of points {2, 4, 6} on a number line.

Solution

Opposite of a number: A number that is the same distance from 0 but on the other side of 0 from the original number on a number line. Opposites are also called **additive inverses.**

Positive numbers: Numbers greater than 0 (to the right of 0 on a number line)

Negative numbers: Numbers smaller than 0 (to the left of 0 on a number line)

Integers: The whole numbers and their opposites

Zero: 0 is a whole number that is neither positive nor negative. The opposite of 0 is 0.

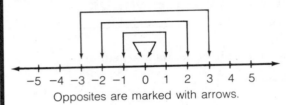

Opposites are marked with arrows.

Integers: ... , −3, −2, −1, 0, 1, 2, 3, ...

Positive integers: 1, 2, 3, 4, 5, ...

Negative integers: ... , −3, −2, −1

Note: The three dots, ... , indicate a continuing pattern without end.

Example 3 Find the opposite of 8.

Solution

−8

Example 4 Find the opposite of −2.

Solution

−(−2)   Read **the opposite of negative 2.**
or +2   Read **positive 2.**
or 2   A number is understood to be positive if there is no sign in front of it.

Example 5 Graph the set $B = \{-2, -1, 3\}$.

Solution

Example 6 Graph the set {−4, −2, 0, 2, 4, . . .}.

Solution

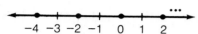

Example 7 Remember:

(a) The opposite of a negative number is a positive number.

$$-(-3) = +3 \text{ and } -(-7) = +7$$

(b) The opposite of 0 is 0.

$$-(0) = 0$$

(c) The opposite of a positive number is a negative number.

$$-(+6) = -6 \text{ and } -(+2) = -2$$

☐ **NOW WORK EXERCISES 1–4 IN THE MARGIN.**

❙ **Absolute value:** The distance a number is from 0 on a number line, regardless of the direction
❙ **Symbol for absolute value:** | |, two vertical lines

Example 8 +5 and −5 are both 5 units from 0 and both have the same absolute value, which is 5.

$$|+5| = 5 \text{ and } |-5| = 5$$

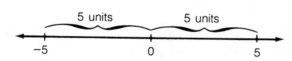

Example 9 $|+3| = 3$ and $|-3| = 3$

Example 10 $|-7| = |+7| = 7$

Example 11 $|0| = 0$ The absolute value of 0 is 0.

1. Find the opposite of −5.

2. Find the opposite of +12.

3. Graph the set {−3, 0, 1}.

4. Graph the set {−5, −4, 1, 2}.

Find each absolute value.

5. $|-10|$

6. $|4|$

7. $|83|$

8. $|-52|$

Example 12 Which number has the greater absolute value, -8 or -6?

Solution

-8 has the greater absolute value since $|-8| = 8$ and $|-6| = 6$, and 8 is greater than 6.

☐ **NOW WORK EXERCISES 5–8 IN THE MARGIN.**

On a horizontal number line, smaller numbers are to the left of larger numbers.

$$a < b: \text{read } a \text{ is less than } b.$$

$$c > d: \text{read } c \text{ is greater than } d.$$

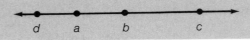

$$\begin{array}{cccc} d & a & b & c \end{array}$$

Example 13 $8 < 10$ 8 is less than 10.

Example 14 $-5 < 0$ -5 is less than 0.

Example 15 $2 > -6$ 2 is greater than -6.

Example 16 $5 > -3$ 5 is greater than -3.

Example 17 For the two numbers -12 and -7, which number is greater and which number has the greater absolute value?

Solution

$-7 > -12$ since -7 is to the right of -12 on a number line.

$|-12| > |-7|$ since $|-12| = 12$ and $|-7| = 7$ and 12 is greater than 7.

Answers to Marginal Exercises

1. $-(-5) = 5$ **2.** $-(+12) = -12$

3.
$$\begin{array}{cccccc} & -3 & -2 & -1 & 0 & 1 \end{array}$$

4.
$$\begin{array}{cccccccc} & -5 & -4 & -3 & -2 & -1 & 0 & 1 & 2 \end{array}$$

5. 10   **6.** 4   **7.** 83   **8.** 52

NAME _____   SECTION _____   DATE _____

EXERCISES **12.1** _____

In each of the following exercises, draw a number line and graph the given set of integers.

1. {0, 1, 2}

2. {0, 2, 4}

3. {−3, −1, 1}

4. {−3, −2, 0}

5. {−10, −9, −8, −7}

6. {−5, −4, −2, −1}

7. {−5, 0, 5}

8. {−3, −1, 0, 1, 3}

9. {1, 2, 3, . . . , 10}

10. {−2, −1, 0, . . . , 5}

1. _____

2. _____

3. _____

4. _____

5. _____

6. _____

7. _____

8. _____

9. _____

10. _____

Answers

11. _____

12. _____

13. _____

14. _____

15. _____

16. _____

17. _____

18. _____

19. _____

20. _____

11. $\{1, 3, 5, 7, \ldots\}$

12. $\{0, 2, 4, 6, \ldots\}$

13. $\{\ldots, -5, -4, -3\}$

14. $\{\ldots, -10, -9, -8\}$

15. $\{-3, 0, 3, 6, 9, \ldots\}$

16. $\{\ldots, -8, -4, 0, 4\}$

17. $\{|-7|, |-2|, 0\}$

18. $\{|-3|, 0, 1\}$

19. $\{-3, 0, |-3|, |-5|\}$

20. $\{-2, -1, |-1|, |-2|\}$

NAME _____ SECTION _____ DATE _____

Find the opposite of each number.

21. −10   **22.** −9   **23.** 14   **24.** 12   **25.** −6

26. −3   **27.** 30   **28.** 40   **29.** 0   **30.** −7

Find each absolute value as indicated.

31. |−6| **32.** |−10| **33.** |+24| **34.** |+16|

35. |−20| **36.** |−50| **37.** |0| **38.** |27|

Answers

21. _____

22. _____

23. _____

24. _____

25. _____

26. _____

27. _____

28. _____

29. _____

30. _____

31. _____

32. _____

33. _____

34. _____

35. _____

36. _____

37. _____

38. _____

Answers

State which number in each pair is greater and then determine which number has the larger absolute value.

39. _____

39. 10, 13

40. 16, 18

40. _____

41. _____

41. −13, −10

42. −18, −16

42. _____

43. _____

44. _____

43. 20, 30

44. 10, 15

45. _____

46. _____

45. −9, 5

46. −6, 3

47. _____

48. _____

47. −8, 9

48. −3, 7

49. _____

50. _____

49. −11, 7

50. −12, 10

12.2

ADDING INTEGERS

The sum of two integers is indicated with a plus sign (+) just as with adding whole numbers.

$$(+2) \quad + \quad (+7)$$

positive 2   plus   positive 7

An intuitive idea of how to add integers using number lines:

Start at the first number to be added, then

1. move **right** if the second number is positive, or
2. move **left** if the second number is negative.

The distance moved is the absolute value of the second number.

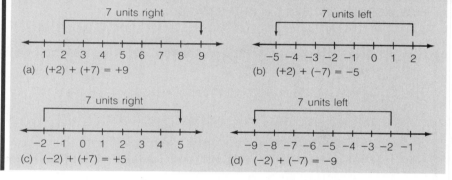

(a) $(+2) + (+7) = +9$

(b) $(+2) + (-7) = -5$

(c) $(-2) + (+7) = +5$

(d) $(-2) + (-7) = -9$

Example 1 $(-5) + (+4) = ?$

     Solution

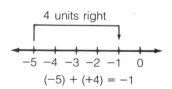

$$(-5) + (+4) = -1$$

Example 2 $(-3) + (-8) = ?$

     Solution

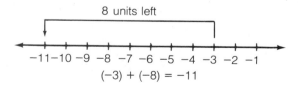

$$(-3) + (-8) = -11$$

Find each sum. Notice that the numbers can be written vertically as well as horizontally.

1. $(-6) + (-10) =$

2. $13 + (-3) =$

3. $\begin{array}{r} -21 \\ +\ 8 \\ \hline \end{array}$

4. $\begin{array}{r} -30 \\ -\ 5 \\ \hline \end{array}$

Example 3 $(+6) + (-10) = ?$

Solution

$(+6) + (-10) = -4$

Example 4 $(-7) + (+7) = ?$

Solution

$(-7) + (+7) = 0$

Note: The sum of two opposites will always be 0. $a + (-a) = 0.$

☐ **NOW WORK EXERCISES 1–4 IN THE MARGIN.**

> 1. **To add two integers with like signs,** add their absolute values and use the common sign.
>
> $(+6) + (+5) = +[\,|+6| + |+5|\,] = +[6 + 5] = +11$
>
> $(-2) + (-8) = -[\,|-2| + |-8|\,] = -[2 + 8] = -10$
>
> 2. **To add two integers with unlike signs,** subtract their absolute values (smaller from larger) and use the sign of the number with the larger absolute value.
>
> $(+10) + (-15) = -[\,|-15| - |+10|\,] = -[15 - 10] = -5$
>
> $(+12) + (-9) = +[\,|+12| - |-9|\,] = +[12 - 9] = +3$

Example 5 $(+4) + (-1) = +(4 - 1) = +3$ (unlike signs)

Example 6 $(-5) + (-3) = -(5 + 3) = -8$ (like signs)

Example 7 $(+7) + (+10) = +(7 + 10) = +17$ (like signs)

Example 8 $(+6) + (-19) = -(19 - 6) = -13$ (unlike signs)

Example 9 $(+30) + (-30) = (30 - 30) = 0$ (opposites)

☐ **NOW WORK EXERCISES 5–8 IN THE MARGIN.**

> To add more than two numbers, add any two, then add this sum to one of the other numbers until all numbers have been added. A common practice is to add all positives and all negatives separately, then add these two sums.

Example 10 $(+9) + (+2) + (-6) = +[9 + 2] + (-6)$

$$= +[11] + (-6)$$
$$= +5$$

Example 11 $\begin{array}{r} -12 \\ +8 \\ \hline -4 \end{array}$

Example 12 $\begin{array}{r} -5 \\ +6 \\ -14 \\ \hline -13 \end{array}$

Example 13 $\begin{array}{r} -3 \\ +1 \\ +7 \\ -5 \\ \hline 0 \end{array}$

Find each sum.

5. $(-10) + (+3) =$

6. $(-5) + (+9) =$

7. $(-2) + (-3) =$

8. $(+4) + (-6) =$

Answers to Marginal Exercises
1. −16 **2.** 10 **3.** −13 **4.** −35 **5.** −7 **6.** 4 **7.** −5 **8.** −2

NAME _____   SECTION _____   DATE _____

EXERCISES **12.2** _____

Find each sum.

1. $(+6) + (-4)$ 2. $(+8) + (-7)$

3. $(+4) + (+6)$ 4. $(+5) + (-8)$

5. $(+16) + (+3)$ 6. $(-8) + (-2)$

7. $(-3) + (-6)$ 8. $(-2) + (+2)$

9. $(+4) + (-4)$ 10. $(+13) + (+12)$

11. $(+6) + (-10)$ 12. $(+14) + (-17)$

13. $(+5) + (-3)$ 14. $(+15) + (-18)$

15. $(-4) + (-12)$ 16. $(-8) + (+8)$

17. $(+2) + (-6)$ 18. $(-9) + (+5)$

Answers

1. _____

2. _____

3. _____

4. _____

5. _____

6. _____

7. _____

8. _____

9. _____

10. _____

11. _____

12. _____

13. _____

14. _____

15. _____

16. _____

17. _____

18. _____

Answers

19. _____

20. _____

21. _____

22. _____

23. _____

24. _____

25. _____

26. _____

27. _____

28. _____

29. _____

30. _____

31. _____

32. _____

33. _____

34. _____

35. _____

36. _____

37. _____

38. _____

39. _____

40. _____

19. $(-16) + (+3) + (+13)$

20. $(-5) + (+5) + (14)$

21. $(-1) + (-2) + (+7)$

22. $(+3) + (-4) + (-5)$

23. $(+6) + (+3) + (+5)$

24. $(-18) + (-5) + (-7)$

25. $(-1) + (+2) + (-4) + (+2)$

26. $\begin{array}{r} -4 \\ +8 \\ \hline \end{array}$

27. $\begin{array}{r} -5 \\ -10 \\ \hline \end{array}$

28. $\begin{array}{r} -13 \\ -6 \\ \hline \end{array}$

29. $\begin{array}{r} +16 \\ +25 \\ \hline \end{array}$

30. $\begin{array}{r} +14 \\ -8 \\ \hline \end{array}$

31. $\begin{array}{r} +20 \\ -7 \\ \hline \end{array}$

32. $\begin{array}{r} +2 \\ -5 \\ -3 \\ \hline \end{array}$

33. $\begin{array}{r} +8 \\ +3 \\ -1 \\ \hline \end{array}$

34. $\begin{array}{r} +10 \\ -4 \\ +2 \\ \hline \end{array}$

35. $\begin{array}{r} -16 \\ -8 \\ +12 \\ \hline \end{array}$

36. $\begin{array}{r} -15 \\ -20 \\ -6 \\ \hline \end{array}$

37. $\begin{array}{r} -4 \\ -17 \\ +11 \\ \hline \end{array}$

38. $\begin{array}{r} +13 \\ -5 \\ +17 \\ -25 \\ \hline \end{array}$

39. $\begin{array}{r} +14 \\ -14 \\ +37 \\ -37 \\ \hline \end{array}$

40. $\begin{array}{r} -8 \\ -5 \\ -13 \\ -22 \\ \hline \end{array}$

12.3
SUBTRACTING INTEGERS

The difference of two integers is indicated with a minus sign (−) just as with subtracting whole numbers.

$$(+2) \quad - \quad (-4)$$

positive 2   minus   negative 4

An intuitive idea of how to subtract integers using number lines: Start at the first number then

1. move **left** if the second number is positive, or
2. move **right** if the second number is negative.

The distance moved is the absolute value of the second number and **the direction moved is the opposite of that number.**

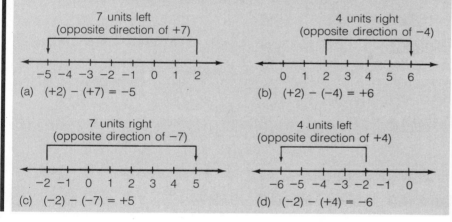

7 units left
(opposite direction of +7)

(a) $(+2) - (+7) = -5$

4 units right
(opposite direction of −4)

(b) $(+2) - (-4) = +6$

7 units right
(opposite direction of −7)

(c) $(-2) - (-7) = +5$

4 units left
(opposite direction of +4)

(d) $(-2) - (+4) = -6$

Find the differences in Examples 1 and 2 by using a number line.

Example 1 $(-5) - (+4) = ?$

Solution

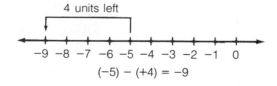

4 units left

$(-5) - (+4) = -9$

Find each difference using a number line as an aid.

1. $2 - (-3) =$

2. $-5 - (+2) =$

Example 2 $(-3) - (-8) = ?$

Solution

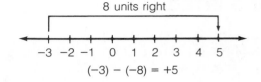

$(-3) - (-8) = +5$

☐ **NOW WORK EXERCISES 1 AND 2 IN THE MARGIN.**

RULE FOR SUBTRACTING INTEGERS

To subtract, add the opposite of the number being subtracted.

$$(a) - (b) = (a) + (-b).$$

Note: This rule means that subtraction is accomplished with a sign change, then **addition**.

Example 3 $(+2) - (-6) = (+2) + (+6) = +8$
                         ↑   ↑   ↑
                       minus   plus   opposite of −6

Example 4 $(-3) - (-7) = (-3) + (+7) = +4$
                                        ↑
                                 opposite of −7

Example 5 $(-5) - (+2) = (-5) + (-2) = -7$
                                        ↑
                                 opposite of +2

Example 6 $(-9) - (-9) = (-9) + (+9) = 0$
                                        ↑
                                 opposite of −9

The numbers may also be written vertically, one underneath the other. In this case, change the sign of the number being subtracted (the bottom number), then add.

Example 7 *TO SUBTRACT:* *ADD:*

$$
\begin{array}{r} -10 \\ \underline{-3} \end{array}
\qquad
\begin{array}{r} -10 \\ \underline{+3} \\ -7 \end{array}
$$

Example 8 *TO SUBTRACT:* *ADD:*

$$
\begin{array}{r} +14 \\ \underline{+9} \end{array}
\qquad
\begin{array}{r} +14 \\ \underline{-9} \\ 5 \end{array}
$$

☐ **NOW WORK EXERCISES 3–7 IN THE MARGIN.**

> **COMMON NOTATION IN ALGEBRA**
>
> 1. Parentheses around the numbers and the + sign for addition and − sign for subtraction are dropped.
> 2. The numbers are written horizontally.
> 3. The problem is considered as **addition** of positive and negative numbers.
> 4. The + sign may be dropped from the first number at the left; otherwise, the signs of all numbers must be written.

Example 9 $9 - 12$ is the same as $(+9) + (-12)$, so

$$9 - 12 = (+9) + (-12) = -3$$

Example 10 $7 - 3 = (+7) + (-3) = 4$
 or $7 - 3 = 4$

Example 11 $-13 + 6 = (-13) + (+6) = -7$
 or $-13 + 6 = -7$

Example 12 $-4 - 8 = (-4) + (-8) = -12$
 or $-4 - 8 = -12$

Example 13 $6 - 9 = -3$

Find each difference.

3. $(+8) - (-3) =$

4. $(-4) - (-5) =$

5. $(-10) - (+3) =$

6. Subtract: $\begin{array}{r} 19 \\ \underline{-13} \end{array}$

7. Subtract: $\begin{array}{r} -2 \\ \underline{-7} \end{array}$

Evaluate each expression.

8. $-10 + 4 =$

9. $-12 - 8 =$

10. $13 - 5 + 6 =$

11. $-2 + 1 - 6 - 5 =$

Example 14 $\quad -20 + 12 = -8$

Example 15 $\quad -3 - 14 = -17$

Example 16 $\quad 17 - 4 - 8 = 13 - 8 = 5$

Example 17 $\quad -14 + 30 - 5 + 1 = 16 - 5 + 1 = 11 + 1 = 12$

☐ **NOW WORK EXERCISES 8–11 IN THE MARGIN.**

Answers to Marginal Exercises

1.

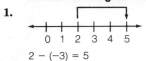

$2 - (-3) = 5$

2.

$-5 - (+2) = -7$

3. 11   **4.** 1   **5.** −13   **6.** 32   **7.** 5   **8.** −6   **9.** −20   **10.** 14   **11.** −12

NAME _____ SECTION _____ DATE _____

EXERCISES **12.3** _____

Find each difference.

1. $(+5) - (+2)$ **2.** $(+16) - (+3)$ **3.** $(+8) - (-3)$

4. $(+12) - (-4)$ **5.** $(-5) - (+2)$ **6.** $(-10) - (+3)$

7. $(-10) - (-1)$ **8.** $(-15) - (-1)$ **9.** $(-3) - (-7)$

10. $(-2) - (-12)$ **11.** $(-4) - (+6)$ **12.** $(-9) - (+13)$

13. $(-13) - (-14)$ **14.** $(-12) - (-15)$ **15.** $(+9) - (-9)$

16. $(+11) - (-11)$ **17.** $(+15) - (-2)$ **18.** $(+20) - (-3)$

1. _____

2. _____

3. _____

4. _____

5. _____

6. _____

7. _____

8. _____

9. _____

10. _____

11. _____

12. _____

13. _____

14. _____

15. _____

16. _____

17. _____

18. _____

19. $(-17) - (+14)$ **20.** $(-16) - (+10)$ **21.** $(+3) - (+8)$

22. $(+1) - (+5)$ **23.** $(-5) - (-5)$ **24.** $(-7) - (-7)$

25. $(+7) - (+12)$

Subtract the second number from the first.

26. 18	**27.** 24	**28.** -8	**29.** -13
-12	$\underline{16}$	$\underline{-12}$	$\underline{-18}$

30. -4	**31.** 32	**32.** -6	**33.** -25
$\underline{+5}$	$\underline{-48}$	$\underline{-30}$	$\underline{-13}$

34. -45	**35.** 28
$\underline{-16}$	$\underline{-15}$

NAME _____ SECTION _____ DATE _____

Evaluate each of the following expressions.

36. $6 + 2$ **37.** $4 + 8$ **38.** $7 - 1$

36. _____

37. _____

38. _____

39. $9 - 4$ **40.** $4 + 6$ **41.** $8 + 9$

39. _____

40. _____

41. _____

42. $-3 - 1$ **43.** $-2 - 6$ **44.** $12 - 6$

42. _____

43. _____

44. _____

45. $9 - 3$ **46.** $-13 + 4$ **47.** $-20 + 14$

45. _____

46. _____

47. _____

48. $-10 + 9$ **49.** $-18 + 3$ **50.** $24 - 32$

48. _____

49. _____

50. _____

51. $14 - 17$ **52.** $-12 - 6$ **53.** $-2 - 8$

51. _____

52. _____

53. _____

Answers

54. _____

55. _____

56. _____

57. _____

58. _____

59. _____

60. _____

61. _____

62. _____

63. _____

64. _____

65. _____

66. _____

67. _____

68. _____

69. _____

70. _____

71. _____

72. _____

73. _____

74. _____

75. _____

54. $-15 + 18$

55. $-25 + 30$

56. $-20 + 21$

57. $-30 + 32$

58. $-7 + 7$

59. $-6 + 6$

60. $18 - 3$

61. $-4 + 16 - 8$

62. $-5 + 12 - 3$

63. $-20 - 2 + 6$

64. $14 - 5 - 12$

65. $13 + 15 - 6$

66. $-6 - 8 - 13$

67. $-4 - 10 - 7$

68. $30 + 12 - 18$

69. $16 + 4 - 20$

70. $15 + 6 - 21$

71. $13 - 4 + 6 - 5$

72. $16 - 3 - 7 - 1$

73. $19 - 5 - 8 - 6$

74. $-4 + 10 - 12 + 1$

75. $-8 + 14 - 10 + 3$

12.4
MULTIPLYING AND DIVIDING INTEGERS

The rules for multiplying integers are stated here without formal development. A more detailed discussion appears in beginning algebra textbooks.

> **RULES FOR MULTIPLYING INTEGERS**
> 1. The product of two positive integers is positive.
> 2. The product of two negative integers is positive.
> 3. The product of a positive integer with a negative integer is negative.

Example 1 $4 \cdot 7 = 28$

Example 2 $5(-2) = -10$

Example 3 $-4(+3) = -12$

Example 4 $-9(-5) = 45$

In Examples 5 and 6, fill in the missing products and study the patterns to help you understand the rules.

Completion Example 5

-5	-5	-5	-5	-5	-5	-5
$+3$	$+2$	$+1$	0	-1	-2	-3
-15	-10	-5	0	___	___	___

Completion Example Answer

5.

-5	-5	-5
-1	-2	-3
5	10	15

Find each product.

1. 6
    −3

2. −8
    −2

3. −5
    4

4. −9(−3) =

5. −14(0) =

Find the following products.

6. (−2)(−2)(−2) =

7. (−3)(−3)(0) =

8. (−6)(−1)(4)(−3) =

Completion Example 6

$(+3)(-2) = -6$
$(+2)(-2) = -4$
$(+1)(-2) = -2$
$(0)(-2) = 0$
$(-1)(-2) = \underline{}$
$(-2)(-2) = \underline{}$
$(-3)(-2) = \underline{}$

□ **NOW WORK EXERCISES 1–5 IN THE MARGIN.**

To multiply more than two integers,

1. Multiply any two integers.
2. Continue to multiply that product by the next integer until all the integers have been multiplied.

Example 7

$$(-3)(4)(-2) = [(-3)(4)](-2)$$
$$= [-12](-2)$$
$$= 24$$

Example 8

$$(-5)(-3)(-10) = [(-5)(-3)](-10)$$
$$= [+15](-10)$$
$$= -150$$

Example 9

$$(-2)(+3)(-2)(-6) = [(-2)(+3)](-2)(-6)$$
$$= [-6](-2)(-6)$$
$$= [(-6)(-2)](-6)$$
$$= [+12](-6)$$
$$= -72$$

□ **NOW WORK EXERCISES 6–8 IN THE MARGIN.**

Completion Example Answer

6. $(-1)(-2) = 2$
$(-2)(-2) = 4$
$(-3)(-2) = 6$

> RULES FOR DIVIDING INTEGERS
>
> 1. The quotient of two positive integers is positive.
> 2. The quotient of two negative integers is positive.
> 3. The quotient of a positive integer and a negative integer is negative.

Division is closely related to multiplication as this definition states.

Division: For integers a and b, and $b \neq 0$,

$$\frac{a}{b} = x \quad \text{means} \quad a = b \cdot x$$

Example 10

$$\frac{+28}{+4}$$

$$\frac{+28}{+4} = +7 \text{ because } +28 = (+4)(+7)$$

Example 11

$$\frac{+16}{-2}$$

$$\frac{+16}{-2} = -8 \text{ because } +16 = (-2)(-8)$$

Example 12

$$\frac{-40}{-4}$$

$$\frac{-40}{-4} = +10 \text{ because } -40 = -4(+10)$$

Example 13

$$\frac{-10}{+2} = -5 \text{ because } -10 = +2(-5)$$

Find the following quotients.

9. $\dfrac{-20}{10}$

10. $\dfrac{-40}{-2}$

11. $\dfrac{0}{-8}$

12. $\dfrac{36}{-2}$

TWO SPECIAL RULES FOR DIVISION

1. $\dfrac{0}{b} = 0$ when $b \neq 0$.

2. $\dfrac{a}{0}$ is undefined. You **cannot** divide by 0.

Example 14

$\dfrac{0}{-5} = 0$ because $0 = -5(0)$

Example 15

$\dfrac{23}{0}$ is undefined. You cannot multiply a number by 0 and get 23.

$23 = 0 \cdot x$ is not possible.

☐ **NOW WORK EXERCISES 9–12 IN THE MARGIN.**

Answers to Marginal Exercises

1. -18 **2.** 16   **3.** -20 **4.** 27   **5.** 0   **6.** -8 **7.** 0   **8.** -72 **9.** -2
10. 20   **11.** 0   **12.** -18

NAME _____ SECTION _____ DATE _____

EXERCISES **12.4** _____

Find the following products.

1. $5(-3)$ **2.** $4(-6)$ **3.** $-6(-4)$

4. $-2(-7)$ **5.** $-5(4)$ **6.** $-8(3)$

7. $14(2)$ **8.** $13(3)$ **9.** $-10(5)$

10. $-11(3)$ **11.** $(-7)3$ **12.** $(-2)9$

13. $6(-8)$ **14.** $9(-4)$ **15.** $-7(-9)$

1. _____

2. _____

3. _____

4. _____

5. _____

6. _____

7. _____

8. _____

9. _____

10. _____

11. _____

12. _____

13. _____

14. _____

15. _____

Answers

16. _____

17. _____

18. _____

19. _____

20. _____

21. _____

22. _____

23. _____

24. _____

25. _____

26. _____

27. _____

28. _____

29. _____

30. _____

31. _____

32. _____

33. _____

34. _____

16. $-8(-9)$ **17.** $0(-6)$ **18.** $0(-4)$

19. $(-6)(-5)(3)$ **20.** $(-2)(-1)(7)$ **21.** $4(-2)(-3)$

22. $5(-6)(-1)$ **23.** $(-5)(3)(-4)$ **24.** $(-3)(7)(-5)$

25. $(-7)(-2)(-3)$ **26.** $(-4)(-4)(-4)$ **27.** $(-3)(-3)(-5)$

28. $(-2)(-2)(-8)$ **29.** $(-3)(-4)(-5)$ **30.** $(-2)(-5)(-7)$

31. $(-5)(0)(-6)$ **32.** $(-6)(0)(-2)$

33. $(-1)(-1)(-1)$ **34.** $(-3)(-3)(-3)$

NAME _____ SECTION _____ DATE _____

35. $(-2)(-2)(-2)(-2)$ **36.** $(-2)(-4)(-4)$

37. $(-1)(-4)(-7)(+3)$ **38.** $(-5)(-2)(-1)(+5)$

39. $(-2)(-3)(-10)(-5)$ **40.** $(-11)(-2)(-4)(-1)$

Find the following quotients.

41. $\dfrac{-12}{4}$ **42.** $\dfrac{-18}{2}$ **43.** $\dfrac{-14}{7}$ **44.** $\dfrac{-28}{7}$

45. $\dfrac{-20}{-5}$ **46.** $\dfrac{-30}{-3}$ **47.** $\dfrac{-50}{-10}$ **48.** $\dfrac{-30}{-5}$

Answers

35. _____

36. _____

37. _____

38. _____

39. _____

40. _____

41. _____

42. _____

43. _____

44. _____

45. _____

46. _____

47. _____

48. _____

Answers

49. _____

50. _____

51. _____

52. _____

53. _____

54. _____

55. _____

56. _____

57. _____

58. _____

59. _____

60. _____

61. _____

62. _____

63. _____

64. _____

65. _____

66. _____

67. _____

68. _____

69. _____

70. _____

49. $\dfrac{30}{-6}$

50. $\dfrac{40}{-8}$

51. $\dfrac{75}{-25}$

52. $\dfrac{80}{-4}$

53. $\dfrac{12}{6}$

54. $\dfrac{24}{8}$

55. $\dfrac{36}{9}$

56. $\dfrac{22}{11}$

57. $\dfrac{-39}{13}$

58. $\dfrac{27}{-9}$

59. $\dfrac{32}{-4}$

60. $\dfrac{23}{-23}$

61. $\dfrac{-34}{-17}$

62. $\dfrac{-60}{-15}$

63. $\dfrac{-8}{-8}$

64. $\dfrac{26}{-13}$

65. $\dfrac{-31}{0}$

66. $\dfrac{17}{0}$

67. $\dfrac{0}{-20}$

68. $\dfrac{0}{-16}$

69. $\dfrac{35}{0}$

70. $\dfrac{0}{25}$

12.5
COMBINING LIKE TERMS AND EVALUATING EXPRESSIONS

Constant: Any one number

-18 is a constant.

Variable: A symbol or letter that can represent more than one number

x is a variable.

Term: An expression that involves only multiplications and/or divisions with constants and/or variables

$5x$, $-6y$, $\dfrac{-15x}{3a}$, and -42 are terms.

Like terms (or similar terms): Terms that are constants or terms that contain exactly the same variables in the numerator or denominator*

Coefficient: The constant in a term

In the term $-6y$, -6 is the coefficient of y.

Example 1 From the following list of terms, pick out the like terms.

$$7, \quad 3x, \quad 32, \quad -7x, \quad 8y, \quad 0, \quad 13y$$

Solution

7, 32, and 0 are like terms. All are constants.

$3x$ and $-7x$ are like terms.

$8y$ and $13y$ are like terms.

Example 2 The terms $\dfrac{-15x}{3a}$ and $\dfrac{4a}{3x}$ are **not** like terms because the variables are not in the same position in each term.

Example 3 The terms $8y$ and $6y^2$ are **not** like terms because $6y^2$ has y twice while $8y$ has y only once.

Example 4 $5x + 2y$ and $8x - 3x$ are not terms at all, because they contain addition and subtraction.

---

*In elementary algebra, you will find that terms such as $5x^2$ and $-2x^2$ are like terms and $3xy^3$ and $4xy^3$ are also like terms.

> **THE DISTRIBUTIVE PROPERTY OF MULTIPLICATION OVER ADDITION**
>
> If a, b, and c are integers, then
>
> $$\text{Form 1: } a(b + c) = ab + ac$$
> $$\text{Form 2: } ba + ca = (b + c)a$$

> **Combine like terms:** Simplify expressions containing like terms using the distributive property

In the expressions in Examples 5–12, Form 2 of the distributive property is used to combine like terms.

Example 5

$5x + 3x$

$ba + ca = (b + c)a$ Using form 2 where x takes the place of a

$5x + 3x = (5 + 3)x$ and 5 and 3 take the place of b and c.

$\qquad\quad = 8x$

Example 6

$-10x - 2x = (-10 - 2)x$

$\qquad\qquad\quad = -12x$

Example 7

$4a - 5a + 7a = (4 - 5 + 7)a$

$\qquad\qquad\qquad = 6a$

Example 8

$2y + y = (2 + 1)y$ **Note:** In the term y, the coefficient

$\qquad\quad = 3y$ is understood to be $+1$.

Example 9

$2y - 3y = (2 - 3)y$ **Note:** In the term $-y$, the coefficient

$\qquad\quad = -1y \text{ or } -y$ is understood to be -1.

Example 10

$-4x + 5x + 3x = (-4 + 5 + 3)x$

$\qquad\qquad\qquad = 4x$

Example 11

$$7a + a + 2 + 3 = (7 + 1)a + 5$$
$$= 8a + 5$$

Note: In the term a, the coefficient is understood to be $+1$.

Example 12

$-6a + 4b - 8$ has no like terms.

☐ **NOW WORK EXERCISES 1–4 IN THE MARGIN.**

> **RULES FOR ORDER OF OPERATIONS IN EVALUATING EXPRESSIONS**
>
> 1. First, simplify within grouping symbols, such as parentheses ( ), brackets [ ], or braces { }.
> 2. Second, find any powers indicated by exponents.
> 3. Third, moving from **left to right,** perform any multiplications or divisions in the order they appear.
> 4. Fourth, moving from **left to right,** perform any additions or subtractions in the order they appear.

> **TO EVALUATE AN ALGEBRAIC EXPRESSION**
>
> 1. Combine like terms.
> 2. Substitute the given value(s) for the variable(s).
> 3. Follow the rules for order of operations.

In Examples 13–16, evaluate each expression for the given values of the variables.

Example 13

$x - 5$ for $x = 4$ and $x = -3$

Solution

For $x = 4$, $x - 5 = 4 - 5 = -1$
For $x = -3$, $x - 5 = -3 - 5 = -8$

Example 14

$3x + 2$ for $x = 4$ and $x = -3$

Solution

For $x = 4$, $3x + 2 = 3 \cdot 4 + 2 = 12 + 2 = 14$
For $x = -3$, $3x + 2 = 3(-3) + 2 = -9 + 2 = -7$

Simplify by combining like terms whenever possible.

1. $5x + 6x$

2. $-3y - 4y$

3. $3x + 4x - 2y$

4. $6x + x + 6 - 8$

Evaluate each expression for $x = 4$ and $a = -5$.

5. $2x - 8 =$

6. $3x + x + 5 =$

7. $2a + 14 =$

8. $(3a + 1) + (x - 7) =$

Example 15

$-2y - 6y - 1$ for $y = -2$ and $y = 0$

Solution

Simplify first:

$$-2y - 6y - 1 = -8y - 1$$

For $y = -2$, $-8y - 1 = -8(-2) - 1 = 16 - 1 = 15$
For $y = 0$, $-8y - 1 = -8(0) - 1 = 0 - 1 = -1$

Example 16

$(2x + 1 - 5x) + (a - 15)$ for $x = 2$ and $a = -3$

Solution

Simplify first.

$$(2x + 1 - 5x) + (a - 15) = (-3x + 1) + (a - 15)$$
$$= -3x + 1 + a - 15$$
$$= -3x + a - 14$$

Substitute $x = 2$ and $a = -3$.

$$-3x + a - 14 = -3(2) + (-3) - 14$$
$$= -6 - 3 - 14$$
$$= -23$$

☐ **NOW WORK EXERCISES 5–8 IN THE MARGIN.**

Answers to Marginal Exercises

1. $11x$ **2.** $-7y$ **3.** $7x - 2y$ **4.** $7x - 2$ **5.** 0   **6.** 21   **7.** 4   **8.** -17

NAME _____   SECTION _____   DATE _____

EXERCISES **12.5** _____

Simplify the following expressions by combining like terms whenever possible.

1. $6x + 2x$

2. $4x - 3x$

3. $5x + x$

4. $7x - 3x$

5. $-10a + 3a$

6. $-11y + 4y$

7. $-18y + 6y$

8. $-2x - 5x$

9. $-5x - 4x$

10. $-x - 2x$

11. $-7x - x$

12. $2x - 2x$

13. $5x - 5x$

14. $16p - 17p$

15. $9c - 10c$

Answers

1. _____

2. _____

3. _____

4. _____

5. _____

6. _____

7. _____

8. _____

9. _____

10. _____

11. _____

12. _____

13. _____

14. _____

15. _____

Answers

16. 3x − 5x + 12x

17. 2a + 14a − 25a

16. _____

17. _____

18. _____

18. 6c − 13c + 5c

19. 40p − 30p − 10p

19. _____

20. _____

21. _____

20. 16x − 15x − 3x

21. 2x + 3x − 7

22. _____

23. _____

24. _____

22. 5x − 6x + 2

23. 7x − 8x + 5

25. _____

26. _____

27. _____

24. −5x − 7x − 4

25. −8a − 3a − 2

26. −4x + x + 1 − 3

27. −2x + 5x + 6 − 5

NAME _____ SECTION _____ DATE _____

28. $4x + 7 - 8 + 3x$

29. $-5x - 1 + 8 + 9x$

30. $10y + 3 - 4 - 6y$

Evaluate each of the following expressions for $x = -3$, $y = 2$, $z = 3$, $a = -1$, and $c = -2$.

31. $x - 2$

32. $y - 2$

33. $z - 3$

34. $2x + z$

35. $3y - x$

36. $x - 4z$

37. $20 - 2a$

38. $10 + 2c$

39. $3c - 5$

40. $2x + 3x - 7$

41. $7a - a + 3$

Answers

28. _____

29. _____

30. _____

31. _____

32. _____

33. _____

34. _____

35. _____

36. _____

37. _____

38. _____

39. _____

40. _____

41. _____

Answers

42. _____

43. _____

44. _____

45. _____

46. _____

47. _____

48. _____

42. $-3y - 4y + 6 - 2$

43. $-2x - 3x + 1 - 4$

44. $5y - 2y - 3y + 4$

45. $2x - 3x + x - 8$

46. $(3a + 2x) + (7x - a + x)$

47. $[6 - (2a + 1)] + 3z$

48. $5y + [14 - (3c + 1)]$

12.6
SOLVING EQUATIONS ($ax + b = c$) _____

> **First-degree equation in x:** Any equation that can be written in the form
>
> $$ax + b = c \quad \text{where } a, b, \text{ and } c \text{ are constants and } a \neq 0.$$
>
> The variable may be something other than x.

Example 1　$2x = 14$ is a first-degree equation in x because we can write the equation in the form

$$2x + 0 = 14$$

where $a = 2$, $b = 0$, and $c = 14$.

Example 2　$x + 4 = 10$
Here $a = 1$, $b = 4$, and $c = 10$.

Example 3　$5y = 2y + 3$ is a first-degree equation in y. (We will see in later examples that the equation can be written in the form $3y = 3$.)

Example 4　$2x = 14$
Show that 7 is a solution and that 5 is **not** a solution.

Solution

Since $2 \cdot 7 = 14$ is true, 7 is a solution.
But　$2 \cdot 5 = 14$ is false. So 5 is not a solution.

Example 5　$y + 11 = 3$
Show that -8 is a solution and that 8 is **not** a solution.

Solution

Since $-8 + 11 = 3$ is true, -8 is a solution.
But　$8 + 11 = 3$ is false. So 8 is not a solution.

Solve the following equations.

1. $x + 5 = 6$

2. $3x = 12$

3. $y - 2 = -5$

4. $9y = -45$

HOW TO SOLVE AN EQUATION

There are two basic principles to use.

1. The **addition principle**
   If $A = B$ is true, then
   $A + C = B + C$ is also true for any number C.
2. The **multiplication principle**
   If $A = B$ is true, then
   $A \cdot C = B \cdot C$ is also true for any number C.
   $\dfrac{A}{B} = \dfrac{B}{C}$ is also true for $C \neq 0$.

Note: Dividing by C is the same as multiplying by the reciprocal $\dfrac{1}{C}$.

Example 6 $\begin{aligned} x + 14 &= 10 \\ x + 14 - 14 &= 10 - 14 \\ x + 0 &= -4 \\ x &= -4 \end{aligned}$ Using the **addition principle**, add -14 to both sides.
Simplify.
The solution.

Example 7 $8y = 72$

$\dfrac{8y}{8} = \dfrac{72}{8}$ Using the **multiplication principle**, divide both sides by 8. This is the same as multiplying both sides by $\dfrac{1}{8}$.

$y = 9$ Simplify to find the solution.

To understand solving equations:

1. If a constant is added to a variable, add its opposite to both sides of the equation.
2. If a constant is multiplied by a variable, divide both sides by that constant.
3. Remember that the object is to isolate the variable on one side of the equation, right side or left side.

☐ **NOW WORK EXERCISES 1–4 IN THE MARGIN.**

More difficult problems involving several steps are illustrated in Examples 8–11. Study each step carefully. Write the equations one under the other. Remember, **the object of the procedures is to get the variable terms on one side and the constant terms on the other side.**

Example 8

$4x + 3 = 11$	Write the equation.
$4x + 3 - 3 = 11 - 3$	Add -3 to both sides.
$4x = 8$	Simplify.
$\dfrac{\cancel{4}x}{\cancel{4}} = \dfrac{8}{4}$	Divide both sides by 4, the coefficient of x.
$x = 2$	Simplify.

Example 9

$2x - 4 + 3x = 26$	Write the equation.
$5x - 4 = 26$	Combine like terms on the left side.
$5x - 4 + 4 = 26 + 4$	Add $+4$ to both sides.
$5x = 30$	Simplify.
$\dfrac{\cancel{5}x}{\cancel{5}} = \dfrac{30}{5}$	Divide both sides by 5.
$x = 6$	Simplify.

Example 10

$5x + 2 = 3x - 8$	Write the equation.
$5x + 2 - 2 = 3x - 8 - 2$	Add -2 to both sides.
$5x = 3x - 10$	Simplify.
$5x - 3x = 3x - 10 - 3x$	Add $-3x$ to both sides.
$2x = -10$	Simplify; now one side has the term with the variable and the other side has the term with a constant.
$\dfrac{\cancel{2}x}{\cancel{2}} = \dfrac{-10}{2}$	Divide both sides by 2.
$x = -5$	Simplify.

Example 11

$3(x + 2) = 18 - x$	Write the equation.
$3 \cdot x + 3 \cdot 2 = 18 - x$	Use the distributive property.
$3x + 6 = 18 - x$	Simplify.
$3x + 6 - 6 = 18 - x - 6$	Add -6 to both sides.
$3x = 12 - x$	Simplify.
$3x + x = 12 - x + x$	Add $+x$ to both sides.
$4x = 12$	Simplify.
$\dfrac{\cancel{4}x}{\cancel{4}} = \dfrac{12}{4}$	Divide both sides by 4.
$x = 3$	Simplify.

Solve each equation.

5. $10x + 4 = 14$

6. $x + 5 - 2x = 7 + x$

7. $2(y - 7) = 5(y + 2)$

CHECKING SOLUTIONS TO EQUATIONS

Checking can be done by substituting the solution into the original equation. However, checking can be time-consuming and need not be done for every problem. Particularly on an exam, check only after you have finished the entire exam.

CHECKING FOR EXAMPLE 9:

$$2x - 4 + 3x = 26$$
$$2(6) - 4 + 3(6) \stackrel{?}{=} 26$$
$$12 - 4 + 18 \stackrel{?}{=} 26$$
$$26 = 26$$

CHECKING FOR EXAMPLE 10:

$$5x + 2 = 3x - 8$$
$$5(-5) + 2 \stackrel{?}{=} 3(-5) - 8$$
$$-25 + 2 \stackrel{?}{=} -15 - 8$$
$$-23 = -23$$

Completion Example 12 Solve the following equation.

$2(x - 5) = x - 1$	Write the equation.
$2x - 10 = x - 1$	Use the distributive property.
$2x - 10 \; + 10 \; = x - 1 \; + 10$	Add $+10$ to both sides.
_____	Simplify.
_____	Add $-x$ to both sides.
_____	Simplify.

☐ **NOW WORK EXERCISES 5–7 IN THE MARGIN.**

Completion Example Answer

12.
$$2(x - 5) = x - 1$$
$$2x - 10 = x - 1$$
$$2x - 10 + 10 = x - 1 + 10$$
$$2x = x + 9$$
$$2x - x = x + 9 - x$$
$$x = 9$$

Answers to Marginal Exercises

1. $x = 1$ **2.** $x = 4$ **3.** $y = -3$ **4.** $y = -5$ **5.** $x = 1$ **6.** $x = -1$
7. $y = -8$

EXERCISES **12.6** _____

Solve the following equations.

1. $x + 4 = 10$

2. $x + 13 = 20$

3. $y - 5 = 17$

4. $y - 12 = 4$

5. $y + 8 = 3$

6. $x + 10 = 7$

7. $x - 5 = -7$

8. $x - 14 = -10$

9. $y - 8 = -6$

10. $x - 12 = -5$

11. $5x = 30$

12. $3y = 15$

13. $10y = -40$

14. $6x = -48$

15. $-2x = 12$

16. $-4x = 24$

17. $-8y = -40$

18. $-12y = -36$

Answers

1. _____

2. _____

3. _____

4. _____

5. _____

6. _____

7. _____

8. _____

9. _____

10. _____

11. _____

12. _____

13. _____

14. _____

15. _____

16. _____

17. _____

18. _____

Answers

19. _____

20. _____

21. _____

22. _____

23. _____

24. _____

25. _____

26. _____

27. _____

28. _____

29. _____

30. _____

31. _____

32. _____

33. _____

34. _____

35. _____

36. _____

19. $16 = x + 3$

20. $25 = x + 14$

21. $2x + 3 = 5$

22. $3x - 4 = 8$

23. $4y + 1 = 9$

24. $3x - 10 = 11$

25. $6x + 4 = -14$

26. $7y - 8 = -1$

27. $3 + 6y = 15$

28. $6 + 5y = 21$

29. $2x + 3 = -9$

30. $3x - 1 = -4$

31. $5y + 12 = -3$

32. $10y + 3 = -17$

33. $15 = 2x - 3$

34. $20 = 3x - 1$

35. $-17 = 5y - 2$

36. $30 = 4y + 6$

37. $4 = 5x + 9$ **38.** $28 = 10x - 2$ **39.** $-24 = 7x - 3$

40. $96 = 25y - 4$ **41.** $3x = x - 10$ **42.** $5y = 2y + 12$

43. $7y = 6y + 5$ **44.** $6x = 2x + 20$ **45.** $5x = 2x$

46. $4x = 3x$ **47.** $4x + 3 = 2x + 9$

48. $5y - 2 = 4y - 6$ **49.** $7x + 14 = 10x + 5$

Answers

37. _____

38. _____

39. _____

40. _____

41. _____

42. _____

43. _____

44. _____

45. _____

46. _____

47. _____

48. _____

49. _____

Answers

50. _____

51. _____

52. _____

53. _____

54. _____

55. _____

56. _____

57. _____

58. _____

59. _____

60. _____

50. $5x + 20 = 8x - 4$

51. $5(x - 2) = 3(x - 8)$

52. $2(y + 1) = 3y + 3$

53. $4(x - 1) = 2x + 6$

54. $6y - 3 = 3(y + 2)$

55. $7y - 6y + 12 = 4y$

56. $6x + 5 + 3x = 3x - 13$

57. $5x - 2x + 4 = 3x + x - 1$

58. $7x + x - 6 = 2(x + 9)$

59. $x - 5 + 4x = 4(x - 3)$

60. $3(-x + 6) = 3x + 2(x + 1)$

12.7

SOLVING WORD PROBLEMS

Look for key words in word problems:

Key Words

Addition	Subtraction	Multiplication	Division
add	subtract	multiply	divide
sum	difference	product	quotient
plus	minus	times	
more than	less than	twice	
increased by	decreased by	of (with fraction)	

First we will translate some English phrases into algebraic expressions using the meanings of key words.

	ENGLISH PHRASE	ALGEBRAIC EXPRESSION
Example 1	7 **multiplied by** the variable x the **product** of 7 and x 7 **times** x	$7x$
Example 2	5 **added to** the unknown y the **sum** of 5 and y 5 **plus** y	$5 + y$
Example 3	8 **subtracted from** a number **times** 6 the **difference** between $6x$ and 8 $6x$ **minus** 8	$6x - 8$
Example 4	3 **more than twice** a number 3 **added to** 2 **times** a number $2x$ **increased by** 3	$2x + 3$
Example 5	three **times** the **quantity** found by **adding** 1 to a number the **product** of 3 with the **quantity** $x + 1$ three **times** the **sum** of a number and 1	$3(x + 1)$

☐ **NOW WORK EXERCISES 1–4 IN THE MARGIN.**

Translate each phrase into an algebraic expression.

1. 9 subtracted from a number

2. twenty plus twice a number

3. a number plus 7

4. 6 less than 5 times a number

> **STEPS IN SOLVING WORD PROBLEMS**
>
> 1. Read the problem carefully. Read the problem a second time.
> 2. Decide what is unknown and represent it with a letter. Draw a picture, if possible, to illustrate the problem.
> 3. Translate the English phrases into mathematical phrases and form an equation indicated by the problem.
> 4. Solve the equation.
> 5. Check to see that the solution of the equation makes sense in the problem.

Example 6   Five times a number is increased by 3, and the result is 38. What is the number?

Solution

Let n = the number

Translate "five times a number is increased by 3" to "$5n + 3$."

Translate "the result is" to "=."

The equation to be solved is

$$5n + 3 = 38$$

$$5n + 3 - 3 = 38 - 3$$

$$5n = 35$$

$$\frac{5n}{5} = \frac{35}{5}$$

$$n = 7$$

CHECK:

"Five times a number is increased by 3" translates to $(5 \cdot 7) + 3$, and

$$(5 \cdot 7) + 3 = 35 + 3 = 38$$

Example 7   Seven less than four times a number is equal to five more than twice the number. Find the number.

Solution

Let n = the number.
Translate "seven less than four times a number" to $4n - 7$.
Translate "five more than twice the number" to $2n + 5$.
The equation to be solved is

$$4n - 7 = 2n + 5$$

$$4n - 7 \;-\; 2n = 2n + 5 \;-\; 2n$$

$$2n - 7 = +5$$

$$2n - 7 \;+\; 7 = 5 \;+\; 7$$

$$2n = 12$$

$$\frac{2n}{2} = \frac{12}{2}$$

$$n = 6$$

CHECK: $4(6) - 7 \stackrel{?}{=} 2(6) + 5$

$$24 - 7 \stackrel{?}{=} 12 + 5$$

$$17 = 17$$

Example 8 A rectangular swimming pool has a perimeter of 160 meters and a length of 50 meters. How wide is the pool?

Solution

Use the formula for the perimeter of a rectangle, $P = 2l + 2w$.

Draw a picture.

Let w = width.

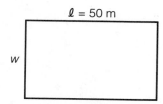

$\ell = 50$ m

w

Substituting $P = 160$ and $l = 50$ we get

$$160 = 2 \cdot 50 + 2w$$

$$160 = 100 + 2w$$

$$160 \;-\; 100 = 100 + 2w \;-\; 100$$

$$60 = 2w$$

$$\frac{60}{2} = \frac{2w}{2}$$

$$30 = w$$

The width of the pool is 30 meters.

Consecutive integers: Each integer is one more than the previous integer

18, 19, 20 are three consecutive integers and can be represented by n, $n + 1$, $n + 2$.

Consecutive odd integers: Each integer is odd and two more than the previous integer

31, 33, 35 are three consecutive odd integers and can be represented by n, $n + 2$, $n + 4$.

Consecutive even integers: Each integer is even and two more than the previous integer

-8, -6, -4 are three consecutive even integers and can be represented by n, $n + 2$, $n + 4$.

Example 9   The sum of three consecutive odd integers is -57. What are the integers?

Solution

Let n = first odd integer

$n + 2$ = second consecutive odd integer

$n + 4$ = third consecutive odd integer

The equation to be solved is

$$n + (n + 2) + (n + 4) = -57$$
$$n + n + 2 + n + 4 = -57$$
$$3n + 6 = -57$$
$$3n + 6 - 6 = -57 - 6$$
$$3n = -63$$
$$\frac{3n}{3} = \frac{-63}{3}$$
$$n = -21$$
$$n + 2 = -19$$
$$n + 4 = -17$$

CHECK: $(-21) + (-19) + (-17) = -57$

Answers to Marginal Exercises

1. $n - 9$ **2.** $20 + 2x$ **3.** $y + 7$ **4.** $5n - 6$

NAME _____ SECTION _____ DATE _____

EXERCISES **12.7** _____

Solve each of the following problems.

1. Find a number whose product with 3 is 57.

2. Find a number that when multiplied by 7 gives 84.

3. The sum of a number and 32 is 86. What is the number?

4. The difference between a number and 16 is −48. What is the number?

Answers

1. _____

2. _____

3. _____

4. _____

5. _____

6. _____

7. _____

8. _____

9. _____

5. If the product of a number and 4 is decreased by 10, the result is 50. Find the number.

6. If the product of a number and 5 is added to 12, the result is 7. Find the number.

7. If the product of a number and 8 is increased by 24, the result is twice the number. What is the number?

8. The sum of a number and 2 is equal to three times the number. What is the number?

9. If twice a number is decreased by 4, the result is the number. Find the number.

10. Three times the sum of a number and 4 is −60. What is the number?

Answers

10. _____

11. _____

12. _____

11. Five more than twice a number is equal to 20 more than the number. What is the number?

13. _____

12. If 7 is subtracted from a number, the result is 8 times the number. Find the number.

13. Twenty plus a number is equal to the sum of twice the number and three times the same number. What is the number?

Answers

14. _____

15. _____

16. _____

17. _____

18. _____

14. The sum of two consecutive integers is 37. Find the two integers.

15. The sum of three consecutive integers is -42. Find the three integers.

16. The sum of three consecutive odd integers is 27. Find the three integers.

17. Find three consecutive odd integers whose sum is 81.

18. Find three consecutive even integers whose sum is 30.

NAME _____ SECTION _____ DATE _____

19. Find four consecutive even integers whose sum is 54 more than the smallest of the four integers.

20. If the product of 4 with the sum of a number and 3 is diminished by 6, the result is 26. Find the number.

21. What is the number whose sum with 18 is ten times the number?

22. If 7 times a number is decreased by 4 times the number, the difference is equal to the sum of the number and 10. What is the number?

Answers

19. _____

20. _____

21. _____

22. _____

23. _____

24. _____

25. _____

26. _____

27. _____

28. _____

23. What is the number whose product with 6 is equal to 12 less than twice the number?

24. The difference of a number and 3 is equal to the difference of five times the number and 15. What is the number?

25. If the sum of two consecutive integers is multiplied by 3, the result is −15. What are the two integers?

26. The perimeter of a rectangular garden is 240 feet. If the width is 40 feet, what is the length of the garden?

27. The perimeter of a triangle is 18 centimeters. If two sides are equal and the third side is 4 centimeters long, what is the length of each of the other two sides?

28. A length of wire is bent to form a triangle with two sides equal. If the wire is 30 cm long and the two equal sides are each 8 cm long, how long is the third side?

NAME _____ SECTION _____ DATE _____

CHAPTER **12** TEST _____

Answers

1. Graph the following set of integers on the number line.

$\{-3, -1, 4, |-1|, |2|\}$

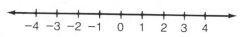

1. _____

2. _____

3. _____

4. _____

2. Find the opposite of -11.

3. Which of the following numbers is greater? -15 or -12

4. Which of the following numbers is greater in absolute value? -15 or -12

Answers

5. _____

6. _____

7. _____

8. _____

9. _____

10. _____

5. Find the sum of 14 and −6.

6. Add: $(-3) + (+7) + (-4) + (-9)$

7. Subtract 3 from −9.

8. Evaluate: $17 - 4 - 12 + 15$

9. Find the product of −4 and −7.

10. Multiply: $(-3)(-5)(2)(-3)$

NAME _____ SECTION _____ DATE _____

For Problems 11–13, find each quotient.

11. $\dfrac{-56}{-7}$

11. _____

12. _____

13. _____

12. $\dfrac{-3}{0}$

14. _____

15. _____

13. $\dfrac{65}{-13}$

14. Simplify the following expression by combining like terms:

$$4x - 5 - x + 10$$

15. Evaluate the following expression if $x = 2$ and $y = -1$:

$$[2x - (y + 3) - (3y + 4)]$$

Answers

16. _____

17. _____

18. _____

19. _____

20. _____

For Problems 16–18, solve each equation.

16. $3x - 4 = -19$

17. $x - 8 = 5x - 32$

18. $3(x + 3) = 2(7 - x)$

19. The difference of twice a number and -3 is equal to 5 more than the number. What is the number?

20. The sum of three consecutive integers is -36. Find the three integers.

Appendix

Powers, Roots, and Prime Factorizations

Number	Square	Square Root	Cube	Cube Root	Prime Factorization
1	1	1.0000	1	1.0000	—
2	4	1.4142	8	1.2599	prime
3	9	1.7321	27	1.4423	prime
4	16	2.0000	64	1.5874	2·2
5	25	2.2361	125	1.7100	prime
6	36	2.4495	216	1.8171	2·3
7	49	2.6458	343	1.9129	prime
8	64	2.8284	512	2.0000	2·2·2
9	81	3.0000	729	2.0801	3·3
10	100	3.1623	1000	2.1544	2·5
11	121	3.3166	1331	2.2240	prime
12	144	3.4641	1728	2.2894	2·2·3
13	169	3.6056	2197	2.3513	prime
14	196	3.7417	2744	2.4101	2·7
15	225	3.8730	3375	2.4662	3·5
16	256	4.0000	4096	2.5198	2·2·2·2
17	289	4.1231	4913	2.5713	prime
18	324	4.2426	5832	2.6207	2·3·3
19	361	4.3589	6859	2.6684	prime
20	400	4.4721	8000	2.7144	2·2·5
21	441	4.5826	9261	2.7589	3·7
22	484	4.6904	10,648	2.8020	2·11
23	529	4.7958	12,167	2.8439	prime
24	576	4.8990	13,824	2.8845	2·2·2·3
25	625	5.0000	15,625	2.9240	5·5
26	676	5.0990	17,576	2.9625	2·13
27	729	5.1962	19,683	3.0000	3·3·3
28	784	5.2915	21,952	3.0366	2·2·7
29	841	5.3852	24,389	3.0723	prime
30	900	5.4772	27,000	3.1072	2·3·5
31	961	5.5678	29,791	3.1414	prime
32	1024	5.6569	32,768	3.1748	2·2·2·2·2
33	1089	5.7446	35,937	3.2075	3·11
34	1156	5.8310	39,304	3.2396	2·17
35	1225	5.9161	42,875	3.2711	5·7
36	1296	6.0000	46,656	3.3019	2·2·3·3
37	1369	6.0828	50,653	3.3322	prime
38	1444	6.1644	54,872	3.3620	2·19
39	1521	6.2450	59,319	3.3912	3·13
40	1600	6.3246	64,000	3.4200	2·2·2·5
41	1681	6.4031	68,921	3.4482	prime
42	1764	6.4807	74,088	3.4760	2·3·7
43	1849	6.5574	79,507	3.5034	prime
44	1936	6.6333	85,184	3.5303	2·2·11
45	2025	6.7082	91,125	3.5569	3·3·5
46	2116	6.7823	97,336	3.5830	2·23
47	2209	6.8557	103,823	3.6088	prime
48	2304	6.9282	110,592	3.6342	2·2·2·2·3
49	2401	7.0000	117,649	3.6593	7·7
50	2500	7.0711	125,000	3.6840	2·5·5

Powers, Roots, and Prime Factorizations

Number	Square	Square Root	Cube	Cube Root	Prime Factorization
51	2601	7.1414	132,651	3.7084	3·17
52	2704	7.2111	140,608	3.7325	2·2·13
53	2809	7.2801	148,877	3.7563	prime
54	2916	7.3485	157,464	3.7798	2·3·3·3
55	3025	7.4162	166,375	3.8030	5·11
56	3136	7.4833	175,616	3.8259	2·2·2·7
57	3249	7.5498	185,193	3.8485	3·19
58	3364	7.6158	195,112	3.8709	2·29
59	3481	7.6811	205,379	3.8930	prime
60	3600	7.7460	216,000	3.9149	2·2·3·5
61	3721	7.8103	226,981	3.9365	prime
62	3844	7.8740	238,328	3.9579	2·31
63	3969	7.9373	250,047	3.9791	3·3·7
64	4096	8.0000	262,144	4.0000	2·2·2·2·2·2
65	4225	8.0623	274,625	4.0207	5·13
66	4356	8.1240	287,496	4.0412	2·3·11
67	4489	8.1854	300,763	4.0615	prime
68	4624	8.2462	314,432	4.0817	2·2·17
69	4761	8.3066	328,509	7.1016	3·23
70	4900	8.3666	343,000	7.1213	2·5·7
71	5041	8.4262	357,911	4.408	prime
72	5184	8.4853	373,248	4.1602	2·2·2·3·3
73	5329	8.5440	389,017	4.1793	prime
74	5476	8.6023	405,224	4.1983	2·37
75	5625	8.6603	421,875	4.2172	3·5·5
76	5776	8.7178	438,976	4.2358	2·2·19
77	5929	8.7750	456,533	4.2543	7·11
78	6084	8.8318	474,552	4.2727	2·3·13
79	6241	8.8882	493,039	4.2908	prime
80	6400	8.9443	512,000	4.3089	2·2·2·2·5
81	6561	9.0000	531,441	4.3267	3·3·3·3
82	6724	9.0554	551,368	4.3445	2·41
83	6889	9.1104	571,787	4.3621	prime
84	7056	9.1652	592,704	4.3795	2·2·3·7
85	7225	9.2195	614,125	4.3968	5·17
86	7396	9.2736	636,056	4.4140	2·43
87	7569	9.3274	658,503	4.4310	3·29
88	7744	9.3808	681,472	4.4480	2·2·2·11
89	7921	9.4340	704,969	4.4647	prime
90	8100	9.4868	729,000	4.4814	2·3·3·5
91	8281	9.5394	753,571	4.4979	7·13
92	8464	9.5917	778,688	4.5144	2·2·23
93	8649	9.6437	804,357	4.5307	3·31
94	8836	9.6954	830,584	4.5468	2·47
95	9025	9.7468	857,375	4.5629	5·19
96	9216	9.7980	884,736	4.5789	2·2·2·2·2·3
97	9409	9.8489	912,673	4.5947	prime
98	9604	9.8995	941,192	4.6104	2·7·7
99	9801	9.9499	970,299	4.6261	3·3·11
100	10,000	10.0000	1,000,000	4.6416	2·2·5·5

Answer Key

CHAPTER 1

Exercises 1.1   (page 7)

1. 0, 1, 2, 3, 4, 5, 6, 7, 8, 9   **3.** 1, 10, 100, 1000, 10,000, and so on   **5.** four   **7.** 30 + 7   **9.** 50 + 6
11. 1000 + 800 + 90 + 2   **13.** 20,000 + 5000 + 600 + 50 + 8   **15.** eighty-three   **17.** ten thousand, five hundred
19. five hundred ninety-two thousand, three hundred   **21.** seventy-one million, five hundred thousand   **23.** 98   **25.** 573
27. 10,011   **29.** 537,082
31. 2—billions
4—hundred millions
3—millions
1—hundred thousand
8—ten thousands
9—thousands
5—hundreds
33. ninety-three million

Exercises 1.2   (page 15)

1. 16   **3.** 13   **5.** 17   **7.** 17   **9.** 15   **11.** 12, commutative property   **13.** 12, associative property
15. 9, associative property   **17.** 9, additive identity   **19.** 16, associative property

21.

+	5	8	7	9
3	8	11	10	12
6	11	14	13	15
5	10	13	12	14
2	7	10	9	11

23. 125   **25.** 1552   **27.** 13,324   **29.** 197   **31.** 835   **33.** 553   **35.** 4168   **37.** 9064   **39.** 1,463,930
41. 2,817,126   **43.** $6,313,323   **45.** $493,154

Exercises 1.3   (page 25)

1. 6   **3.** 4   **5.** 5   **7.** 1   **9.** 8   **11.** 0   **13.** 7   **15.** 9   **17.** 8   **19.** 7   **21.** 11   **23.** 20   **25.** 5   **27.** 126
29. 376   **31.** 395   **33.** 966   **35.** 1424   **37.** 5871   **39.** 3,800,559   **41.** 5,671,011   **43.** 398   **45.** $815

Exercises 1.4 *(page 33)*

1. 760   **3.** 80   **5.** 300   **7.** 990   **9.** 4200   **11.** 500   **13.** 600   **15.** 76,500   **17.** 7000   **19.** 8000
21. 13,000   **23.** 62,000   **25.** 80,000   **27.** 260,000   **29.** 120,000   **31.** 180,000   **33.** 100   **35.** 1000
37. 168 (estimate 180)   **39.** 1173 (estimate 1200)   **41.** 1881 (estimate 1900)   **43.** 9224 (estimate 9500)
45. 5467 (estimate 6000)   **47.** 5931 (estimate 5000)   **49.** 20,804 (estimate 20,000)

Exercises 1.5 *(page 43)*

1. 72   **3.** 48   **5.** 45   **7.** 24   **9.** 56   **11.** 0   **13.** 24   **15.** 32   **17.** 1
19. 27, commutative property of multiplication   **21.** 7, multiplicative identity

23.

•	5	8	7	9
3	15	24	21	27
6	30	48	42	54
5	25	40	35	45
2	10	16	14	18

25. $8 \cdot 10$ **27.** $6 \cdot 1000$ **29.** $54 \cdot 100$ **31.** $7 \cdot 1,000,000$ **33.** 47,000   **35.** 13   **37.** 400   **39.** 5600   **41.** 4000
43. 3600   **45.** 16,000   **47.** 5000   **49.** 800,000   **51.** 3,200,000   **53.** 240   **55.** 36,000   **57.** 150,000

Exercises 1.6 *(page 51)*

1. 224 (estimate 240)   **3.** 432 (estimate 450)   **5.** 344 (estimate 320)   **7.** 455 (estimate 450)   **9.** 252 (estimate 240)
11. 2352 (estimate 2400)   **13.** 330 (estimate 400)   **15.** 2412 (estimate 2800)   **17.** 960 (estimate 1000)
19. 7055 (estimate 7200)   **21.** 544 (estimate 600)   **23.** 880 (estimate 800)   **25.** 375 (estimate 600)
27. 2064 (estimate 1800)   **29.** 156 (estimate 100)   **31.** 5166 (estimate 4000)   **33.** 2850 (estimate 3000)
35. 29,601 (estimate 20,000)   **37.** 9800 (estimate 10,000)   **39.** 125,178 (estimate 140,000)   **41.** 31,200   **43.** 380,000
45. 496,400   **47.** 217,300   **49.** $32,760; $93,600   **51.** 275 miles; 200 miles   **53.** 36,400 sq ft

Exercises 1.7 *(page 65)*

1. 4   **3.** 9   **5.** 8   **7.** 5   **9.** 5   **11.** 5   **13.** 8   **15.** 9   **17.** 9   **19.** 8   **21.** 0   **23.** 1
25. 30 R 0; 7 and 30 are factors of 210.   **27.** 12 R 0; 11 and 12 are factors of 132.   **29.** 17 R 0; 3 and 17 are factors of 51.
31. 20 R 13; divisor and quotient are not factors of 413.   **33.** 14 R 0; 13 and 14 are factors of 182.
35. 3 R 10; divisor and quotient are not factors of 52.   **37.** 38 R 6; divisor and quotient are not factors of 424.   **39.** 15 R 5
41. 42 R 3   **43.** 20 R 0   **45.** 400 R 3   **47.** 301 R 4   **49.** 3 R 3   **51.** 54 R 3   **53.** 22 R 74   **55.** 7 R 358
57. $1610 \div 35 = 46$ (or $1610 \div 46 = 35$)   **59.** 203

Exercises 1.8 *(page 73)*

1. 16   **3.** 921   **5.** $262   **7.** The blue car would be cheaper by $33.   **9.** $85   **11.** 114 cm   **13.** 168 sq in.
15. $90   **17.** $485   **19.** $316

Chapter 1 Test   *(page 77)*

1. $8000 + 900 + 50 + 2$; eight thousand nine hundred fifty-two   **2.** identity
3. $a \cdot b = b \cdot a$ where a and b are whole numbers   **4.** 1000   **5.** 140,000   **6.** 12,009   **7.** 1735   **8.** 13,781,661   **9.** 488
10. 1229   **11.** 5707   **12.** 2584   **13.** 220,405   **14.** 210,938   **15.** 403 R 0   **16.** 172 R 388   **17.** 2005 R 0   **18.** 74
19. 54   **20.** \$306; \$224

CHAPTER 2 _____

Exercises 2.1   *(page 83)*

	Exponent	Base	Power
1.	3	2	8
3.	2	5	25
5.	0	7	1
7.	4	1	1
9.	0	4	1
11.	2	3	9
13.	0	5	1
15.	1	62	62
17.	2	10	100
19.	4	10	10,000

21. 2^2 **23.** 2^4 or 4^2 **25.** 2^5 **27.** 7^2 **29.** 3^2 **31.** 5^3 **33.** 8^2 or 4^3 or 2^6
35. 10^3 **37.** 6^5 **39.** $2^2 \cdot 7^2$ **41.** $2^2 \cdot 3^3$ **43.** $3^2 \cdot 5^3$ **45.** 11^3 **47.** $3 \cdot 5^3$
49. 64   **51.** 144   **53.** 225   **55.** 324

Exercises 2.2   *(page 89)*

1. 3   **3.** 7   **5.** 22   **7.** 3   **9.** 5   **11.** 5   **13.** 5   **15.** 3   **17.** 7   **19.** 0   **21.** 6   **23.** 3   **25.** 27   **27.** 26
29. 0   **31.** 69   **33.** 68   **35.** 140   **37.** 5   **39.** 0   **41.** 9   **43.** 12   **45.** 24   **47.** 118   **49.** 0

Exercises 2.3   *(page 99)*

1. 2, 3, 4, 9   **3.** 3, 5   **5.** 2, 3, 5, 10   **7.** 2, 4   **9.** 2, 3, 4   **11.** none   **13.** 3, 9   **15.** none   **17.** none   **19.** 3
21. 3, 5   **23.** 2, 3, 4, 5, 9, 10   **25.** 2, 3   **27.** none   **29.** 2   **31.** 3, 5   **33.** 3, 9   **35.** 2, 3, 9   **37.** 2, 4, 5, 10
39. 3, 9   **41.** 2, 3, 4, 5, 10   **43.** 2, 4   **45.** 2, 3, 4, 5, 10   **47.** 3, 5   **49.** none   **51.** 2, 3, 9   **53.** 2, 3, 4
55. 2, 3, 4, 9   **57.** 3, 9   **59.** 2, 3, 4   **61.** yes; 18, 36, 54, 72, 90 (Other answers are possible.)

Exercises 2.4   *(page 107)*

1. 5, 10, 15, 20, 25, 30, 35, 40, . . .   **3.** 11, 22, 33, 44, 55, 66, 77, 88, . . .   **5.** 12, 24, 36, 48, 60, 72, 84, 96, . . .
7. 20, 40, 60, 80, 100, 120, 140, 160, . . .   **9.** 16, 32, 48, 64, 80, 96, 112, 128, . . .
11. 2, 3, 5, 7, 11, 13, 17, 19, 23, 29, 31, 37, 41, 43, 47, 53, 59, 61, 67, 71, 73, 79, 83, 89, 97   **13.** Prime
15. Composite: 2 and 16; 4 and 8   **17.** Prime   **19.** Composite: 3 and 21; 9 and 7   **21.** Composite: 3 and 17   **23.** Prime
25. Prime   **27.** Composite: 3 and 19   **29.** Composite: 2 and 43   **31.** Prime   **33.** 3 and 4   **35.** 1 and 12
37. 2 and 25   **39.** 3 and 8   **41.** 3 and 12   **43.** 3 and 21   **45.** 5 and 5   **47.** 5 and 12   **49.** 3 and 9   **51.** 2

Exercises 2.5 *(page 113)*

1. $24 = 2^3 \cdot 3$ **3.** $27 = 3^3$ **5.** $36 = 2^2 \cdot 3^2$ **7.** $72 = 2^3 \cdot 3^2$ **9.** $81 = 3^4$ **11.** $125 = 5^3$ **13.** $75 = 3 \cdot 5^2$
15. $210 = 2 \cdot 3 \cdot 5 \cdot 7$ **17.** $250 = 2 \cdot 5^3$ **19.** $168 = 2^3 \cdot 3 \cdot 7$ **21.** $126 = 2 \cdot 3^2 \cdot 7$ **23.** Prime   **25.** $51 = 3 \cdot 17$
27. $121 = 11^2$ **29.** $225 = 3^2 \cdot 5^2$ **31.** $32 = 2^5$ **33.** $108 = 2^2 \cdot 3^3$ **35.** Prime   **37.** $78 = 2 \cdot 3 \cdot 13$ **39.** $10{,}000 = 2^4 \cdot 5^4$
41. 1, 12, 2, 3, 4, 6   **43.** 1, 28, 2, 4, 7, 14   **45.** 1, 121, 11   **47.** 1, 105, 3, 5, 7, 15, 21, 35   **49.** 1, 97

Exercises 2.6 *(page 121)*

1. $2 \cdot 2 \cdot 2 \cdot 3 = 2^3 \cdot 3 = 24$ **3.** $2 \cdot 2 \cdot 3 \cdot 3 = 2^2 \cdot 3^2 = 36$ **5.** $2 \cdot 5 \cdot 11 = 110$ **7.** $2 \cdot 2 \cdot 3 \cdot 5 = 2^2 \cdot 3 \cdot 5 = 60$
9. $2 \cdot 2 \cdot 2 \cdot 5 \cdot 5 = 2^3 \cdot 5^2 = 200$ **11.** $2 \cdot 2 \cdot 7 \cdot 7 = 2^2 \cdot 7^2 = 196$ **13.** $2 \cdot 2 \cdot 2 \cdot 2 \cdot 3 \cdot 5 = 2^4 \cdot 3 \cdot 5 = 240$
15. $2 \cdot 2 \cdot 5 \cdot 5 = 2^2 \cdot 5^2 = 100$ **17.** $2 \cdot 2 \cdot 5 \cdot 5 \cdot 7 = 2^2 \cdot 5^2 \cdot 7 = 700$ **19.** $2 \cdot 2 \cdot 3 \cdot 3 \cdot 7 = 2^2 \cdot 3^2 \cdot 7 = 252$ **21.** $2 \cdot 2 \cdot 2 = 2^3 = 8$
23. $2 \cdot 2 \cdot 2 \cdot 3 \cdot 5 \cdot 13 = 2^3 \cdot 3 \cdot 5 \cdot 13 = 1560$ **25.** $2 \cdot 2 \cdot 3 \cdot 5 = 2^2 \cdot 3 \cdot 5 = 60$ **27.** $2 \cdot 2 \cdot 2 \cdot 2 \cdot 3 \cdot 5 = 2^4 \cdot 3 \cdot 5 = 240$
29. $2 \cdot 3 \cdot 3 \cdot 5 \cdot 5 \cdot 5 = 2 \cdot 3^2 \cdot 5^3 = 2250$ **31.** $2 \cdot 3 \cdot 11 \cdot 11 = 2 \cdot 3 \cdot 11^2 = 726$ **33.** $2 \cdot 3 \cdot 3 \cdot 5 \cdot 29 = 2 \cdot 3^2 \cdot 5 \cdot 29 = 2610$
35. $3 \cdot 3 \cdot 3 \cdot 5 \cdot 5 = 3^3 \cdot 5^2 = 675$ **37.** $2 \cdot 2 \cdot 2 \cdot 3 \cdot 5 = 2^3 \cdot 3 \cdot 5 = 120$ **39.** $2 \cdot 2 \cdot 2 \cdot 2 \cdot 3 \cdot 5 \cdot 7 = 2^4 \cdot 3 \cdot 5 \cdot 7 = 1680$

41. (a) LCM $= 2^3 \cdot 3 \cdot 5 = 120$;
    (b) $120 = 8 \cdot 15$
           $120 = 10 \cdot 12$

43. (a) LCM $= 2^3 \cdot 3 \cdot 5 = 120$;
    (b) $120 = 10 \cdot 12$
           $120 = 15 \cdot 8$
           $120 = 24 \cdot 5$

45. (a) LCM $= 2 \cdot 3^3 \cdot 5 = 270$;
    (b) $270 = 6 \cdot 45$
           $270 = 18 \cdot 15$
           $270 = 27 \cdot 10$

47. (a) LCM $= 2 \cdot 3^2 \cdot 5 \cdot 7^2 = 4410$;
    (b) $4410 = 45 \cdot 98$
        $4410 = 63 \cdot 70$

49. (a)   LCM $= 3^2 \cdot 11^2 \cdot 13 = 14{,}157$;
    (b) $14{,}157 = 99 \cdot 143$
        $14{,}157 = 363 \cdot 39$

51. (a) 70 min; (b) 5 laps and 7 laps
53. (a) 48 hr; (b) 3 orbits and 4 orbits
55. (a) 180 days; (b) 18, 15, 12, and 10 trips

Chapter 2 Test *(page 125)*

1. 13   **2.** 48   **3.** 0   **4.** 72   **5.** 2, 3, 5, 9, 10   **6.** none   **7.** 2, 3, 4, 9   **8.** 2, 4, 5, 10   **9.** 13, 26, 39, 52, 65, 78, 91
10. one   **11.** 2, 7, 23, 47   **12.** $2 \cdot 5 \cdot 7$ **13.** $2 \cdot 2 \cdot 31$ **14.** $2 \cdot 5 \cdot 5 \cdot 13$ **15.** 1, 60, 2, 3, 4, 5, 6, 10, 12, 15, 30
16. 216   **17.** 84   **18.** 60   **19.** 420   **20.** Every 36 days; salesman A: 6; salesman B: 4; salesman C: 3

CHAPTER 3

Exercises 3.1 *(page 133)*

1. 0   **3.** $\dfrac{1}{3}$ **5.** $\dfrac{1}{4}$ **7.** $\dfrac{1}{4}$ **9.**

$$\frac{1}{5} \cdot \frac{3}{4} = \frac{3}{20}$$

11.

$$\frac{7}{8} \cdot \frac{1}{4} = \frac{7}{32}$$

13.

$$\frac{2}{9} \cdot \frac{2}{3} = \frac{4}{27}$$

15. $\dfrac{1}{16}$ **17.** $\dfrac{12}{35}$ **19.** $\dfrac{1}{9}$ **21.** $\dfrac{3}{32}$ **23.** $\dfrac{9}{49}$ **25.** $\dfrac{15}{32}$ **27.** 0 **29.** $\dfrac{35}{12}$ **31.** $\dfrac{10}{1}$ **33.** $\dfrac{45}{2}$ **35.** $\dfrac{32}{15}$ **37.** $\dfrac{99}{20}$

39. $\dfrac{48}{455}$ **41.** $\dfrac{1}{360}$ **43.** $\dfrac{840}{1}$ **45.** 0 **47.** 0 **49.** $\dfrac{728}{45}$ **51.** commutative **53.** associative

Exercises 3.2 *(page 143)*

1. 21 **3.** 10 **5.** 45 **7.** 16 **9.** 42 **11.** 54 **13.** 40 **15.** 44 **17.** 10 **19.** 45 **21.** 6 **23.** 9 **25.** 6

27. 1 **29.** 11 **31.** 30 **33.** $\dfrac{1}{3}$ **35.** $\dfrac{3}{4}$ **37.** $\dfrac{2}{5}$ **39.** $\dfrac{7}{18}$ **41.** 0 **43.** $\dfrac{2}{5}$ **45.** $\dfrac{5}{6}$ **47.** $\dfrac{2}{3}$ **49.** $\dfrac{2}{3}$ **51.** $\dfrac{6}{25}$

53. $\dfrac{2}{3}$ **55.** $\dfrac{12}{35}$ **57.** $\dfrac{2}{9}$ **59.** $\dfrac{25}{76}$ **61.** $\dfrac{1}{2}$ **63.** $\dfrac{6}{1}$ **65.** $\dfrac{2}{17}$

Exercises 3.3 *(page 151)*

1. $\dfrac{3}{8}$ **3.** $\dfrac{12}{13}$ **5.** $\dfrac{11}{15}$ **7.** $\dfrac{8}{9}$ **9.** $\dfrac{2}{1}$ **11.** $\dfrac{8}{9}$ **13.** $\dfrac{5}{7}$ **15.** $\dfrac{1}{3}$ **17.** 1 **19.** $\dfrac{1}{6}$ **21.** $\dfrac{1}{12}$ **23.** $\dfrac{2}{25}$ **25.** $\dfrac{10}{3}$

27. $\dfrac{4}{15}$ **29.** $\dfrac{5}{18}$ **31.** $\dfrac{1}{4}$ **33.** $\dfrac{21}{4}$ **35.** $\dfrac{77}{4}$ **37.** $\dfrac{8}{5}$ **39.** $\dfrac{2}{7}$

Exercises 3.4 *(page 157)*

1. $\dfrac{8}{9}$ **3.** $\dfrac{5}{7}$ **5.** $\dfrac{7}{5}$ **7.** $\dfrac{1}{3}$ **9.** $\dfrac{3}{4}$ **11.** $\dfrac{4}{9}$ **13.** $\dfrac{5}{8}$ **15.** $\dfrac{39}{32}$ **17.** $\dfrac{9}{10}$ **19.** 1 **21.** $\dfrac{32}{21}$ **23.** $\dfrac{10}{3}$ **25.** $\dfrac{16}{21}$

27. $\dfrac{7}{60}$ **29.** $\dfrac{63}{50}$ **31.** $\dfrac{7}{5}$ **33.** $\dfrac{5}{7}$ **35.** $\dfrac{8}{3}$ **37.** $\dfrac{3}{8}$ **39.** $\dfrac{3}{2}$ **41.** $\dfrac{1}{4}$ **43.** $\dfrac{22}{7}$ **45.** 37 **47.** $\dfrac{62}{43}$ **49.** 225

Exercises 3.5 *(page 167)*

1. 1 **3.** $\dfrac{1}{5}$ **5.** $\dfrac{3}{2}$ **7.** $\dfrac{6}{5}$ **9.** $\dfrac{3}{5}$ **11.** $\dfrac{17}{15}$ **13.** $\dfrac{23}{20}$ **15.** $\dfrac{43}{200}$ **17.** $\dfrac{11}{16}$ **19.** 1 **21.** $\dfrac{3}{4}$ **23.** $\dfrac{8}{5}$ **25.** $\dfrac{151}{140}$

27. $\dfrac{377}{280}$ **29.** $\dfrac{3}{5}$ **31.** $\dfrac{11}{20}$ **33.** $\dfrac{49}{60}$ **35.** $\dfrac{157}{100}$ **37.** $\dfrac{31}{96}$ **39.** $\dfrac{106}{945}$ **41.** $\dfrac{103}{90}$ **43.** $\dfrac{171}{100}$ **45.** $\dfrac{631}{100}$ **47.** $\dfrac{49}{100}$

49. $\dfrac{63}{50}$ **51.** $\dfrac{181}{10,000}$ **53.** $\dfrac{7683}{10,000}$ **55.** $\dfrac{1331}{1,000,000}$

Exercises 3.6 *(page 175)*

1. $\dfrac{3}{7}$ 3. $\dfrac{3}{5}$ 5. $\dfrac{1}{2}$ 7. $\dfrac{1}{3}$ 9. $\dfrac{3}{5}$ 11. $\dfrac{1}{2}$ 13. $\dfrac{13}{30}$ 15. $\dfrac{1}{12}$ 17. $\dfrac{9}{32}$ 19. $\dfrac{13}{20}$ 21. $\dfrac{7}{54}$ 23. $\dfrac{1}{40}$ 25. $\dfrac{7}{60}$

27. $\dfrac{27}{8}$ 29. $\dfrac{3}{16}$ 31. $\dfrac{87}{100}$ 33. $\dfrac{3}{50}$ 35. $\dfrac{1}{25}$ 37. $\dfrac{3}{20}$ 39. 0   41. $\dfrac{11}{48}$ 43. $\dfrac{3}{32}$

Exercises 3.7 *(page 183)*

1. $\dfrac{3}{4}$ is larger by $\dfrac{1}{12}$ 3. $\dfrac{17}{20}$ is larger is $\dfrac{1}{20}$ 5. $\dfrac{13}{20}$ is larger by $\dfrac{1}{40}$ 7. equivalent fractions   9. $\dfrac{11}{48}$ is larger by $\dfrac{1}{60}$

11. $\dfrac{3}{5}, \dfrac{2}{3}, \dfrac{7}{10}$ 13. $\dfrac{11}{12}, \dfrac{19}{20}, \dfrac{7}{6}$ 15. $\dfrac{1}{4}, \dfrac{1}{3}, \dfrac{1}{2}$ 17. $\dfrac{13}{18}, \dfrac{7}{9}, \dfrac{31}{36}$ 19. $\dfrac{20}{10,000}, \dfrac{3}{1000}, \dfrac{1}{100}$ 21. $\dfrac{2}{3}$ 23. $\dfrac{13}{9}$ 25. $\dfrac{187}{32}$

27. $\dfrac{1}{4}$ 29. $\dfrac{8}{39}$ 31. $\dfrac{15}{64}$ 33. $\dfrac{29}{36}$

Chapter 3 Test *(page 185)*

1. $\dfrac{1}{9}$ 2. $\dfrac{35}{80}$ 3. $\dfrac{5}{6}$ 4. $\dfrac{13}{51}$ 5. $\dfrac{19}{24}$ 6. $\dfrac{6}{35}$ 7. $\dfrac{7}{18}$ 8. $\dfrac{1}{9}$ 9. $\dfrac{2}{9}$ 10. $\dfrac{17}{19}$ 11. 25   12. $\dfrac{3}{8}$ 13. $\dfrac{9}{70}$ 14. $\dfrac{1}{2}$

15. $\dfrac{25}{22}$ 16. $\dfrac{3}{16}$ 17. $\dfrac{1}{6}$ 18. 3   19. $\dfrac{2}{3}$ 20. 2   21. $\dfrac{13}{20}$ 22. $\dfrac{1}{3}$ 23. $\dfrac{11}{30}$ 24. $\dfrac{1}{16}$ 25. $\dfrac{2}{3}, \dfrac{3}{4}, \dfrac{7}{8}$

CHAPTER 4

Exercises 4.1 *(page 191)*

1. $\dfrac{4}{3}$ 3. $\dfrac{4}{3}$ 5. $\dfrac{3}{2}$ 7. $\dfrac{7}{5}$ 9. $\dfrac{5}{4}$ 11. $4\dfrac{1}{6}$ 13. $1\dfrac{1}{3}$ 15. $1\dfrac{1}{2}$ 17. $6\dfrac{1}{7}$ 19. $7\dfrac{1}{2}$ 21. $3\dfrac{1}{9}$ 23. 3   25. $4\dfrac{1}{2}$

27. 3   29. $1\dfrac{3}{4}$ 31. $\dfrac{37}{8}$ 33. $\dfrac{76}{15}$ 35. $\dfrac{46}{11}$ 37. $\dfrac{7}{3}$ 39. $\dfrac{32}{3}$ 41. $\dfrac{34}{5}$ 43. $\dfrac{50}{3}$ 45. $\dfrac{101}{5}$ 47. $\dfrac{92}{7}$ 49. $\dfrac{51}{3}$

Exercises 4.2 *(page 195)*

1. $14\dfrac{5}{7}$ 3. 8   5. $12\dfrac{3}{4}$ 7. $8\dfrac{2}{3}$ 9. $34\dfrac{7}{18}$ 11. $7\dfrac{2}{3}$ 13. $42\dfrac{7}{20}$ 15. $10\dfrac{3}{4}$ 17. $12\dfrac{7}{27}$ 19. $18\dfrac{23}{45}$ 21. $28\dfrac{1}{2}$

23. $9\dfrac{29}{40}$ 25. $12\dfrac{47}{60}$ 27. $63\dfrac{11}{24}$ 29. $53\dfrac{83}{192}$ 31. $18\dfrac{7}{8}$ 33. $36\dfrac{23}{24}$ 35. $90\dfrac{1}{8}$ 37. $35\dfrac{7}{10}$ km   39. $12\dfrac{1}{8}$ in.

Exercises 4.3 *(page 203)*

1. $4\dfrac{1}{2}$ 3. $1\dfrac{5}{12}$ 5. 4   7. $3\dfrac{1}{2}$ 9. $3\dfrac{1}{4}$ 11. $3\dfrac{5}{8}$ 13. $4\dfrac{2}{3}$ 15. $6\dfrac{7}{12}$ 17. $10\dfrac{4}{5}$ 19. $5\dfrac{7}{12}$ 21. $1\dfrac{7}{16}$ 23. $\dfrac{1}{3}$

25. $\dfrac{5}{8}$ 27. $1\dfrac{11}{16}$ 29. $1\dfrac{13}{16}$ 31. 5   33. $7\dfrac{3}{40}$ 35. $7\dfrac{4}{5}$ 37. $1\dfrac{3}{20}$ hr   39. $6\dfrac{3}{5}$ min   41. $3\dfrac{1}{4}$ lb   43. $3\dfrac{1}{4}$ ft

Exercises 4.4 *(page 211)*

1. $\frac{91}{12}$ or $7\frac{7}{12}$ 3. $\frac{21}{2}$ or $10\frac{1}{2}$ 5. $\frac{45}{2}$ or $22\frac{1}{2}$ 7. $\frac{187}{6}$ or $31\frac{1}{6}$ 9. 8 11. 30 13. $\frac{87}{4}$ or $21\frac{3}{4}$ 15. 1 17. $\frac{55}{2}$ or $27\frac{1}{2}$

19. $\frac{77}{4}$ or $19\frac{1}{4}$ 21. $\frac{1}{7}$ 23. $\frac{1}{10}$ 25. $\frac{36}{385}$ 27. 11 29. 242 31. $\frac{91}{8}$ or $11\frac{3}{8}$ 33. 17 35. $\frac{21}{10}$ or $2\frac{1}{10}$

37. $\frac{18,271}{10}$ or $1827\frac{1}{10}$ 39. $\frac{226,226}{75}$ or $3016\frac{26}{75}$ 41. 40 43. 20 45. 75 47. $\frac{5}{16}$ 49. $\frac{9}{14}$ 51. 10 ft; 22 ft

53. 177 mi 55. 90 pages, 18 hr

Exercises 4.5 *(page 221)*

1. $\frac{1}{3}$ 3. $\frac{5}{9}$ 5. $\frac{10}{39}$ 7. $\frac{29}{155}$ 9. $\frac{28}{17}$ or $1\frac{11}{17}$ 11. $\frac{200}{301}$ 13. $\frac{7}{5}$ or $1\frac{2}{5}$ 15. $\frac{41}{3}$ or $13\frac{2}{3}$ 17. $\frac{63}{5}$ or $12\frac{3}{5}$

19. $\frac{20}{11}$ or $1\frac{9}{11}$ 21. $\frac{1}{5}$ 23. $\frac{172}{9}$ or $19\frac{1}{9}$ 25. $\frac{791}{120}$ or $6\frac{71}{120}$ 27. $\frac{3}{5}$ 29. $\frac{5}{2}$ or $2\frac{1}{2}$ 31. $\frac{141}{40}$ or $3\frac{21}{40}$ 33. $\frac{50}{27}$ or $1\frac{23}{27}$

35. 175 passengers 37. $\frac{543}{40}$ or $13\frac{23}{40}$ 39. $\frac{42}{5}$ or $8\frac{2}{5}$

Chapter 4 Test *(page 225)*

1. $3\frac{2}{5}$ 2. $\frac{61}{8}$ 3. $6\frac{7}{12}$ 4. $2\frac{5}{6}$ 5. 24 6. $\frac{1}{6}$ 7. $10\frac{5}{24}$ 8. $1\frac{2}{5}$ 9. $7\frac{1}{20}$ 10. $1\frac{37}{42}$ 11. $4\frac{17}{40}$ 12. $\frac{8}{9}$

13. $1\frac{2}{3}$ 14. $\frac{1}{3}$ 15. 20 16. $1\frac{1}{2}$ 17. $11\frac{2}{3}$ 18. $\frac{29}{42}$ 19. $\frac{33}{8}$ or $4\frac{1}{8}$ 20. $7\frac{1}{2}$ shelves

CHAPTER **5**

Exercises 5.1 *(page 231)*

1. 37.498 3. 4.11 5. 87.003 7. 62.7 9. 100.38 11. $82\frac{56}{100}$ 13. $10\frac{576}{1000}$ 15. $65\frac{3}{1000}$ 17. 0.014

19. 6.28 21. 72.392 23. 700.77 25. 600,500.402 27. ninety-three hundredths

29. thirty-two and fifty-eight hundredths 31. thirty-five and seventy-eight thousandths

33. eighteen and one hundred two thousandths 35. six hundred seven and six hundred seven thousandths

37. eight hundred six thousandths 39. five thousand and five thousandths

41. nine hundred and four thousand six hundred thirty-eight ten-thousandths

43.

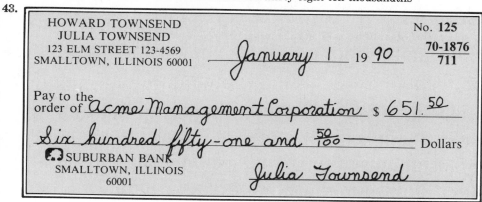

Exercises 5.2 *(page 239)*

1. 4.8   **3.** 76.3   **5.** 89.0   **7.** 18.0   **9.** 14.3   **11.** 0.39   **13.** 5.72   **15.** 7.00   **17.** 0.08   **19.** 5.71   **21.** 0.067
23. 0.634   **25.** 32.479   **27.** 17.364   **29.** 0.002   **31.** 479   **33.** 18   **35.** 382   **37.** 440   **39.** 6333   **41.** 5160
43. 500   **45.** 1000   **47.** 5480   **49.** 92,540   **51.** 7000   **53.** 48,000   **55.** 217,000   **57.** 380,000   **59.** 7,305,000
61. 0.00058   **63.** 470   **65.** 500   **67.** 3.230   **69.** 80,000

Exercises 5.3 *(page 247)*

1. 2.3   **3.** 7.55   **5.** 72.31   **7.** 276.096   **9.** 93.877   **11.** 103.429   **13.** 15.89   **15.** 64.947   **17.** 30.475
19. 34.186   **21.** 0.25 in.   **23.** 12.28 in.

Exercises 5.4 *(page 255)*

1. 0.42   **3.** 0.04   **5.** 21.6   **7.** 0.42   **9.** 0.004   **11.** 0.108   **13.** 0.0276   **15.** 0.0486   **17.** 0.0006   **19.** 0.375
21. 1.4   **23.** 1.725   **25.** 5.063   **27.** 0.08   **29.** 0.0329   **31.** 0.000045   **33.** 0.00822   **35.** 0.535   **37.** 0.0009222
39. 9.224655   **41.** 4.9077   **43.** 54.5034   **45.** 2.32986   **47.** 2057   **49.** 693   **51.** 3820   **53.** 7.19   **55.** 4178.2
57. 470   **59.** 61.5   **61.** 37.125 ft   **63.** (a) $4400 approximately; (b) $4209.50; (c) $190.50
65. (a) $636 approximately; (b) $617.98   **67.** 7.8; 8.0; The answers are not the same.   **69.** $418.75

Exercises 5.5 *(page 269)*

1. 2.34   **3.** 0.99   **5.** 0.08   **7.** 2056   **9.** 20   **11.** 56.9   •**13.** 0.7   **15.** 0.1   **17.** 283.6   **19.** 56.4   **21.** 12.12
23. 0.96   **25.** 0.22   **27.** 5.04   **29.** 3.087   **31.** 0.285   **33.** 0.079   **35.** 0.007   **37.** 196.613   **39.** 0.164963
41. 0.045621   **43.** 0.0186   **45.** 13.81   **47.** 0.000154   **49.** 0.0010413   **51.** (a) 400 mi approximately; (b) 442.8 mi
53. (a) 15 mi/gal approximately; (b) 18.75 mi/gal   **55.** (a) about $1/lb; (b) $1.25/lb   **57.** $239.56   **59.** $295

Exercises 5.6 *(page 281)*

1. $\dfrac{9}{10}$ **3.** $\dfrac{5}{10}$ **5.** $\dfrac{62}{100}$ **7.** $\dfrac{57}{100}$ **9.** $\dfrac{526}{1000}$ **11.** $\dfrac{16}{1000}$ **13.** $\dfrac{51}{10}$ **15.** $\dfrac{815}{100}$ **17.** $\dfrac{1}{8}$ **19.** $\dfrac{9}{50}$ **21.** $\dfrac{9}{40}$ **23.** $\dfrac{17}{100}$

25. $3\dfrac{1}{5}$ **27.** $6\dfrac{1}{4}$ **29.** $0.\overline{6}$ **31.** $0.\overline{45}$ **33.** 0.6875   **35.** $0.\overline{428571}$ **37.** $0.1\overline{6}$ **39.** $0.\overline{5}$ **41.** 0.292   **43.** 0.417

45. 0.031   **47.** 1.231   **49.** 0.938   **51.** 0.70   **53.** 1.635   **55.** 14.98   **57.** 1.125   **59.** 13.51   **61.** 0.2555
63. 11.0825   **65.** 2.638   **67.** 27.3   **69.** 2.3 is larger; 0.05   **71.** 0.28 is larger; $0.00\overline{72}$

Chapter 5 Test *(page 285)*

1. $\dfrac{7}{125}$ **2.** twenty-one and twenty-one thousandths   **3.** 3005.0203   **4.** 203.02   **5.** 2000   **6.** 0.1   **7.** 92.76
8. 762.59   **9.** 122.11   **10.** 18.107   **11.** 37.313   **12.** 0.90395   **13.** 0.294   **14.** 12.80335   **15.** 1920   **16.** 17.83
17. 66.28   **18.** 0.04   **19.** $\dfrac{3}{8}$ **20.** 0.313   **21.** (a) about 400 mi; (b) 392 mi   **22.** $544

CHAPTER **6**

Exercises 6.1   *(page 293)*

1. $\dfrac{2 \text{ nickels}}{3 \text{ nickels}}$ 3. $\dfrac{25 \text{ miles}}{1 \text{ gallon}}$ 5. $\dfrac{50 \text{ miles}}{1 \text{ hour}}$ 7. $\dfrac{6 \text{ chairs}}{5 \text{ people}}$ 9. $\dfrac{3 \text{ inches}}{4 \text{ inches}}$ 11. $\dfrac{1 \text{ cm}}{10 \text{ cm}}$ 13. $\dfrac{8 \text{ days}}{7 \text{ days}}$ 15. $\dfrac{\$1 \text{ profit}}{\$5 \text{ invested}}$

17. true   19. true   21. true   23. true   25. true   27. false   29. true   31. true   33. true   35. false
37. true

Exercises 6.2   *(page 299)*

1. $x = 12$ 3. $x = 20$ 5. $B = 40$ 7. $x = 50$ 9. $A = \dfrac{21}{2}$ 11. $D = 100$ 13. $x = 1$ 15. $x = \dfrac{3}{5}$ 17. $w = 24$

19. $y = 6$ 21. $x = \dfrac{3}{2}$ 23. $R = 60$ 25. $A = 2$ 27. $B = 120$ 29. $R = \dfrac{100}{3}$ 31. $x = 22$ 33. $x = 1$

35. $x = 15$ 37. $B = 7.8$ 39. $x = 1.56$ 41. $R = 50$ 43. $B = 48$ 45. $A = 27.3$ 47. $A = 255$ 49. $B = 5800$

Exercises 6.3   *(page 309)*

1. 45 mi   3. $3.27   5. $120   7. $400   9. $160,000   11. 200 ft   13. 7.5   15. $11\dfrac{1}{4}$ hr   17. $26.25

19. 259,200 revolutions   21. 2100 min (or 35 hr)   23. 30.48 cm   25. 180 min (or 3 hr)   27. 18 biscuits   29. 2.7 kg

31. 500 mg   33. 4 drams   35. 1920 minims   37. 60 g   39. $\dfrac{1}{2}$ oz

Chapter 6 Test   *(page 315)*

1. $\dfrac{7}{6}$ 2. extremes   3. means   4. false   5. $\dfrac{5 \text{ inches}}{12 \text{ inches}}$ 6. $\dfrac{55 \text{ miles}}{1 \text{ hour}}$ 7. $\dfrac{1 \text{ inch}}{2.54 \text{ cm}}$ 8. $x = 36$ 9. $x = 45$

10. $x = 0.625$ 11. $x = \dfrac{1}{90}$ 12. $x = 3.4$ 13. $x = \dfrac{5}{6}$ 14. $x = 15$ 15. $x = 4\dfrac{1}{2}$ 16. 171.4 mi   17. 24 mi

18. $295   19. 45 mph   20. 27 boys   21. 4700 g   22. 12 weeks   23. 437.5 g   24. $7.50

CHAPTER **7**

Exercises 7.1   *(page 323)*

1. 60%   3. 65%   5. 3.5%   7. 110%   9. 99%   11. 30%   13. 40%   15. 7%   17. 90%   19. 25%   21. 45%
23. 75%   25. 53%   27. 77%   29. 125%   31. 150%   33. 200%   35. 236%   37. 16.3%   39. 13.4%

41. 20.25%   43. 0.5%   45. 0.25%   47. $10\dfrac{1}{4}\%$ 49. $24\dfrac{1}{2}\%$ 51. a   53. both are 10%   55. a

Exercises 7.2   *(page 331)*

1. 2%   3. 10%   5. 70%   7. 36%   9. 83%   11. 25%   13. 40%   15. 2.5%   17. 4.6%   19. 0.3%   21. 110%
23. 125%   25. 200%   27. 105%   29. 230%   31. 0.02   33. 0.1   35. 0.15   37. 0.25   39. 0.35   41. 0.101

43. 0.132　　**45.** 0.0525　　**47.** 0.1375　　**49.** 0.2025　　**51.** 0.0025　　**53.** 0.0017　　**55.** 1.25　　**57.** 1.3　　**59.** 2.22
61. 0.06　　**63.** 0.085　　**65.** 0.45　　**67.** 12%　　**69.** 2%

Exercises 7.3　(page 341)

1. 3%　　**3.** 7%　　**5.** 50%　　**7.** 25%　　**9.** 55%　　**11.** 30%　　**13.** 20%　　**15.** 80%　　**17.** 26%　　**19.** 48%　　**21.** 12.5%
23. 87.5%　　**25.** 55.6%　　**27.** 42.9%　　**29.** 63.6%　　**31.** 107.1%　　**33.** 105%　　**35.** 175%　　**37.** 137.5%　　**39.** 210%
41. $\dfrac{1}{10}$　　**43.** $\dfrac{3}{20}$　　**45.** $\dfrac{1}{4}$　　**47.** $\dfrac{1}{2}$　　**49.** $\dfrac{3}{8}$　　**51.** $\dfrac{1}{3}$　　**53.** $\dfrac{33}{100}$　　**55.** $\dfrac{1}{400}$　　**57.** 1　　**59.** $1\dfrac{1}{5}$　　**61.** $\dfrac{3}{1000}$　　**63.** $\dfrac{5}{8}$
65. $\dfrac{3}{400}$

Exercises 7.4　(page 351)

1. $A = 7$　　**3.** $A = 9$　　**5.** $A = 9$　　**7.** $A = 36$　　**9.** $B = 150$　　**11.** $B = 700$　　**13.** $B = 75$　　**15.** $B = 42$
17. $R = 150$　　**19.** $R = 20$　　**21.** $R = 50$　　**23.** $R = 33.3$　　**25.** $A = 12.5$　　**27.** $B = 110$　　**29.** $R = 180$　　**31.** $B = 72$
33. $B = 520$　　**35.** $R = 200$　　**37.** $A = 16.32$　　**39.** $A = 58.5$　　**41.** 43.92　　**43.** 72　　**45.** 80　　**47.** 40　　**49.** 10
51. 37.5　　**53.** 70　　**55.** 76.3

Exercises 7.5　(page 363)

1. $5700　　**3.** $710　　**5.** $12,000　　**7.** $4.30　　**9.** 20 problems　　**11.** (a) 40%; (b) 60%　　**13.** (a) $1.82; (b) $32.02
15. (a) $250; (b) 20%; (c) $33\dfrac{1}{3}$%; (d) 25%　　**17.** (a) $10.53; (b) 60%　　**19.** (a) $1875; (b) $2025　　**21.** $656.40

Chapter 7 Test　(page 371)

1. 101%　　**2.** 0.3%　　**3.** 17.3%　　**4.** 16%　　**5.** 112.5%　　**6.** 0.3　　**7.** 3.25　　**8.** 0.045　　**9.** $1\dfrac{2}{5}$　　**10.** $\dfrac{49}{200}$　　**11.** $\dfrac{7}{125}$
12. $A = 3.6$　　**13.** $R = 50$　　**14.** $B = 30$　　**15.** $A = 146.58$　　**16.** $B = 20.3$　　**17.** $R = 33.1$　　**18.** $15　　**19.** $1335
20. 58.3%　　**21.** 32% on rent; 35% on food; 12% on taxes

CHAPTER **8**

Exercises 8.1　(page 379)

1. $48　　**3.** $31.50　　**5.** $11.25　　**7.** $48　　**9.** $730　　**11.** $463.50; $36.50 savings　　**13.** $37,500
15. 10%　　**17.** interest: $7.50　　**19.** 19.2%; $562.50　　**21.** $712,500　　**23.** 6 mo
　　　　　　time: 60 days or 2 mo
　　　　　　rate: 18%
　　　　　　principal: $100

Exercises 8.2 *(page 389)*

1. (a) $13,791.70; (b) $14,631.61   **3.** (a) $306.82; (b) $5306.82; (c) $6.82   **5.** (a) $370.80; (b) $376.53   **7.** $800
9. $720.74   **11.** In 7 years, the value will be $19,487.18.
13. (a) $466.56; (b) $832; (c) $970.45; (d) The principal was larger; (e) $2934.39; (f) $634.76

Exercises 8.3 *(page 397)*

1.

Check No.	Date	Transaction Description	Payment (−)	(✓)	Deposit (+)	Balance
		Balance brought forward				─0─
	7-15	Deposit ( )		✓	700.00	+700.00 / 700.00
1	7-15	Quiet Town Apt. (rent/deposit)	520.00	✓		−520.00 / 180.00
2	7-15	Pa Bell Telephone (phone installation)	32.16	✓		−32.16 / 147.84
3	7-15	XYZ Power Co. (gas/elect. hookup)	46.49	✓		−46.49 / 101.35
4	7-16	Foodway Stores (groceries)	51.90	✓		−51.90 / 49.45
	7-20	Deposit ( )		✓	350.00	+350.00 / 399.45
5	7-23	Comfy Furniture (sofa, chair)	300.50			−300.50 / 98.95
	8-1	Deposit ( )			350.00	+350.00 / 448.95
	7-31	Interest ( )		✓	2.50	+2.50 / 451.45
	7-31	Service Charge ( )	2.00	✓		−2.00 / 449.45
		True Balance				449.45

BANK STATEMENT — Checking Account Activity

Transaction Description	Amount	(✓)	Running Balance	Date
Beginning Balance		✓	0.00	7-15
Deposit	700.00	✓	700.00	7-15
Check #1	520.00	✓	180.00	7-16
Check #4	51.90	✓	128.10	7-17
Check #2	32.16	✓	95.94	7-18
Check #3	46.49	✓	49.45	7-18
Deposit	350.00	✓	399.45	7-20
Interest	2.50	✓	401.95	7-31
Service Charge	2.00		399.95	7-31
Ending Balance			399.95	

RECONCILIATION SHEET

A. First, mark ✓ beside each check and deposit listed in both your checkbook register and on the bank statement.
B. Second, in your checkbook register, add any interest paid and subtract any service charge listed on the bank statement.
C. Third, find the total of all outstanding checks.

Outstanding Checks

No.	Amount
5	300.50
Total	300.50

Statement Balance	399.95
Add deposits not credited	+350.00
Total	749.95
Subtract total amount of checks outstanding	−300.50
True Balance	449.45

3.

YOUR CHECKBOOK REGISTER						
Check No.	Date	Transaction Description	Payment (−)	(✓)	Deposit (+)	Balance
			Balance brought forward			756.14
271	6-15	Parts, Parts, Parts (spark plugs)	12.72	✓		−12.72 743.42
272	6-24	Firerock Tire Co. (2 tires)	121.40	✓		−121.40 622.02
273	6-30	Gus' Gas Station (tune-up)	75.68			− 75.68 546.34
	7-1	Deposit ( )			250.00	+250.00 796.34
274	7-1	Prudent Ins. Co. (car insurance)	300.00			−300.00 496.34
	6-30	Service Charge ( )	1.00	✓		− 1.00 495.34
		( )				
		( )				
		( )				
		( )				
		True Balance				495.34

BANK STATEMENT
Checking Account Activity

Transaction Description	Amount	(✓)	Running Balance	Date
Beginning Balance			756.14	6-01
Check #271	12.72	✓	743.42	6-16
Check #272	121.40	✓	622.02	6-26
Service Charge	1.00	✓	621.02	6-30
Ending Balance			621.02	

RECONCILIATION SHEET

A. First, mark ✓ beside each check and deposit listed in both your checkbook register and on the bank statement.

B. Second, in your checkbook register, add any interest paid and subtract any service charge listed on the bank statement.

C. Third, find the total of all outstanding checks.

Outstanding Checks			
No.	Amount		
273	75.68	Statement Balance	621.02
274	300.00	Add deposits not credited	+250.00
		Total	871.02
		Subtract total amount of checks outstanding	−375.68
Total	375.68	True Balance	495.34

5.

		YOUR CHECKBOOK REGISTER				
Check No.	Date	Transaction Description	Payment (−)	(✓)	Deposit (+)	Balance
					Balance brought forward	967.22
772	4-13	C.P. Hay (accountant)	85.00	✓		− 85.00 \n 882.22
	4-14	Deposit ( )		✓	1200.00	+1200.00 \n 2082.22
773	4-14	E.Z. Pharmacy (aspirin)	4.71	✓		− 4.71 \n 2077.51
774	4-15	I.R.S. (income tax)	2000.00			−2000.00 \n 77.51
775	4-30	Heavy Finance Co. (loan payment)	52.50			− 52.50 \n 25.01
	5-1	Deposit ( )			600.00	+600.00 \n 625.01
	4-30	Interest ( )		✓	2.82	+ 2.82 \n 627.83
	4-30	Service Charge ( )	4.00	✓		− 4.00 \n 623.83
		( )				
		( )				
					True Balance	623.83

BANK STATEMENT
Checking Account Activity

Transaction Description	Amount	(✓)	Running Balance	Date
Beginning Balance			967.22	4-01
Deposit	1200.00	✓	2167.22	4-14
Check #772	85.00	✓	2082.22	4-15
Check #773	4.71	✓	2077.51	4-15
Interest	2.82	✓	2080.33	4-30
Service Charge	4.00	✓	2076.33	4-30
Ending Balance			2076.33	

RECONCILIATION SHEET

A. First, mark ✓ beside each check and deposit listed in both your checkbook register and on the bank statement.

B. Second, in your checkbook register, add any interest paid and subtract any service charge listed on the bank statement.

C. Third, find the total of all outstanding checks.

Outstanding Checks			
No.	Amount	Statement Balance	2076.33
774	2000.00	Add deposits not credited	+ 600.00
775	52.50	Total	2676.33
		Subtract total amount of checks outstanding	−2052.50
Total	2052.50	True Balance	623.83

7.

		YOUR CHECKBOOK REGISTER				
Check No.	Date	Transaction Description	Payment (−)	(✓)	Deposit (+)	Balance
			Balance brought forward			602.82
14	6-20	Aisle Bridal (flowers)	402.40	✓		−402.40 / 200.42
	6-22	Deposit ( )		✓	1000.00	+1000.00 / 1200.42
15	6-24	Tuxedo Junction (tux)	155.65	✓		−155.65 / 1044.77
16	6-28	D. Lohengrin (organist)	55.00			−55.00 / 989.77
17	6-28	D-Lux Limo (limo rental)	125.00			−125.00 / 864.77
18	6-30	C. C. Catering (food caterer)	700.00			−700.00 / 164.77
19	7-1	Luv-Lee Stationers (thank-you cards)	35.20			−35.20 / 129.57
	6-30	Service Charge ( )	1.00	✓		− 1.00 / 128.57
		( )				———
		( )				———
		True Balance				128.57

BANK STATEMENT Checking Account Activity				
Transaction Description	Amount	(✓)	Running Balance	Date
Beginning Balance			602.82	6-01
Deposit	1000.00	✓	1602.82	6-22
Check #14	402.40	✓	1200.42	6-22
Check #15	155.65	✓	1044.77	6-26
Service Charge	1.00	✓	1043.77	6-30
Ending Balance			1043.77	

RECONCILIATION SHEET

A. First, mark ✓ beside each check and deposit listed in both your checkbook register and on the bank statement.

B. Second, in your checkbook register, add any interest paid and subtract any service charge listed on the bank statement.

C. Third, find the total of all outstanding checks.

Outstanding Checks	
No.	Amount
16	55.00
17	125.00
18	700.00
19	35.20
Total	915.20

Statement Balance	1043.77
Add deposits not credited	+ 0
Total	1043.77
Subtract total amount of checks outstanding	− 915.20
True Balance	128.57

9.

YOUR CHECKBOOK REGISTER						
Check No.	Date	Transaction Description	Payment (−)	(✔)	Deposit (+)	Balance
					Balance brought forward	147.02
203	2-3	Food Stoppe (groceries)	26.90	✔		−26.90 / 120.12
204	2-8	Ekkon Oil (gasoline bill)	71.45	✔		−71.45 / 48.67
205	2-14	Rose's Roses (flowers)	25.00	✔		−25.00 / 23.67
206	2-14	L.M.R.U. (alumni dues)	20.00			−20.00 / 3.67
	2-15	Deposit ( )		✔	600.00	+600.00 / 603.67
207	2-26	SRO (theater tickets)	52.50			−52.50 / 551.17
208	2-28	MPG Mtg. (house payment)	500.00			−500.00 / 51.17
	2-28	Service Charge ( )	3.00	✔		−3.00 / 48.17
		( )				
		( )				
		( )				
					True Balance	48.17

BANK STATEMENT — Checking Account Activity				
Transaction Description	Amount	(✔)	Running Balance	Date
Beginning Balance			147.02	2-01
Check #203	26.90	✔	120.12	2-04
Check #204	71.45	✔	48.67	2-14
Deposit	600.00	✔	648.67	2-15
Check #205	25.00	✔	623.67	2-15
Service Charge	3.00	✔	620.67	2-28
Ending Balance			620.67	

RECONCILIATION SHEET

A. First, mark ✔ beside each check and deposit listed in both your checkbook register and on the bank statement.

B. Second, in your checkbook register, add any interest paid and subtract any service charge listed on the bank statement.

C. Third, find the total of all outstanding checks.

Outstanding Checks	
No.	Amount
206	20.00
207	52.50
208	500.00
Total	572.50

Statement Balance	620.67
Add deposits not credited	+ 0
Total	620.67
Subtract total amount of checks outstanding	−572.50
True Balance	48.17

Exercises 8.4 (page 409)

1. (a) $40,625; (b) $121,875; (c) $44,462.50   **3.** (a) 2%; (b) $19,010   **5.** (a) $24,000; (b) $72,000; (c) $1080; (d) $25,915

7. (a) $1405; (b) about 30.4%   **9.** (a) $738; (b) 41%

Exercises 8.5 (page 415)

1. $1548   **3.** $5780   **5.** $1637   **7.** $842

9. No. Her car expenses would be $493 and she could only afford to spend $420.

Chapter 8 Test (page 417)

1. $97.50   **2.** $800   **3.** 2 years   **4.** 14%   **5.** $1126   **6.** $6   **7.** $7563   **8.** $1512.60   **9.** (a) $24,000; (b) $960

10. $25,810

11.

YOUR CHECKBOOK REGISTER						
Check No.	Date	Transaction Description	Payment (−)	(✓)	Deposit (+)	Balance
					Balance brought forward	1201.40
203	2-3	Food Mart (groceries)	49.12	✓		-49.12 / 1152.28
204	2-8	Enco Oil (gasoline bill)	82.50	✓		-82.50 / 1069.78
205	2-14	Top Drawer (blouse)	25.00	✓		-25.00 / 1044.78
206	2-14	I.O. Dyne (medicine)	20.00			-20.00 / 1024.78
	2-15	Deposit ( )		✓	800.00	+800.00 / 1824.78
207	2-26	Stop 'n Shop (groceries)	52.50			-52.50 / 1772.28
208	2-28	Mayne Mtg. (house payment)	900.00			-900.00 / 872.28
	2-28	Service Charge ( )	3.00	✓		-3.00 / 869.28
		( )				
		( )				
		( )				
					True Balance	869.28

BANK STATEMENT
Checking Account Activity

Transaction Description	Amount	(✓)	Running Balance	Date
Beginning Balance		✓	1201.40	2-01
Check #203	49.12	✓	1152.28	2-04
Check #204	82.50	✓	1069.78	2-14
Deposit	800.00	✓	1869.78	2-15
Check #205	25.00	✓	1844.78	2-15
Service Charge	3.00	✓	1841.78	2-28
Ending Balance			1841.78	

RECONCILIATION SHEET

A. First, mark ✓ beside each check and deposit listed in both your checkbook register and on the bank statement.
B. Second, in your checkbook register, add any interest paid and subtract any service charge listed on the bank statement.
C. Third, find the total of all outstanding checks.

Outstanding Checks	
No.	Amount
206	20.00
207	52.50
208	900.00
Total	972.50

Statement Balance	1841.78
Add deposits not credited	+ -0-
Total	1841.78
Subtract total amount of checks outstanding	-972.50
True Balance	869.28

12. (a) $1090; (b) about 36.3%   **13.** $34,070

CHAPTER 9

Exercises 9.1 (page 429)

1. (a) 79; (b) 84; (c) 38   **3.** (a) 6.2; (b) 6; (c) 6   **5.** (a) $400; (b) $375; (c) $225   **7.** (a) 81; (b) 79; (c) 26
9. (a) 22; (b) 19; (c) 21   **11.** (a) 18.5 in.; (b) 14.9 in.; (c) 22.0 in.   **13.** 85   **15.** $24,432

Exercises 9.2 *(page 437)*

1. (a) Faculty Salaries—$5,625,000
   Administration Salaries—$2,500,000
   Student Program—$625,000
   Savings—$500,000
   Non-Teaching Salaries—$1,625,000
   Maintenance—$1,250,000
   Supplies—$375,000
   (b) $3,125,000
   (c) 22%
   (d) $2,750,000

3. (a) $500
   (b) Loan—60%
       Insurance—14%
       Gas—10%
       Repairs—16%
   (c) $\dfrac{7}{5}$

Exercises 9.3 *(page 441)*

1. (a) social science
   (b) chemistry and physics; humanities
   (c) 3300
   (d) 21.2%

3. (a) The scales represent different items.
   (b) Sue
   (c) Bob and Sue
   (d) 54.5%
   (e) 75%
   (f) GPA seems to be higher for those who work fewer job hours per week.

5. (a) Wheat
   (b) $1500
   (c) Steel
   (d) $1000
   (e) Steel
   (f) $2000
   (g) Wheat
   (h) $1500

7. (a) August
   (b) 3.5 million dollars
   (c) 0
   (d) 5.5 million dollars
   (e) 29.4%
   (f) 33.3%

Exercises 9.4 *(page 449)*

1. (a) 5
   (b) 5000
   (c) 3rd class
   (d) 40
   (e) 15,000.5 and 20,000.5
   (f) 100
   (g) 5%
   (h) 35%

3. (a) 3rd
   (b) 15
   (c) 120
   (d) 50
   (e) 35
   (f) 54%

Chapter 9 Test *(page 453)*

1. (a) 2; (b) 2; (c) 4   2. (a) 10.16; (b) 8.89; (c) 17.78   3. $9900   4. $33   5. No; they will be short by $228.
6. $1167   7. 62.5%   8. 49.1%   9. (a) 5; (b) 10; (c) 12   10. 52%   11. 60%

CHAPTER 10

Exercises 10.1 *(page 467)*

1. milli-, centi-, deci-, deka-, hecto-, kilo-   3. 500   5. 600   7. 300   9. 1400   11. 18   13. 350   15. 160   17. 21
19. 5000   21. 6400   23. 2.6   25. 4.8   27. 0.12   29. 0.03   31. 0.256   33. 0.32   35. 1.7   37. 2.4   39. 0.4
41. 462   43. 0.052   45. 6410   47. 0.5   49. 0.057   51. 35   53. 2300   55. 300   57. 870   59. 6   61. 14
63. 400; 40,000   65. 5700; 570,000   67. 1700; 170,000; 17,000,000   69. 3; 300; 30,000   71. 1.42   73. 2   75. 780
77. 4   79. 869; 86,900   81. 16; 1600   83. 0.15 ha   85. 47,600 a; 476 ha

Exercises 10.2 *(page 481)*

1. 7000 **3.** 0.0345 **5.** 4 **7.** 73,000,000 **9.** 540 **11.** 5000 **13.** 2000 **15.** 896,000 **17.** 75 **19.** 7,000,000
21. 0.00034 **23.** 0.016 **25.** 3940 **27.** 5600 **29.** 3.547 **31.** 1000; 1000; 1000; 1,000,000,000
33. 10; 100; 100; 10,000; 1000; 1,000,000 **35.** 73,000 **37.** 0.4 **39.** 63,000 **41.** 0.5 **43.** 19 **45.** 2000
47. 5300 **49.** 0.03 **51.** 48,000 **53.** 0.32 **55.** 0.29

Exercises 10.3 *(page 491)*

1. 77 **3.** 50 **5.** 10 **7.** 0 **9.** 157.48 **11.** 2.742 **13.** 96.6 **15.** 644 **17.** 124 **19.** 21.7 **21.** 19.7
23. 19.35 **25.** 55.8 **27.** 83.6 **29.** 405 **31.** 741 **33.** 53.82 **35.** 4.65 **37.** 9.46 **39.** 94.625 **41.** 10.6
43. 11.088 **45.** 3277.4 **47.** 4.54 **49.** 1102.5

Exercises 10.4 *(page 501)*

1. $\frac{3}{10}$ **3.** 60 **5.** $4\frac{1}{2}$ **7.** 3 **9.** 20 **11.** $1\frac{1}{2}$ **13.** 5 **15.** 6 **17.** 3 **19.** 150 **21.** 15 mL Ampicillin

23. 2 mL castor oil **25.** 4 mL Robitussin® **27.** $\frac{2}{3}$ mL Polaramine® **29.** $7\frac{1}{2}$ mL castor oil

Chapter 10 Test *(page 505)*

1. 0.07 **2.** 1500 **3.** 0.173 **4.** 300 **5.** 0.017 **6.** 10,300 **7.** 0.042 **8.** 6000 **9.** 40,000 **10.** 0.0016
11. 2.7 **12.** 2 **13.** 2000 **14.** 0.175 **15.** 33 **16.** $\frac{39}{2} = 19\frac{1}{2}$ **17.** $\frac{37}{6} = 6\frac{1}{6}$ **18.** $\frac{2}{3}$ **19.** 76 **20.** 2600
21. $\frac{27}{8} = 3\frac{3}{8}$ **22.** $0.4 = \frac{2}{5}$ **23.** 192 **24.** 58 **25.** $2.8 = 2\frac{4}{5}$ **26.** $3.5 = 3\frac{1}{2}$ **27.** $25.6 = 25\frac{3}{5}$ **28.** $\frac{19}{2} = 9\frac{1}{2}$
29. $\frac{13}{16}$ **30.** $\frac{11}{8} = 1\frac{3}{8}$ **31.** 75 mL

CHAPTER **11**

Exercises 11.1 *(page 511)*

1. (a) 2; (b) 3; (c) 4; (d) 1; (e) 6; (f) 5 **3.** 104 mm **5.** 31.4 ft **7.** 19.468 yd **9.** 13.5 cm **11.** 16,000 m **13.** 20.8 in.
15. 11.9634 cm

Exercises 11.2 *(page 519)*

1. (a) 3; (b) 2; (c) 4; (d) 1; (e) 6; (f) 5 **3.** 6 cm² **5.** 78.5 yd² **7.** 227.5 in.² **9.** 500 mm² (or 5 cm²) **11.** 28.26 ft²
13. 21.195 yd² **15.** 106 cm² **17.** 158,000 a **19.** 0.785 ft² **21.** 2500 cm² and 0.25 m²

Exercises 11.3 *(page 527)*

1. (a) 3; (b) 5; (c) 4; (d) 2; (e) 1 **3.** 1695.6 in.³ **5.** 904.32 ft³ **7.** 80 cm³ **9.** 9106 dm³ **11.** 113.04 in.³
13. 3456 in.³

Exercises 11.4 *(page 537)*

1. (a) m∠2 = 75°; (b) m∠2 = 87°; (c) m∠2 = 45°; (d) m∠2 = 15° **3.** (a) a right angle; (b) an acute angle; (c) an acute angle
5. 25° **7.** 40° **9.** 55° **11.** 110° **13.** m∠A = 55°; m∠B = 45°; m∠C = 80°
15. m∠L = 115°; m∠M = 90°; m∠N = 105°; m∠O = 110°; m∠P = 120°
17. (a) straight angle; (b) right angle; (c) acute angle; (d) obtuse angle
19. (a) 150°; (b) Yes; ∠3 is supplementary to ∠2; (c) ∠1 and ∠2; ∠2 and ∠3; ∠3 and ∠4; ∠4 and ∠1
21. m∠2 = 150°; m∠3 = 30°; m∠4 = 150° **23.** m∠3 = m∠5; m∠2 = m∠5; m∠1 = m∠4; m∠1 = m∠6; m∠4 = m∠6

Exercises 11.5 *(page 547)*

1. scalene (and obtuse) **3.** scalene (and right) **5.** isosceles (and acute) **7.** isosceles (and right)
9. scalene (and acute)
11. Yes, the triangles are similar.
∠RPQ = ∠TPS because vertical angles are equal.
∠R = ∠S because the sum of the measures of the angles in both triangles is 180°. Since two of the angles are the same in both triangles, the third angles must be the same.
13. Yes, because the sum of any two sides is longer than the third side. The triangles would be scalene.

Exercises 11.6 *(page 555)*

1. Yes $(144 = 12^2)$ **3.** Yes $(81 = 9^2)$ **5.** Yes $(400 = 20^2)$ **7.** No **9.** No
11. $(1.732)^2 = 2.999824$ and $(1.733)^2 = 3.003289$ **13.** $2\sqrt{3}$ **15.** $2\sqrt{6}$ **17.** $4\sqrt{3}$ **19.** $11\sqrt{3}$ **21.** $10\sqrt{5}$
23. $8\sqrt{2}$ **25.** $6\sqrt{2}$ **27.** $11\sqrt{5}$ **29.** 13 **31.** $20\sqrt{2}$ **33.** $3\sqrt{10}$ **35.** 16 **37.** Yes; $6^2 + 8^2 = 10^2$
39. $c = \sqrt{5}$ **41.** $x = 5\sqrt{2}$ **43.** $c = 2\sqrt{29}$

Chapter 11 Test *(page 559)*

1. 18.84 in. **2.** 28.26 in.² **3.** 16.8 m **4.** 12 m² **5.** 0.785 L **6.** 16 L
7. (a) right angle; (b) obtuse; (c) ∠BOA and ∠DOE (or ∠AOE and ∠BOD); (d) ∠BOA and ∠COD (or ∠COD and ∠DOE);
(e) ∠AOB and ∠BOD, ∠BOD and ∠DOE, ∠DOE and ∠AOE (or ∠AOE and ∠AOB)
8. (a) isosceles; (b) obtuse; (c) scalene **9.** x = 2.5 **10.** $2\sqrt{7}$ **11.** $3\sqrt{5}$ **12.** $4\sqrt{3}$ **13.** $10\sqrt{3}$ **14.** $3\sqrt{10}$
15. $13\sqrt{2}$ **16.** Yes; $12^2 + 16^2 = 20^2$. **17.** $2\sqrt{5}$

CHAPTER **12**

Exercises 12.1 *(page 567)*

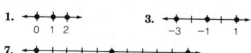
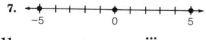
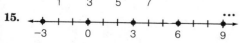

21. +10 or 10 **23.** −14 **25.** +6 or 6 **27.** −30 **29.** 0

31. 6   **33.** 24   **35.** 20   **37.** 0   **39.** $13 > 10$ and $|13| > |10|$ **41.** $-10 > -13$ and $|-13| > |-10|$
43. $30 > 20$ and $|30| > |20|$ **45.** $5 > -9$ and $|-9| > |5|$ **47.** $9 > -8$ and $|9| > |-8|$ **49.** $7 > -11$ and $|-11| > |7|$

Exercises 12.2   *(page 575)*

1. 2   **3.** 10   **5.** 19   **7.** -9 **9.** 0   **11.** -4 **13.** 2   **15.** -16 **17.** -4 **19.** 0   **21.** 4   **23.** 14   **25.** -1
27. -15 **29.** 41   **31.** 13   **33.** 10   **35.** -12 **37.** -10 **39.** 0

Exercises 12.3   *(page 581)*

1. 3   **3.** 11   **5.** -7 **7.** -9 **9.** 4   **11.** -10 **13.** 1   **15.** 18   **17.** 17   **19.** -31 **21.** -5 **23.** 0
25. -5 **27.** 8   **29.** 5   **31.** 80   **33.** -12 **35.** 43   **37.** 12   **39.** 5   **41.** 17   **43.** -8 **45.** 6   **47.** -6
49. -15 **51.** -3 **53.** -10 **55.** 5   **57.** 2   **59.** 0   **61.** 4   **63.** -16 **65.** 22   **67.** -21 **69.** 0   **71.** 10
73. 0   **75.** -1

Exercises 12.4   *(page 589)*

1. -15 **3.** 24   **5.** -20 **7.** 28   **9.** -50 **11.** -21 **13.** -48 **15.** 63   **17.** 0   **19.** 90   **21.** 24   **23.** 60
25. -42 **27.** -45 **29.** -60 **31.** 0   **33.** -1 **35.** 16   **37.** -84 **39.** -300 **41.** -3 **43.** -2 **45.** 4
47. 5   **49.** -5 **51.** -3 **53.** 2   **55.** 4   **57.** -3 **59.** -8 **61.** 2   **63.** 1   **65.** undefined   **67.** 0
69. undefined

Exercises 12.5   *(page 597)*

1. $8x$ **3.** $6x$ **5.** $-7a$ **7.** $-12y$ **9.** $-9x$ **11.** $-8x$ **13.** 0   **15.** $-c$ **17.** $-9a$ **19.** 0   **21.** $5x - 7$
23. $-x + 5$ **25.** $-11a - 2$ **27.** $3x + 1$ **29.** $4x + 7$ **31.** -5 **33.** 0   **35.** 9   **37.** 22   **39.** -11
41. -3 **43.** 12   **45.** -8 **47.** 16

Exercises 12.6   *(page 605)*

1. $x = 6$ **3.** $y = 22$ **5.** $y = -5$ **7.** $x = -2$ **9.** $y = 2$ **11.** $x = 6$ **13.** $y = -4$ **15.** $x = -6$ **17.** $y = 5$
19. $x = 13$ **21.** $x = 1$ **23.** $y = 2$ **25.** $x = -3$ **27.** $y = 2$ **29.** $x = -6$ **31.** $y = -3$ **33.** $x = 9$ **35.** $y = -3$
37. $x = -1$ **39.** $x = -3$ **41.** $x = -5$ **43.** $y = 5$ **45.** $x = 0$ **47.** $x = 3$ **49.** $x = 3$ **51.** $x = -7$ **53.** $x = 5$
55. $y = 4$ **57.** $x = 5$ **59.** $x = -7$

Exercises 12.7   *(page 613)*

1. 19   **3.** 54   **5.** 15   **7.** -4 **9.** 4   **11.** 15   **13.** 5   **15.** $-15, -14, -13$ **17.** 25, 27. 29   **19.** 14, 16, 18, 20
21. 2   **23.** -3 **25.** $-2, -3$ **27.** 7 cm

Chapter 12 Test   *(page 619)*

1.   **2.** 11   **3.** -12 **4.** -15 **5.** 8   **6.** -9 **7.** -12 **8.** 16   **9.** 28   **10.** -90
11. 8   **12.** undefined   **13.** -5 **14.** $3x + 5$ **15.** 1   **16.** $x = -5$ **17.** $x = 6$ **18.** $x = 1$ **19.** 2
20. $-13, -12, -11$

Index

recommended approach, 69
solving, 609–612
Writing expressions, 609

Y

Young's rule, 499

Z

z-scores, 423
Zero, 564
  as additive identity, 11
  as numerator of fraction,
    137

undefined in denominator of
  fraction, 131
Zero property of multiplication,
  37